New Directions in Terahertz Technology

NATO ASI Series

Advanced Science Institutes Series

*A Series presenting the results of activities sponsored by the NATO Science Committee,
which aims at the dissemination of advanced scientific and technological knowledge,
with a view to strengthening links between scientific communities.*

The Series is published by an international board of publishers in conjunction with the NATO
Scientific Affairs Division

A Life Sciences	Plenum Publishing Corporation
B Physics	London and New York
C Mathematical and Physical Sciences	Kluwer Academic Publishers
D Behavioural and Social Sciences	Dordrecht, Boston and London
E Applied Sciences	
F Computer and Systems Sciences	Springer-Verlag
G Ecological Sciences	Berlin, Heidelberg, New York, London,
H Cell Biology	Paris and Tokyo
I Global Environmental Change	

PARTNERSHIP SUB-SERIES

1. Disarmament Technologies	Kluwer Academic Publishers
2. Environment	Springer-Verlag / Kluwer Academic Publishers
3. High Technology	Kluwer Academic Publishers
4. Science and Technology Policy	Kluwer Academic Publishers
5. Computer Networking	Kluwer Academic Publishers

*The Partnership Sub-Series incorporates activities undertaken in collaboration with NATO's
Cooperation Partners, the countries of the CIS and Central and Eastern Europe, in Priority Areas of
concern to those countries.*

NATO-PCO-DATA BASE

The electronic index to the NATO ASI Series provides full bibliographical references (with keywords
and/or abstracts) to more than 50000 contributions from international scientists published in all
sections of the NATO ASI Series.
Access to the NATO-PCO-DATA BASE is possible in two ways:

– via online FILE 128 (NATO-PCO-DATA BASE) hosted by ESRIN,
Via Galileo Galilei, I-00044 Frascati, Italy.

– via CD-ROM "NATO-PCO-DATA BASE" with user-friendly retrieval software in English, French
and German (© WTV GmbH and DATAWARE Technologies Inc. 1989).

The CD-ROM can be ordered through any member of the Board of Publishers or through NATO-
PCO, Overijse, Belgium.

Series E: Applied Sciences - Vol. 334

New Directions in Terahertz Technology

edited by

J. M. Chamberlain

Department of Physics,
The University of Nottingham,
Nottingham, U.K.

and

R. E. Miles

Department of Electronic & Electrical Engineering,
The University of Leeds,
Leeds, U.K.

Kluwer Academic Publishers

Dordrecht / Boston / London

Published in cooperation with NATO Scientific Affairs Division

Proceedings of the NATO Advanced Research Workshop on
New Directions in Terahertz Technology
Château de Bonas, Castéra-Verduzan, France
June 30–July 11, 1996

A C.I.P. Catalogue record for this book is available from the Library of Congress.

ISBN 0-7923-4537-1

Published by Kluwer Academic Publishers,
P.O. Box 17, 3300 AA Dordrecht, The Netherlands.

Kluwer Academic Publishers incorporates the publishing programmes of
D. Reidel, Martinus Nijhoff, Dr W. Junk and MTP Press.

Sold and distributed in the U.S.A. and Canada
by Kluwer Academic Publishers,
101 Philip Drive, Norwell, MA 02061, U.S.A.

In all other countries, sold and distributed
by Kluwer Academic Publishers Group,
P.O. Box 322, 3300 AH Dordrecht, The Netherlands.

Printed on acid-free paper

Contents

PREFACE

This meeting brought together researchers from a wide variety of physical science and engineering disciplines with a common interest in the application of new technologies to the creation of devices and systems operating at Terahertz Frequencies. This region, defined for the purposes of the Institute as between 100 GHz and 10 THz, lies at the transition from optics to electronics and has traditionally been studied for a limited number of specialist scientific purposes only. However, following a number of recent developments in semiconductor physics and in materials-growth and processing, there is now every prospect of a more widespread application of practical terahertz systems. The Institute placed much emphasis on solid state devices capable of operating in the terahertz region at room temperature, which involved detailed studies of device principles for: fundamental sources of terahertz power, frequency-multiplication from lower frequencies, down-conversion from higher frequencies and the implementation of completely novel device principles. An inescapable conclusion which emerged from the Institute was that the skills of scientists and engineers in the areas of solid state physics, materials science, device fabrication, measurement science and physical modelling must be combined to ensure the successful introduction of terahertz electronic systems.

A significant part of the Institute was also devoted to actual or potential applications of terahertz technology in communications, astronomy, remote sensing and medicine. The ASI was based around lecture/tutorials given by distinguished scientists and engineers from both academic and industrial backgrounds. This format ensured that each topic could be developed from basic principles and placed within the wider context of the Institute. The success of this formula could be judged by the lively discussions, often initiated by the younger delegates, which were often followed up into the early hours of the morning in the convivial atmosphere of the venue. It was quite apparent that the mix of disciplines, and the open accessibility of the senior researchers, was much appreciated and enabled the younger members to appreciate the role of their own efforts within the broader scheme and to acquire an appreciation and understanding of the multidisciplinary nature of this subject area. Apart from their determination to benefit as much as possible from the lectures, the younger members also displayed their own work for critical appraisal in the sixty posters which were presented over two sessions. The ASI also included a Technical Visit to Matra Marconi Space in Toulouse to view state-of-the-art terahertz technology in satellite systems, together with an Open Discussion on the way forward.

The ASI took place in the idyllic surroundings of the Château de Bonas, near the town of Castéra-Verduzan some 100 km from Toulouse in south-west France. More than 80 participants from all over the NATO countries attended. The organisers would particularly like to thank Profs. Tatsuo Itoh and Erik Kollberg, the honorary co-directors of the ASI, for their enthusiastic support and tireless energy in attention to detail, making sure that the meeting ran smoothly and for presenting their own papers. Finally the Organisers would like to express their sincere thanks to the speakers, some of whom volunteered their services at the last minute, for contributing their papers to the course workbook and this volume which we believe will be a useful source book for anyone involved in the field of terahertz technology.

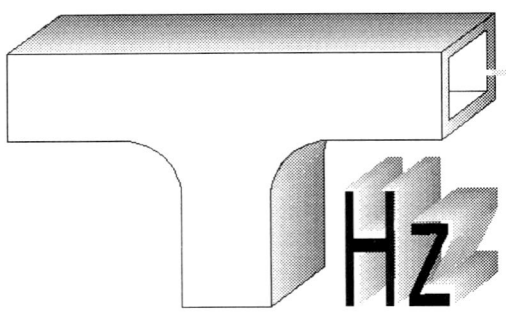

Philip Pieters
IMEC, Leuven, Belgium

"Logo Contest Winner"

NATO ASI: New Directions in Terahertz Technology
Chateau de Bonas, Castera- Verduzan, France, 1996.

ORGANISING COMMITTEE

Co-Directors: T. Itoh (UCLA) & E. Kollberg (Chalmers)

J.Bowen (Reading)   Workbook
J.M.Chamberlain (Nottingham)   Secretary
N.Cronin (Bath)   Sessions Management
J.Leotin (Toulouse)   Local Arrangements
R.E.Miles (Leeds)   Proceedings
R.D.Pollard (Leeds)   Treasurer

SPONSORS

We are most grateful to the following companies and organisations for their generous financial support and sponsorship of this meeting:

North Atlantic Treaty Organisation Science Committee

Engineering and Physical Science Research Council (UK)
European Research Office of the US Army
Hughes Electronics
Matra Marconi Space – France
The University of Leeds
The University of Nottingham
URSI

ACKNOWLEDGEMENTS

The organisers would like to thank: Nathan Mattick for his invaluable work in editing these Proceedings; Terry Davies and Mel Wragg for photographic, artwork and secretarial support; Geoff Parkhurst, Lucas Karatzas and John Digby for poster sessions management and driving duties. The organisers would also like to add their sincere thanks to Mme Simon, Patrice Wozniak and their colleagues at the Château de Bonas for creating a wonderful atmosphere for the meeting. Finally the organisers wish to record their gratitude to all of the speakers for their efforts and interest in this meeting and of course to all of the participants for their attendance.

Theme 1

Background and Introduction

INTRODUCTION TO TERAHERTZ SOLID-STATE DEVICES

J.M.CHAMBERLAIN
Department of Physics, The University of Nottingham, Nottingham NG7 2RD, United Kingdom

R.E.MILES, C.E.COLLINS AND D.P.STEENSON
Department of Electronics and Electrical Engineering, The University of Leeds, Leeds LS2 9JT, United Kingdom

1. Opportunities at Terahertz Frequencies

Terahertz frequencies, somewhat loosely defined as those in the range 100 GHz - 10 THz, form a significant region in the electromagnetic spectrum that has not yet been opened up for commercial exploitation. The main reason for this is the virtual absence in this frequency range of reliable, low cost, miniaturised solid-state power sources. Related to this is the fact that the technology for the fabrication of the necessary passive components is also not well developed. These Proceedings will address these issues in the light of recent developments in solid state physics and engineering which are poised to open up the terahertz frequencies to much wider application than has hitherto been possible. This chapter focuses in particular on the background to the power source problem by surveying the types of device now available, and goes on to indicate which novel (or traditional) approaches may eventually produce viable terahertz frequency sources.

The traditional applications of terahertz systems lie in the fields of astronomy [1], atmospheric studies [2], plasma diagnostics [3] and solid-state physics [4]. At present there are emerging applications in imaging and surveillance [5] and other potential uses of terahertz communication systems have also been suggested [6]. The advantages for systems operation at terahertz frequencies accrue from the high resolution and wide bandwidth afforded, leading to such practical outcomes as smaller antennae (and thus the possibility of imaging arrays), a resolution comparable with the size of everyday objects, the possibility of chemical-specific absorption, transmission through skin and plastic [5] and the ability to carry large amounts of information either directly by free space transmission or on an optical carrier. Limited atmospheric transmission at these frequencies, as noted elsewhere [7], can be turned to advantage to provide security in, for example, an office local area network (LAN) [8,9] and to minimise interference between neighbouring networks.

The design and operation of terahertz detection systems has reached a higher degree of sophistication [10] and, as noted above, one of the key issues in the exploitation of this frequency band in newer areas is the development of compact, efficient solid-state sources for free space transmission and to act as a local oscillator (LO) for heterodyne detection. The specification of power levels required from such a source depends, of course, on many factors such as the noise properties of the associated detection circuitry

3

J.M. Chamberlain and R.E. Miles (eds.), New Directions in Terahertz Technology, 3–27.
© *1997 Kluwer Academic Publishers. Printed in the Netherlands.*

and the precise nature of the application. However, it is useful to note that for a 300 GHz source acting as a LO in a typical office LAN a power of 1-5 mW is required; further consideration of these system requirements is given in the final section of this chapter. Additional characteristics required of such sources are that they should operate at room temperature and be amenable to integration with a transmission line system such as waveguide structures [11] made by micromachining techniques.

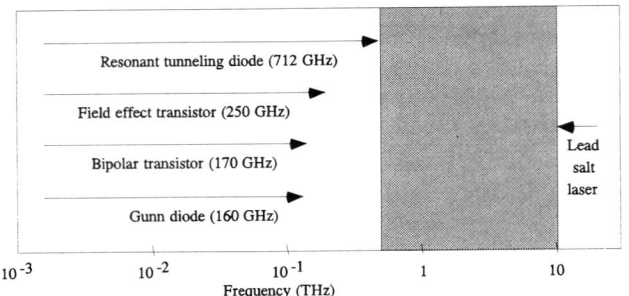

Figure 1. Conventional (fundamental) solid-state sources

Figure 1 is a schematic diagram of the types of source currently available on a commercial basis between 10 GHz and 10 THz, (free-electron lasers [12], molecular-gas lasers [13], travelling wave tubes and related devices [14] are omitted because they are considered too expensive, are bulky, dissipate too much power or are otherwise inappropriate for ·commercial exploitation in a mass market). The paucity of power sources operating in the frequency region 200 GHz to 2 THz, dubbed the "terahertz gap" [15], is immediately conspicuous in the Figure and the reasons for the existence of this gap will now be examined.

2. Terahertz Solid-State Sources: General Comments

2.1. "TRANSIT TIME" AND "TRANSITION" DEVICES

Below the terahertz gap indicated in Figure 1, devices and circuits operating up to 40 GHz are commercially available, with laboratory examples of integrated circuits operating up to 170 GHz [16]. These systems are in general advanced developments of conventional electronics where reduced dimensions or sophisticated layer structures are utilised for operation beyond 100 GHz. Most of these devices may be broadly classified as "transit time devices" and include such familiar examples as Bipolar Junction Transistors (BJTs), Heterojunction Bipolar Transistors (HBTs), Field Effect Transistors (FETs) High Electron Mobility Transistors (HEMTs) and Transferred Electron (Gunn) devices. In all of these, the time taken for carriers (usually electrons) to move a characteristic distance determines the maximum frequency of operation. The characteristic distance is the base width for BJTs, the gate length for FETs and the length of the active region for Gunn devices. An interesting case is the Resonant Tunnelling Diode (RTD), which currently holds the record as "the fastest purely electronic device" [17]; although the ultimate frequency limit is set by the width of the barrier transmission resonance, further practical limits are set by the depletion-region transit time and the RC time constant of the device [18, 19].

Above the terahertz gap, we again have well-developed solid-state sources such as near infra-red lasers and Light Emitting Diodes (LEDs). The interband lead-salt semiconductor lasers [20] represent the long-wavelength limit (34 μm) of such devices. These devices may be classified as "transition devices", since the charge carriers undergo a transition from a higher to a lower energy state with the direct emission of radiation at a frequency f given by E = hf where E is the energy state separation. A schematic illustration of the difference in the two types of device is shown in Figure 2. It is evident that the terahertz gap serves to mark the boundary between "electronic" and

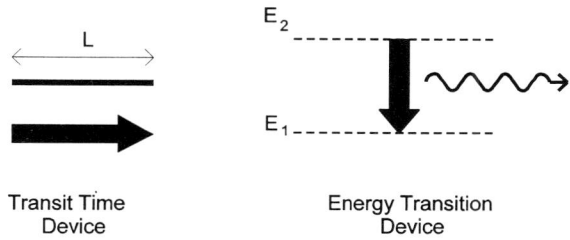

Transit Time Device   Energy Transition Device

Figure 2. Schematic representation of transit and transition devices

"optical" sources. A further discussion of the terahertz region as the "crossover" [21] between regimes, where it is appropriate to consider the boundary between classical electromagnetic/carrier interactions and completely quantum descriptions is given later in this volume [22].

2.2. CLOSING THE GAP

The terahertz gap can be filled in three ways; (i) by developing new devices which can be made to generate power directly at terahertz frequencies i.e. fundamental sources; (ii) multiplying up the lower frequency of an existing device; (iii) mixing down (or downconversion) from a higher frequency. As examples, the RTD operates as a fundamental source, albeit weak, up to 712 GHz [23] whereas Gunn devices may be cascaded with frequency multipliers to provide a practical LO to 1.5 THz [24]. Recently the outputs from a pair of solid state lasers have been mixed to give a difference frequency in the terahertz region [25]. These devices will soon be available in a solid-state "integrated" form as will become clear from further comments to be made later in this Chapter.

2.2.1. "Transit Time" Devices

For a characteristic distance l and carrier velocity v the transit time τ is l/v. The limiting condition is that during this transit time the phase of the generated signal has advanced through only a small fraction of its period T i.e. $\tau << T$. Thus the maximum operation frequency $(= 1/T)$ is therefore $<< v/l$. This is usually written as:

$$f_T \ll \frac{v}{l} \tag{1}$$

where f_T is known as the cut-off frequency.

From equation (1) it may be seen that a high value of f_T requires a high particle velocity and a short transit distance. In all semiconductors, v has an effective maximum (the saturation velocity v_s) of approximately $10^5\,\mathrm{ms^{-1}}$. The length l for generation at 1 THz is therefore given by:

$$l \ll \frac{v}{f_T} = \frac{10^5}{10^{12}} = 10^{-7}m \;\Rightarrow\; 0.1\ \mu m \tag{2}$$

In practice this means that l must be at most $0.01\ \mu m$: not only is this approaching the limit of the lithographer's art, but is also the approximate separation distance of electrons in a typical semiconductor of doping density $10^{24}\mathrm{m^{-3}}$; this condition will mean that significant statistical fluctuations will occur in instantaneous charge flow giving rise to a source of noise.

Transit time devices also exhibit another limitation as a consequence of their small dimensions. In order to generate as much power as possible, and accelerate the carriers up to their maximum velocity, the voltage V developed across the device must be as high as possible. However for a device of length l this is limited by the breakdown field E_B given by V_M/l where V_M is the maximum voltage that can be applied. Therefore

$$V_M = E_B \times l \tag{3}$$

Hence from (1) and (3),

$$V_M f_T = v_S E_B \tag{4}$$

Now, the power P delivered by the device to a load of resistance R is given by $P = V_M^2 /R$ which using (4) may be written as:

$$P = \frac{1}{R} \left(\frac{v_S E_B}{f_T} \right)^2 \tag{5}$$

in essence

$$P \propto \frac{1}{f_T^2} \tag{6}$$

The exact proportionality constant in (6) depends on the structure of the device, the material and the doping level but the consequence is that the power available will decrease relentlessly as one moves up through the terahertz region.

2.2.2. *"Transition" Devices*

The simple equation $E = hf$ indicates that at 1 THz, $E = 4$ meV, which must be compared with the room temperature thermal energy of 25 meV. This indicates that population inversion will never be achieved for a room temperature device unless the relevant energy levels are both high above the ground state. This principle is already employed in the molecular gas laser [13] and analogues must be found in solid state (semiconductor) devices if a compact room temperature terahertz laser is to be constructed. Such analogues may exist in III-V semiconductor hole systems [26] and in quantum dots [27]. Another general issue, which complicates the realisation of terahertz lasers, is that the photon density of states varies as the square of the frequency, hence spontaneous emission rates at a wavelength of 40 μm (or approximately 8.3 THz) will be 100 times slower than at 4 μm where solid-state intersubband sources are now available [28]. In addition, any terahertz source must be constructed to avoid restrahl absorption bands in the device active material. Nevertheless, specific schemes for such terahertz laser sources have been suggested, and this is reviewed elsewhere in these Proceedings [29, 30], together with p-Ge intersubband sources which, although operating at low temperature, is the only solid-state terahertz laser source presently available [31]. As a final comment, it may be noted that if a compact terahertz solid-state laser source is ever realised, attention will have to be paid to mode-confinement, since beyond wavelengths of 30 μm conventional dielectric schemes are impractical (although other methods may be available [32]).

3. Electronic/ "Transit Time" Solid State Terahertz Sources

3.1. GENERAL CONSIDERATIONS AND HIGH FREQUENCY FIGURES OF MERIT

Before listing the current performance figures of electronic (transit time) devices operating as fundamental sources of power at terahertz and sub-terahertz frequencies, it is pertinent to comment on several practical issues which affect their ultimate usefulness.

The first issue is that of device parasitics: in reality the active region of a semiconductor device must be accessed via contacts and other parts of the structure which do not play a part in its operation. Furthermore, the device must be formed on a surrounding substrate which may have its own particular electrical properties. These non-functional but unavoidable regions can be treated as electrical parasitics which act in series or in parallel with the device proper. While the actual magnitudes of parasitic capacitances and inductances are very small (typically fractions of a picofarad or picohenry), at high frequencies the impedance of a parallel capacitor reduces to such an extent that it represents a significant parallel conducting path. Alternatively, the impedance of a series inductor increases such that the potential across it cannot be ignored. Parasitic resistance, although not frequency dependent, may vary with applied bias voltage. The overall effect of these parasitic elements is therefore to reduce bias voltages applied to active regions with a fraction of the current bypassing the active region. Although careful layout and the use of appropriate material systems [33] may reduce such effects, they can never be eliminated.

Another, related, issue is that of commensurability of device and circuit size. At a frequency of 1 THz, for example, a typical waveguide would be 50 μm high and it is clear that the presence of a substrate of high relative permittivity (e.g. GaAs) within the waveguide circuit may have a substantial effect on the mode structure. Under these circumstances, it may no longer be appropriate to separate device and circuit and the use of sophisticated electro-magnetic modelling procedures becomes necessary.

The next issue concerns the historic distinction between passive and active devices: passive devices, while having many kinds of functionality, do not add energy to a signal (rather a fraction of the energy is dissipated in the device). On the other hand, active devices have the ability to convert DC to AC and hence generate or amplify a signal. Historically, passive devices have always been found to operate at higher frequencies than active devices. For example, suitably-designed (passive) Schottky Diodes [34] can operate up to 1.5 THz, whereas the upper limit for HEMTs is currently projected to be about 800 GHz [35]. An interesting exception is the RTD which has been demonstrated to act in both passive mode (as a multiplier) and in active mode (as an oscillator, or a self-oscillating mixer with gain [36]) up to several hundred GHz.

Two figures of merit often quoted to characterise the high frequency performance of a device are:

a) The cut-off frequency f_T: for a **passive device** this is the frequency where the function of the device is swamped by the effect of the parasitics; for an **active device** it is the frequency where the short circuit current gain drops below unity.
b) The maximum frequency of oscillation f_{max} is defined for active devices only as that frequency where the power gain drops below unity. f_{max} is usually greater than f_T because power gain involves voltage as well as current.

3.2. SUMMARY OF DEVICE PERFORMANCE

TABLE 1. Solid state device performance at terahertz frequencies.

DEVICE	Refs	f_T (GHz)	f_{MAX} (GHz)	Power at f_{MAX}	Comments
Gunn	[16]	N/A	160	20 mW (CW)	InP based
	[38]	N/A	200	0.2 mW	3rd harmonic
FET	[35]	~ 350	255 (target 800)	2 μW	InP 0.05 μm
HBT	[39]	80 - 170	100 - 218		
RTD	[23]	N/A	712	< 1 μW	InAs/AlSb

Table 1 gives an indication of the high frequency performance of currently available devices. Further comments on some of these and other devices will now be made.

3.2.1. *Gunn Diodes*

The operational principle of this device is discussed at length elsewhere [40]. The materials used must exhibit the transferred electron effect, and InP is in general preferred owing to the higher saturation velocity (v_s) and faster rate of transfer of electrons between the light and heavy mass conduction bands attainable in this material. In practical terms, this device still represents the "workhorse" source of terahertz power when operated with a multiplier chain [41].

The frequency of operation of a Gunn diode is inversely proportional to the device length which is about 1 μm for the InP diodes referred to in the Table. Reducing the active length to below 1 μm (as would be required for operation above 100 GHz) is not difficult but then one encounters the problem of the time taken for electrons to gain the energy needed to transfer to the higher effective mass conduction band. In essence, the transferred electron effect disappears at frequencies much above 100 GHz - 150 GHz owing to the time taken for interband transfer [42] and the device ceases to exhibit a negative differential resistance. The state-of-the-art performance of Gunn diodes [16] has been achieved using n^+nn^+ structures with a graded doping profile, paying strict attention to heat-sinking. Second and third harmonic operation can also be achieved and powers in the region of 1 mW have been demonstrated at about 200 GHz [38].

3.2.2. *Field Effect Transistors*

Field Effect Transistors (FETs) are a range of devices including MESFETS (Metal Semiconductor FETs), HEMTS (High Electron Mobility Transistors) and pHEMTS (pseudomorphic High Electron Mobility Transistors) which have a broad range of applications in electronics with recent developments in the HEMT type structures indicating that they have the potential of operating in the terahertz region. As mentioned earlier, 140 GHz integrated circuits based on pHEMTS have been demonstrated under laboratory conditions and f_{max} values in the region of 800 GHz are predicted [35]. These performance figures have been achieved through advances in crystal growth in a range of material systems particularly in the control of heterojunctions, improvements in electron-beam lithography and a better understanding of the device physics through computer simulation [43]. As these devices are reduced in size in order to function at ever higher frequencies the scaling rules put great demands on the accuracy of the whole fabrication process but the technology is advancing incrementally to meet these challenges [39].

It is therefore expected that three-terminal device operation well into the terahertz region will soon become a practical reality but it must be remembered that these devices are subject to the $1/f^2$ power scaling factor common to all transit time devices. It may therefore be that in their conventional "electronic" form HEMT type devices will be more suited to high speed signal processing applications rather than power generation.

A recent publication [44] has taken a novel view of HEMTs as "Plasma Wave Electronic" devices. In these devices plasma waves are generated in the two dimensional electron gas and as the velocity of these waves is about ten times higher than the saturation velocity for electrons, the restrictions on device dimensions are somewhat relaxed. These are solid state devices which have features in common with travelling wave tubes but although effects attributed to plasma waves have been observed there is still much development work required to realise a practical device.

3.2.3. *Heterojunction Bipolar Transistors*

The cut-off frequencies (see Table 1) of HBTs, while entering the terahertz region, trail behind those for field effect devices. The main frequency determining parameter of the HBT is the base width with the lowest values reported at the time of writing as low as 5 nm. At these values, the carriers (electrons) undergo ballistic transport across the base and surface recombination, which takes place around the edges, is significantly reduced. However, the measured values of f_{MAX} [39] are still only 170 GHz. The HBT is therefore not likely to compete directly with the field effect devices at terahertz frequencies.

On the other hand the vertical structure of the HBT means that it can be configured as a multiple finger device to increase the area of cross section and hence the power handling ability and it is therefore being considered as a driver for frequency multiplication circuits.

3.2.4. *RTDs*

Applications of this device and its variants have been reviewed extensively elsewhere [18, 19, 45] but to date the power output of the RTD when used as an oscillator is disappointingly low. In spite of this, it may yet prove of practical value in high speed logic [46] and certain niche analogue applications where the low power consumption and versatility of layer structure can be exploited. Significant steps have already been made to increase both the peak-to-valley current ratio and the peak current through judicious choice of heterostructure materials [18, 19] and the optimisation of device design. A variety of schemes for coupling power out of the device have been used, including quasi-optics [47], coplanar waveguide [48] and traditional whisker mounting in rectangular waveguide. In an effort to increase the power available, proof of principle of power combining of RTDs has been demonstrated [49] but the exploitation of this approach requires a more careful control of the circuit environment. Recent work on integrated waveguides [11] suggests that a micromachining approach to power-combining circuits is well suited to the simplicity of RTD construction. In addition, the issue of DC stability in the negative differential resistance region of the RTD is highly relevant to any such power-combining schemes, and a more viable alternative than the use of an RTD in the quasi-sinusoidal mode may be to construct a relaxation oscillator [45]. In this type of oscillator circuit, a multi-stable DC bias circuit may be used which results in an easier "start-up" of oscillation [50] and, in principle, lead to the fabrication of large RTD arrays. In view of the rich harmonic content of such an oscillator, the further development of self-oscillating multipliers (akin to mode-locked lasers) may also be a realistic possibility [45].

In all of the above applications, two-terminal devices have been used. However, the RTD may be inserted into a transistor to create a three-terminal device. Although the majority of such applications reported to date are in the area of high-speed logic [46], multipliers using HEMT [51] structures with an integral RTD have been described and a series RTD-FET combination with an output rich in (terahertz) harmonic content [52] has been simulated. Although three-terminal devices may have intrinsically lower frequency limits than their two-terminal counterparts, it is evident that further work in this area is desirable in view of their greater power-handling capability.

Finally, it should be noted that hybrid opto-terahertz devices [53] based on the RTD may prove of value as modulators in conventional opto-electronic systems and, conversely, as optically-controlled terahertz oscillators [54]. Such modulators may have an eventual role in "photonic oscillators" for terahertz application [55].

3.2.5. *Superconducting Josephson Junction Oscillators*

Although room temperature performance of these devices has not been achieved, their power output does already appear sufficient to drive SIS mixer systems in the 600-700 GHz range [24]. Furthermore, the report of a superconducting Josephson Junction (SJJ) device utilising a high temperature superconductor [57] is encouraging, especially in view of the possibility of array fabrication. A performance figure of 0.36 μW for free-space propagation at 190 GHz [58] puts these devices into a similar performance category as the RTDs, with which they also share digital applications [59].

4. Optical/ "Transition" type solid state Terahertz sources

As noted in section 2 above, there are certain fundamental difficulties associated with the fabrication of a room-temperature compact solid state terahertz source relying on radiation-emitting transitions. In this section, a brief review will be given of bulk semiconductor terahertz lasers operating at low temperature and the difficulties attending the realisation of emitters based on semiconductor systems of reduced dimensionality.

4.1. P-GERMANIUM LASER

The only viable laboratory solid-state laser source of terahertz radiation is the p-Ge laser [15, 31, 59] which operates at low temperature and relies on the different dynamics of holes in the valence bands of bulk semiconductor material in crossed electric and magnetic fields. In this device a one-way pumping process takes place where heavy holes are transformed into light holes. To obtain lasing it is necessary to apply a large electric field (a few kV per centimetre) and to use large germanium crystals of exceptional quality. Output powers of the order of 10 mW are, however, obtainable at 5 K with a pump power of 100 W [60]. Research in this area is presently concerned with the improvement of the time structure of the output radiation to achieve either continuous or short-pulse (<100 ps) operation; recently 11μs pulses with a repetition rate of 6.7 kHz have been achieved [61] which represents a considerable advance. Another important goal is to achieve lasing in a more easily obtainable material than high quality germanium. Possible laser action in silicon is being actively pursued at several laboratories: in this material the different mass ratio of the holes should result in operation around 1 THz (rather than 3 THz for p-Ge) and, arguably, emission at 77K. Despite the problems associated with p-Ge lasers they do offer the advantage of tunability and are, in addition, capable of miniaturisation for space-borne applications [61].

4.2. III-V INTERSUBBAND EMITTERS

An attractive route to the fabrication of a room temperature coherent terahertz source may involve intersubband transition effects in multi-quantum wells [62]. Operation of such devices in the 4-8.5 μm range has already been achieved [28] and an important issue is whether this device principle might be extended to longer wavelengths.

Early efforts to obtain terahertz emission from Quantum Well (QW) structures concentrated on the possibility of obtaining an optical gain in a superlattice experiencing an applied electric field with optical transitions between subband states of adjacent quantum wells [63]. Although Helm *et al.* [64] observed emission due to intrawell transitions, superlattice structures are not ideal owing to fundamental problems associated with charge storage effects leading to de-tuning from resonance. A variety of other lasing or emission schemes have been suggested [65] using asymmetric or double wells, but to date there has been little progress towards the attainment of an electrically-pumped device operating at room temperature. It would appear at first sight that the fabrication of a long-wavelength emitter, utilising the cascade-laser or a related principle, would be easier than at shorter wavelengths: this conclusion follows from the anticipated removal of optical phonon transitions as a non-radiative channel for emission below about 30 meV in GaAs and from the increase of the dipole matrix element for intersubband transitions in the wider QWs necessary for such long-wavelength operation. Unfortunately, recent experimental work [66] has demonstrated that the longitudinal optical phonon scattering is only partially suppressed below 30 meV in n-type QWs and, furthermore, any benefits gained from dipole matrix element increases will be lost due to the general reduction in the photon density of states noted in section 2.2.2. Finally, all realistic emitter schemes must eventually incorporate electrical pumping and thus difficulties related to free-carrier absorption in the contacts and space-charge effects must also be overcome. It is noteworthy that only one terahertz emission result [67] has been reported (at sub-nanowatt levels) and that this involved an optical pumping scheme. Nevertheless, despite the manifold difficulties presented here, this is an area which demands much further work, especially in relation to the modelling of device structures (by self-consistent Schrodinger-Poisson solvers) and carrier dynamics (by Monte Carlo techniques) [68].

A comparatively new area concerns terahertz emission from p-type low dimensional semiconductor structures. Recent advances in growth techniques [69] have resulted in GaAs/AlGaAs two dimensional hole systems of exceptional quality. The exploitation of intersubband effects in such hole systems may have many advantages: the effective masses and band-architecture may be varied by suitable choice of sample design to a much greater extent than in the n-type counterparts [26]; the dispersion-relations exhibit parallelism which can be deployed to provide a very high joint density-of-states [70, 71, 72]; the existence of different "ladders" in the hole subband structure could be deployed to provide lasing states far-removed from the ground state so that room-temperature lasing might be achieved; the breakdown of selection rules implies that out-coupling of radiation could be accomplished without recourse to grid-structures [73] and the longer hole lifetimes [74] may, in certain circumstances, point to easier-to-obtain lasing conditions. To date, there is only a limited literature on terahertz emission from such systems [75], although further work in the area of hot-hole emission is clearly indicated.

5. Frequency Multiplied Sources

Frequency (or harmonic) multiplication requires a circuit element that has a non-linear current voltage-characteristic which for illustrative purposes could be of the form

$$I = a_0 + a_1 V + a_2 V^2 + a_3 V^3 \qquad (7)$$

where the coefficients a_0, a_1, a_n are constants. For an applied voltage V, the V^2 term will give a current through the device at twice the applied frequency, V^3 at three times the frequency, and so on. The amplitudes of the various harmonics depend on the values of the coefficients a_n but generally reduce rapidly with increasing harmonic number. Non-linearity can be achieved in a number of ways and all electronic devices exhibit this behaviour in some part of their characteristic.

The principle method of generating power above 100 GHz is through harmonic multiplication in one form or another with its popularity largely due to the maturity of the technique and previous space qualification of the component parts. However, present methods, based on machined waveguide, are costly because the cavities on which they depend require watchmaker levels of precision and "hand-tuning" to give the desired performance. Unique designs of multiplier using "hand-crafted" techniques, have exhibited efficiencies of 12% with output powers of 5 mW at 270 GHz [76]. However, as the frequency increases the degree of manufacturing difficulty (and therefore cost) becomes increasingly high. The solution to these fabrication difficulties may be to use micromachining based on photolithographic techniques and to increase the level of integration.

In harmonic multiplication, the high powers generated from a fundamental mode oscillator are converted to higher harmonic frequencies using varactor diodes [77, 78]. Traditionally, these varactor diodes are of the Schottky contact type, where even in the ideal case, the efficiency of the process drops off inversely with the square of the harmonic number [79]. This behaviour is common to all resistive-type multipliers. In theory, a purely capacitive nonlinear reactance can convert all the input power to any higher harmonic. In practice, the multiplication efficiency is greatly reduced because of parasitic losses in the idling circuits at undesired harmonic frequencies. This is especially true when higher harmonic frequencies are sought. In this case the power at the unwanted lower harmonic frequencies (where the majority of the power is generated), must be "recycled" back into the varactor in order not to be lost. This recycling requires idling circuits (which must appear wholly reactive at these frequencies) positioned to reflect all of this power back to the varactor for conversion to the desired frequency. Subsequently, even in well designed cases, the efficiency of these variable reactive-type multipliers remains relatively low and drops dramatically for the higher harmonic frequencies. This is why such multipliers or combinations thereof are usually restricted to generating harmonic frequencies below the sixth. Double barrier structures have been investigated as harmonic up-converters [36, 80] and show promise for replacing Schottky based multipliers.

The phase noise of multiplied sources depends on the stability and noise of the input signal at the fundamental (or pump) frequency. Subsequently, the phase noise at a predefined offset from the up-converted pump frequency will be degraded compared to that for a similar offset from the pump frequency. Hence, much tighter frequency stability criteria are required for the pump source in harmonic up-converters. Frequency stabilisation of the fundamental mode device is achieved through phase locking to a reference oscillator, either through injection locking to a harmonic of the reference oscillator or through suitable feedback using the phase difference between the reference oscillator and a down-converted component of the fundamental oscillator. Such phase locking can improve the close-to-carrier noise by factors of 50 or more. Feedback can be easily achieved using three terminal devices, but combined arrangements of two-terminal devices such as Gunn diodes and varactors can also satisfy feedback control of the oscillation frequency.

The following sections will discuss briefly three devices that have been shown to have useful non-linear characteristics in the terahertz frequency range.

5.1. SCHOTTKY DIODES

Schottky diodes in the form of the metal/semiconductor contact "cats whisker" used as a detector in the early days of radio were perhaps the first solid state electronic devices. Remarkably they are still in regular use (albeit in a refined form) operating as nonlinear devices over all frequency ranges up to and including terahertz frequencies. In fact Schottky diodes have always been amongst the first devices to be used whenever there has been a need to move up in frequency. Perhaps the reason for this is that they have a simple structure with only two terminals and improvements in technology have made it possible to shrink their dimensions to keep pace with the requirements of ever higher frequencies.

The functionality of the Schottky diode resides in its well known rectifying I/V characteristic which is usually written as:-

$$I = I_o(e^{\frac{qV}{nK_BT}} - 1) \tag{8}$$

where q is the charge on the electron, K_B is Boltzmann's constant and n is the ideality factor representing departures from simple behaviour. I_o is the reverse bias saturation current. In forward bias where $qV > nK_BT$ the characteristic takes an exponential form which can be written as a power series similar to that in equation (7)

Recent improvements in the nanofabrication of Schottky Barrier diodes discussed in [34] have lead to a better understanding of their operation at terahertz frequencies. Practical measurements indicate that their optimum operation as detectors, regardless of frequency, occurs for a fixed number of electrons flowing in and out of the device during one cycle. The device dimensions and doping levels are such that a relatively small number of electrons are involved, typically 2000 to 5000, and the distances they must travel are smaller than the mean free path even at room temperature. The electrons therefore undergo ballistic transport and, owing to the small dimensions of the device, maintain phase coherence. This behaviour is observed experimentally in that the performance of these devices does not change much as they are cooled below room temperature. The importance of the Schottky diode as a room temperature, high frequency detector looks certain to be maintained for the foreseeable future.

5.2. RESONANT TUNNELLING DIODES

The potential of the double barrier Resonant Tunnelling Diode (RTD) as a power source has been discussed in section 3.2.4 where it is pointed out that, although this device can oscillate at frequencies throughout the terahertz range, the power developed is so far too small to be of use in practical applications other than to drive an SIS mixer. Nevertheless the RTD exhibits a high frequency, non-linear current/voltage characteristic which is being investigated for use as a high frequency multiplier [36]. Significantly, the negative differential resistance part of the characteristic is not needed for frequency conversion so the limitations of low power generation do not apply for this mode of operation.

As outlined above, multipliers generally result in power being produced at all multiples of the input frequency (i.e. at both odd and even harmonics) and this can lead to only a limited amount of power at the desired frequency. However, in the case of the RTD the I/V characteristic is anti-symmetric which means that only odd harmonics of the input frequency are generated. This immediately means that there is no power at the second harmonic and therefore more in the third. Similarly no power is converted into the higher order even harmonics. In addition, the RTD has a versatile layer structure and can be implemented in many different material systems; there is therefore great scope in engineering the electrical characteristics to favour the production of a limited number (or just one) of the harmonics. Investigations are currently [81] in progress to integrate a RTD as a non-linear load in HEMT oscillator circuits. These two devices are technologically compatible and the combination has the potential of multiplying the frequency of a HEMT oscillator by a factor of 3 or 5 to the 400-700 GHz range.

Sub-millimetre wave fundamental mode oscillators in high-Q resonant circuits are usually preferable to harmonic up-converted sources for minimising close-to-carrier noise, hence there is still a case for improving present RTD oscillators for fundamental mode sub-millimetre wave power generation.

5.3. QUANTUM BARRIER VARACTOR

When used as a mixer the RTD is in fact a "varistor", with the mixing action being dependent on its non-linear resistance. This process dissipates power in the mixing device resulting in a lowered efficiency. Developments in Sweden and the USA of a new type of varactor multiplier, the Quantum Barrier Varactor (QBV) [82], have suggested that efficiencies in excess of 60% for third harmonic generation are possible. The advantage of using QBVs as multipliers is twofold: firstly, the anti-symmetric current-voltage and symmetric capacitance-voltage characteristics of structurally symmetric quantum barrier devices, result in an intrinsic suppression of even mode harmonics. This leads to improved efficiency, making the QBV diode especially useful as a third harmonic generator at millimetre wave frequencies [83]. Secondly, there is the possibility of "tailoring" the current-voltage and capacitance-voltage characteristics to favour particular harmonics and, more importantly, to produce series stacks of Quantum Barrier Varactor (QBV) diodes to decrease the capacitance per unit area and therefore increase the frequency and power handling capabilities for a given size of device. A stack of three QBV layers in a single device used as a tripler has been reported with an output power of 20 mW m at 186 GHz, an efficiency of 46% and a cut-off frequency of 1.4 THz for a 40 μm diameter device [84].

6. New Device Principles

6.1. HETERODYNE CONVERSION

The principle of difference-frequency mixing to produce a tunable coherent output in the mid-infrared has recently been extended to the far-infrared regime. The attraction of this technique lies in the possibility, now realised [25], of using solid-state lasers as sources. Although quantum wells have been used as the photomixing element [85], the most promising results have been obtained using low temperature grown GaAs epitaxial films with interdigitated electrodes fabricated on the top surface at the centre of a log spiral antenna. Tunability of output can be achieved with relative ease and speed, and useable power levels (approximately $1\mu W$ at 1 THz [86]) have been reported for spectroscopic work [87]. The crucial issue in the further development of this type of source centres on the line width and stability which are achievable; for dye-laser pumps the line width is around 2 MHz which is governed by laser jitter. For solid state laser diodes operating at 850 nm a linewidth of 100 MHz has been obtained with 100 kHz predicted [25]. Further work in this area is likely to centre on the development of more effective photomixers and integration of lasers, perhaps using epitaxial lift-off techniques.

6.2. TRAVELLING WAVE DEVICES

The travelling-wave (TW) tube principle for generation of electromagnetic radiation has a long tradition of application at terahertz frequencies [14] and has recently been applied to terahertz radiation detection [88]. In a TW device, a finely-constructed periodic structure is used to slow down an electromagnetic wave so that energy can be exchanged with an electron beam, usually within a cavity and useful power extracted. Although there do not appear to be any direct implementations of this principle using artificial semiconductor structures, an analogous mechanism has been utilised in a semiconductor patterned with a periodic grating arrangement, so that terahertz radiation can be extracted from a hot-carrier distribution moving beneath the grating [89]. The output power is extremely weak, however, and the device is unlikely to have much technological relevance. In the present Proceedings, a novel application of the TW concept is reported [90] in which power-transfer occurs in an efficient heterojunction phototransistor. This power-transfer takes place along a polyimide channel waveguide and the device itself can, in principle, operate up to hundreds of GHz [91]. At present, operation as an efficient convertor from optical to terahertz frequencies has been demonstrated up to 60 GHz and improvements in coupling efficiency, power saturation and optical gain are now being pursued to higher frequencies. An important issue in this area is the development of non-linear electro-optic materials which may yield TW devices with intrinsically higher bandwidths than other systems. Finally, it should be noted that the TW principle is also applied extensively in four-wave mixing (FWM) arrangements. Very recently, frequency-conversion in FWM in a Fabry-Perot oscillator has been modelled which suggests the possibility of conversion gain up to terahertz frequencies [92].

6.3. OPTOELECTRONIC OSCILLATORS AND MILLIMETER-WAVE MODULATED OPTICAL SYSTEMS

An important principle was introduced by Yao and Maleki [93] who converted light energy into stable and spectrally pure microwave signals. Their device relied on a feed-back loop, in which the output of a pump laser is intensity-modulated with a modulator driven by a photodetector monitoring the laser beam. Fetterman [90] describes developments of this concept using a polymer modulator and HBT phototransistors and there is now a genuine prospect of this type of system operating beyond 100GHz with an unprecedented degree of signal purity. Furthermore, the use of diode lasers may also lead to an all-solid-state terahertz emitter with adequate power for point-to-point operation in LANs.

There is clearly great potential in the application of optically controlled high frequency systems, and within these Proceedings [90], a single tunable laser system with one optical modulator configured as a beam-steered radar system which is capable of extension to achieve passive imaging is also described. Furthermore, efforts are now underway to harness the concept of phase conjugation (in which both the direction and phase factors of an incoming wave are reversed) to millimetre/submillimetre frequencies. Phase conjugation has previously only been used at visible wavelengths.

Further developments in these areas will rely on advances in single HBTs [94] as phototransistors and mixers and on progress in the development of polymer materials for modulator application [95].

6.4. TERAHERTZ TRANSIENT DEVICES

The concept of the Hertzian oscillating dipole as an emitter of terahertz radiation was first demonstrated by Auston *et al.* [96]; subsequent developments of this idea have led to terahertz imaging systems [5]. Common to these arrangements is the use of a fast-pulse optical stimulus which generates a transient current in an appropriate photoconductor, usually decorated with an antenna pattern to enable the terahertz component to radiate into free-space. Using optical techniques, the radiation pattern is focused onto a device under test which may itself be a receiver or a device through which the radiation is transmitted to a photoconductive or electro-optic receiver. These systems offer the possibility of imaging the chemical compositions of solid and liquid samples at a scale of a few hundred micrometres. Since fatty tissue is essentially transparent to terahertz radiation, this technique may offer considerable advantages for the diagnosis of skin cancers and monitoring the healing of burns. One additional advantage of the transient terahertz concept is that powerful signal analysis techniques can be implemented to improve data quality and to provide both amplitude and phase information. Another related area of interest, reviewed in the present Proceedings [97] is that of spatial electric-field mapping of integrated circuits for failure-analysis using freely-positionable electro-optic probes of terahertz radiation. Frequency-response information may also be obtained using such techniques, for example the behaviour of the effective permittivity of coplanar waveguide up to several-hundred GHz.

A wider exploitation of these transient techniques clearly depends on improvements in the efficiency of the optically-gated terahertz transmitter. Low-temperature grown GaAs [98], decorated with a suitable antenna arrangements [99] is at present widely used.

However, there now appear to be advantages to be gained from the use of organic poled polymers which have low-loss and dispersion and also large non-linear coefficients [100]. Another development of vital importance in the use of this transient technology is likely to be the replacement of titanium-sapphire lasers with suitable solid-state sources. Very recently, Ironside [101] has described the fabrication of multiple colliding-pulse modelocking solid-state lasers which may be suitable for this application in view of their high efficiency, short pulse-length and compact size. The integration of such lasers with appropriate transmitter structures may offer the best route to an all solid state compact terahertz imaging system.

The technique of time-resolved terahertz spectroscopy, discussed above, has also been used to investigate [102] the coherent emission of radiation from Bloch oscillation (BO) processes in semiconductors. Following recent observations of room-temperature BO emission, the possibility of a useful technological implementation of this concept now appears likely. The search for BO has a long history in solid-state physics, but the realisation of the effect awaited the development of artificial superlattice structures over the last two decades. In a Bloch oscillator an electron in a uniform electric field E undergoes periodic oscillations with a frequency $f_B = qEd/h$, where d is the (super)lattice constant, but to date, the only measurable output has been obtained with optical excitation. In this process, a superlattice is designed and biassed so that the energy difference between the lowest conduction band state in a given well and the lowest states in adjacent wells is qEd. At least two such conduction band states are populated by ultra-short optical pulses and the different evolution of the wave-functions leads to temporal and spatial oscillations at the BO frequency f_B. Of particular relevance to technological development has been the recent discovery of superradiance [103], i.e. the charge oscillations are fully phase-locked . This discovery indicates that the use of a cavity may be advantageous so that a solid-state terahertz amplifier might result. It is noteworthy that, to date, the best obtainable power levels for the equivalent of continuous operation are around $1\mu W$; clearly the use of a solid-state maser technique would lead to larger output powers. Again, it is appropriate to remark that electrical device pumping, or perhaps integration with a solid-state laser source, is desirable for any commercial exploitation of this most venerable of solid-state concepts.

6.5. OTHER NEGATIVE RESISTANCE DEVICES

Gunn diodes derive their negative resistance behaviour from the slowing down of the conduction electrons as they transfer from the light to the heavy mass conduction band and, as mentioned previously, the upper frequency of operation is limited by the time taken for these electrons to gain sufficient energy to make the transfer [42]. There are however other ways of slowing down the electrons and one of these is to make use of velocity overshoot. In this case, hot electrons injected into a semiconductor will have a velocity which is not in equilibrium with the lattice and will in fact be higher than the equilibrium value. As the electrons equilibrate with the lattice they therefore loose speed giving rise to a negative resistance effect. Theoretical calculations [104] suggest that $0.25~\mu$m length InP devices could be expected to oscillate in the 350 - 700 GHz range with powers up to 100 mW at 500 GHz. This performance has not yet been demonstrated experimentally. If this mechanism for negative resistance can be shown to be feasible then it is quite possible that terahertz emitters could be made from silicon since the light and heavy mass conduction band structure of the III-V semiconductors will not be necessary.

7. Terahertz Systems

7 1. POWER REQUIREMENTS OF TERAHERTZ DEVICES

So far, this chapter has looked at the possibilities and prospects of producing terahertz oscillations in solid state devices. It is now pertinent to examine how much power might be required in terahertz electronic systems. Two major cases arise; local oscillators in heterodyne receivers and free space power transmission.

7.1.1. *Local Oscillator Power Required for Mixing at Terahertz Frequencies*
The low power levels obtained from solid-state devices at terahertz frequencies become a particular problem when trying to find one suitable to provide a local oscillator (LO) signal to pump a mixer. All diode mixer performance factors are sensitive to the local oscillator power level with, for example, a maximum junction conductance variation required for minimum conversion loss (linked strongly to noise figure) and input and output impedances dependent on junction conductance and capacitance and hence LO power. The variation in diode junction capacitance drops rapidly if the peak LO voltage (V_{LO}) is below the voltage knee of the device I-V characteristic (approximately 0.65 V for a GaAs Schottky diode). This causes the conversion loss (and also the noise figure) to rise rapidly as the LO level is reduced [105].

At microwave frequencies, little effort is usually made to match the LO port to the mixer diode as the LO level can be increased to overcome any inefficiency and make sure the device is sufficiently pumped. However at terahertz frequencies this option is not available and so local oscillator matching becomes important to minimise reflections and hence maximise the power delivered to the device. Matching occurs when the impedance presented by the mixer diode is the complex conjugate of the LO source impedance. Under this condition the power delivered to the device is the same as that available from the local oscillator (P_{LO}):

$$P_{LO} = \frac{V_{LO}^2}{8Re(Z_S)} \tag{9}$$

In equation (9), V_{LO} is the voltage generated by the local oscillator and $Re(Z_S)$ is the real part of its impedance. In practice, mismatch does occur and minimum conversion loss is usually achieved for LO powers of around 10 mW. The deterioration in conversion loss with reduced LO power is generally less rapid if dc bias is applied to the mixing device as this moves the operating point closer to the voltage knee. Under these conditions LO powers of a few mW are generally required.

7.1.2. *Transmitter Power Required for Terahertz Communications Link*
The transmitter power (P_T) required to form a terahertz communications link depends on the power required at the receiver (P_R) as well as a number of other factors [106] and is

given by:-

$$P_T = \frac{P_R \, \lambda_0^2 \, d^2}{A_T \, A_R \, L_0 \, L_r \, L_e} \tag{10}$$

A_T and A_R are the effective apertures of the transmitting and receiving antennas, d is the distance between the antennas, λ_0 represents wavelength, and the attenuation per km due to atmospheric gases, rain, and equipment losses are given by L_0, L_r and L_e respectively. The minimum power necessary at the receiver is determined by noise present in the atmosphere (P_A), the detector noise figure (F_R), and the signal-to-noise ratio desired (SNR_{MIN}) such that:

$$P_R = P_A \, F_R \, SNR_{MIN} \tag{11}$$

If the noise present in the atmosphere is assumed to be simply thermal, then P_A is given by $K_B T B$, where T is the absolute temperature and B is the detector bandwidth. The signal-to-noise ratio depends on the bit error rate (BER) required and on the type of modulation used, but a for a variety of modulation techniques and a BER of 10^{-5} a desirable signal-to-noise ratio is 10 dB [106]. So for example, if we assume a temperature of 290K, a bandwidth of 4 MHz and a receiver noise figure of 15 dB, this gives a required receiver power of -93 dBm.

If we further assume matching 10 cm diameter paraboloidal reflector antennas (using corrugated or other types of feed horn), which have effective apertures equal to πr^2, this gives A_T and A_R to be 7.854×10^{-3}. Taking a frequency of 200 GHz ($\lambda_0 = 1.5$ mm), the attenuation due to oxygen and water vapour is approximately 2dB/km [106]. Assuming an indoor communication link (i.e. $L_r = 0$ dB) and 100% efficient antenna but taking equipment losses to be 1 dB, the transmitter power required for a distance of 100 m works out to be -58 dBm or 1.5μW. Obviously for a system with different parameters (e.g. larger bandwidth, lower antenna efficiency) this figure could be substantially higher.

7.2. CONCLUSION

In order to develop commercially viable terahertz systems the basic building blocks familiar in conventional electronic circuits need to be implemented. In addition to interconnects and resistive loads these include detectors, multipliers, mixers, sources, amplifiers and switches. The first three are passive devices and, as discussed in this chapter, can already be realised in solid state from making use of the non-linear electrical properties of either Schottky diodes or QBVs. Detectors, mixers and sources are at the heart of heterodyne systems, and fast (pico second) switches are required to increase the speed of information transfer in high speed digital systems.

At present there is no single device that could be used as a terahertz source across the whole frequency range. Electronic devices with performance targets up to 800 GHz are closing the gap from below and at the same time intensive development work in solid state laser technology using new multilayer structures and nonlinear optical materials is pushing the output of these devices to longer and longer wavelengths. Similar comments apply to amplifiers which have many principles in common with sources.

The outlook for terahertz systems, depending as it does on the development of active devices, therefore looks very promising at this time. It is an area of new developments in solid state and optical physics, materials technology and engineering and electronics. The necessary developments are on the verge of practical realisation and once such systems are given a practical demonstration then the applications mentioned at the beginning of this chapter will be implemented and no doubt others not yet thought of. The remainder of this book will examine the whole terahertz area including basic physics, materials technology, engineering and different areas of application.

8. References

1. Phillips, T.G. and Keene, J. (1992) Submillimetre Astronomy, *Proc. IEEE* **80**, 1662-1678.
2. Waters, J.W. (1993) Microwave Limb Sounding, in M.A. Janssen (ed.), *Atmospheric Remote Sensing*, John Wiley and Sons Inc., New York, pp. 383-395.
3. Stott, P.E., Gorina, G., and Sindoni, E. (1996) Diagnostics for Experimental Thermonuclear Fusion Reactors, Plenum Press, New York.
4. For reviews of recent work, consult the annual conference series: *International Conference on Infrared and Millimeter Waves*.
5. Hu, B.B. and Nuss, M.C. (1995) Imaging with Terahertz Waves, *Optics Letters* **20**, 1716-1720.
6. Cronin, N.J. (1996) Terahertz systems: the demands on devices, *Phil. Trans. R. Soc. Lond.* **A354**, 2425-2433.
7. Carli, B. (1997) Submillimetre Measurements from Satellites, *this Volume*.
8. Takimoto, Y. (1985) Recent activities on millimeter-wave indoor LAN system development, IEEE NTC '95-The Microwave Systems Conference, Orlando (USA), pp 7-10.
9. Meinel, H.H. (1995) Commercial Applications of Millimeter Waves: History, Present Status and Future Trends, *IEEE Trans. MTT.* **43**, 1639-1653.
10. Röser, H.P. (1991) Heterodyne Spectroscopy for Submillimeter and FIR Wavelengths from 100 μm to 500 μm, *IR Phys.* **32**, 385-407.
11. Brown, D.A., Treen, A.S. and Cronin, N.J. (1994) Micromachining of Terahertz Waveguide Components with Integrated Active Devices, *Proc. 19th Intl. Conf. IR and MM Waves,* Sendai, Japan, 359-360.
12. Van Amersfoort, P.W., Bakker, R.J., Bekkers, J.B. *et al.* (1992) First Lasing with FELIX, *Nucl. Instrum. Methods* **A318**, 42-50.
13. Chantry, G.W. (1971) *Submillimetre Spectroscopy*, Academic Press, London.
14. Benford, J. and Swegle, J. (1992) *High-Power Microwaves*, Artech House, Boston.
15. Andronov, A.A. (1987) Hot Electrons in Semiconductors and Submillimeter

22

Waves, *Sov. Phys. Semicond.* **21,** 701-721.

16. Eisele, H. and Haddad, G.I. (1995) High Performance InP Gunn devices for fundamental mode operation at D-band, *IEEE Microwave and Guided Wave Letters* **5**, 385-387.

17. Syme, R., Kelly, M. and Higgs, A. (1993) New Day Dawning for Tunnelling Devices?, *Physics World* **5**, 22-23.

18. Liu, H.C. and Sollner, T.C.L.G. (1994) High Frequency Resonant-Tunneling Devices, in R.A.Kiehl and T.C.L.G Sollner (eds.), *Semiconductors and Semimetals Vol. 41*, Academic Press, New York, pp. 359-418.

19. Sollner, T.C.L.G. (1994) High-Speed Resonant-Tunneling Diodes, in N.G. Einspruch and W.R. Frensley (eds.), *Heterostructures and Quantum Devices*, Academic Press, New York, pp. 305-337.

20. Horikoshi, Y.(1985) Semiconductor Lasers with Wavelengths Exceeding 2 μm , in W.T. Tsang (ed.), *Semiconductors and Semimetals Vol 22C*, Academic Press, New York, pp. 93-151.

21. Zeuner, S., Keay, B.J., Allen, S.J., Maranowski, K.D., Gossard, A.C., Bhattacharya, U. and Rodwell, M.J.W. (1996) Transition from Classical to Quantum Response in Semiconductor Superlattices at THz Frequencies, *Phys. Rev. B* **53**, R1717-R1720.

22. Truscott, W.S. (1977) From Quantum Mechanics to S-Parameters, *this Volume*.

23. Brown, E.R., Söderstrom, J.R., Parker, C.D., Mahoney, L.J., Molvar, K.M. and McGill, T.C. (1991) Oscillations up to 712 GHz in InAs/AlSb Resonant Tunneling Diodes, *Appl. Phys. Lett.* **58**, 2291-2293.

24. Beaudin, G. and Encrenaz, P.J. (1997) Fundamentals of Receivers for Terahertz Systems, *this Volume*.

25. Brown, E.R., McIntosh, K.A., Nichols, K.B. and Dennis, C.L. (1996) Photomixing up to 3.8 THz in low temperature-grown GaAs, *Appl. Phys.Lett.* **66**, 285-287.

26. Cole, B.E., Batty, W., Imanaka, Y., Shimamoto, Y., Singleton, J., Chamberlain, J.M., Miura, N., Henini, M. and Cheng., T. (1995) Effective Mass anisotropy in GaAs-(Ga,Al)As two dimensional hole systems, *J.Phys.:Condens. Matter.* **7**, L675-681.

27. Darnhofe, T., Roessler, U. and Broido, D. (1995) Far Infrared Response of Holes in Quantum Dots, *Phys. Rev. B,* **52**, R14376-R14379.

28. Faist, J., Capasso, F., Sivco, D.L., Sirtori, C., Hutchinson, A.L. and Cho, A. Y. (1994) Quantum Cascade Laser, *Science* **264**, 553-556.

29. Smet, J.H., Fonstad, C.G. and Hu,Q. (1996) Intrawell and Interwell Intersubband transitions in MQWs for Far Infrared Sources, *J. Appl. Phys.* **79,** 9305-9320.

30. Berger, V. (1994) Three-level laser based on intersubband transitions in asymmetric quantum wells: a theoretical study, *Semicond. Sci. Technol.* **9**, 1493-1499.

31. Gornik, E. and Andronov, A.A. (1991) Far-infrared semiconductor lasers, *Optical and Quantum Electronics* **23**, S111-S349.

32. Hu., Q. and Feng, S. (1991) Feasibility of Far-Infra Lasers using Multiple Semiconductor Quantum Wells, *Appl. Phys. Lett.* **59**, 2923- 2925

33. Lippens, D. (1997) Materials Issues for New Devices, *this Volume.*

34. Röser, H. P. (1997) Schottky Barrier Devices for Terahertz Applications, *this Volume.*

35. Rosenbaum, S.E., Kormanyos, B.K., Jelloin, L.M., Matloubian, M., Brown, A.S., Larson, L.E., Nguyen, L.D., Thompson, M.A., Katehi, L.P.B. and Rebeiz, G.M. (1995) 155 and 213 GHz AlInAs/GaInAs/InP HEMT MMIC Oscillators, *IEEE Trans. MTT.* **43**, 927-932.

36. Sammut, C.V. and Cronin, N.J. (1992) Comparison of measured and computed conversion loss from a resonant tunnelling device multiplier, *IEEE. Microwave and Guided Wave Letters* **2**, 486-488.

37. Millington, G., Miles, R.E., Pollard, R.D., Steenson, D.P. and Chamberlain, J.M. (1993) A Resonant Tunnelling Diode Self-oscillating Mixer with Conversion Gain, *IEEE Microwave and Guided Wave Letters* **1**, 320-322.

38. Rydberg, A. (1990) High Efficiency and Output Power from Second and Third Harmonic Oscillators at Frequencies above 170 GHz, *IEEE Electron Device Letters* **EDL-11,** 439-441.

39. Ito, H., Yamahata, S., Shigekama, N. and Kurishina, K. (1996) High F_{MAX} Carbon-Doped Base InP/InGaAs Heterjunction Bipolar Transistors Grown by MOCVD, *Electronics Letters* **32**, 1415-1416.

40. Monaco, F. (1989) *Introduction to Microwave Technology*, Merril Publishing Company, Columbus, Ohio.

41. Crowe, T.W., Grein, T.C., Zimmerman, R. and Zimmerman, P. (1996) Progress Toward Solid-State Local Oscillators at 1 THz, *IEEE Microwave and Guided Wave Letters* **6**, 207-208.

42. Friscourt, M-R., Rolland, P-A., Cappy, A., Constant, E., and Salmer, G. (1983) Theoretical contribution to the Design of Millimeter-Wave TEO's *IEEE Transactions on Electron Devices* **ED-30,** 223-229.

43. Steenson, D.P. (1996) *Phil. Trans. R. Soc. Lond.* **A354**, 2435-2446.

44. Dyakonov, M.I. and Shur, M.S. (1996) Plasma Wave Electronics: Novel Terahertz Devices using a Two Dimensional Electron Fluid, *IEEE Electron Devices* **43,** 1640-1645.

45. Brown, E.R. and Parker, C.D. (1996) Resonant Tunnel Diodes as Submillimetre -Wave Sources, *Phil. Trans. R. Soc. Lond.* **A354**, 2365-2381.

46. Yokohama, N., Iamura, K., Takatsu, M., Mori, T., Adachihara, T., Sugiyama, Y., Sakuma, Y., Tackeuch, A. and Muto, S. (1996) Resonant Tunnelling Hot Electron Transistors: present status and future prospects, *Phil. Trans. R. Soc. Lond.* **A354**, 2399-2411.

47. Brown, E.R., Parker, C.D., Molvar, K.M. and Stephan, K.D. (1992) A Quasioptically Stabilized Resonant-Tunneling-Diode Oscillator for the Millimeter-Wave and Submillimeter-Wave Regions, *IEEE Trans. MTT.* **40**, 846-850.

48. Lheurette, E., Grimbert, B., Francois, M., Tilmant, P., Lippens, D., Nagle, J. and Vinter B. (1992) InGaAs/GaAs/AlAs Pseudomorphic Resonant Tunnelling Diodes integrated with an Airbridge *Electronics Letters* **28**, 937-938.

49. Miles, R.E., Steenson, D.P., Pollard, R.D., Chamberlain, J.M. and Henini, M. (1994) Power Combining of Double Barrier Resonant Tunnelling Diodes at W-

Band, in H. Goronkin and U. Mishra (eds.), *Proc. 21st Symposium on Compound Semiconductors: IOP Conf. Ser.141* , IOP Publishing, Bristol, pp. 679-684.

50. Boric-Lubecke, O., Pan, D-S. and Itoh, T. (1995) Fundamental and Subharmonic Excitation for an Oscillator with Several Tunneling Diodes in Series, *IEEE Trans. MTT*. **43**, 969-975.

51. Chen, K. J. and Yamamoto, M. (1996) Frequency Multipliers using InP-Based Resonant-Tunneling High Electron Mobility Transistors, *IEEE Electron Device Letters* **EDL-17**, 235 - 238.

52. Janssen, G., Prost, W., Reuter, U. K., Auer, U., Schröder, W. and Tegude, F.J. (1996) Modelling the Properties of a novel 3D integrated RTD/HFET frequency multiplier, International Workshop on Millimeter Waves, Orvieto, Italy.

53. McMeekin, S.G., Taylor, M.R.S., Vögele, B., Stanley, C.R. and Ironside, C.N.C. (1994) Franz-Keldysh effect in an optical waveguide containing a resonant tunneling diode, *Appl. Phys. Lett.* **65**, 1076-1078.

54. Kazemi, H. and Miles, R.E. (1996) Proceedings International Symposium on Quantum Devices and Circuits, Alexandria, Egypt. (In Press)

55. Neyer, A. and Voges, E. (1982) Dynamics of Electrooptic Bistable Devices with Delayed Feedback, *IEEE J. Quantum Electronics* **QE-18**, 2009-2015.

56. Yao, X.S. and Maleki, L.O. (1996) Converting light into spectrally pure microwave oscillation, *Optics Letters* **21**, 483-485.

57. Martens, J.S., Pance, A., Char, K., Lee, L. and Whiteley, S. (1993) Superconducting Josephson Arrays as tunable microwave sources operating at 77K, *Appl. Phys. Lett.* **63**, 1681-1683.

58. Wengler, M.J., Guan, B. and Track, E.K. (1995) 190-GHz Radiation from a Quasioptical Josephson Junction Array, *IEEE Trans. MTT*. **43**, 984-988.

59. Wenckebach, W.Th. (1997) Hot Hole Lasers, *this Volume*.

60. Kimmitt, M. F. (1996) Private Communication.

61. Bründermann, E., Linhart, A.M., Röser, H.P., Dubon, O.D., Hansen, W.L. and Haller, E.E. (1996) Miniaturisation of p-Ge Lasers: Progress towards continuous wave operation, *Appl. Phys. Lett.* **68**, 1359-1361.

62. Harrison, P. (1997) Device Physics of Intersubband Lasers, *this Volume*.

63. Allen, S.J., Brozak, G., Colas, E., DeRosa, F., England, P., Harbison, J., Helm., M., Florez, L. and Leadbeater, M. (1992) Far-infrared emission and absorption by hot carriers in superlattices, *Semicond. Sci. Technol.* **7**, B1-B5.

64. Helm, M., England, P., Colas, E., DeRosa, F., and Allen, S.J. (1989) Intersubband Emission from Semiconductor Superlattices Excited by Sequential Resonant Tunnelling, *Phys. Rev. Lett.* **63**, 74-77.

65. Helm, M. (1995) Infrared-Spectroscopy and Transport of Electrons in Semiconductor Superlattices, *Semicond. Sci. Technol.* **10**, 557-575.

66. Murdin, B.N., Langerak, C.J.G.M., Helm, M., Kruck, P., Heiss, W., Rosskopf, G., Strasser, G., Gornik, E., Dür, M., Goodnick, S.M., Lee, S-C., Galbraith, I. and Pidgeon, C. (1996) Time resolved studies of intersubband relaxation in GaAs/AlGaAs quantum wells below the optical phonon energy using a free electron laser, *Superlattices and Microstructures* **19**, 17-24.

67. Bales, J.W., McIntosh, K.A., Sollner, T.C.L.G., Goodhue, W.D. and Brown,

E.R. (1990) Observation of Optically Pumped Intersubband Emission from Quantum Wells, *SPIE Quantum Wells and Superlattices III* **1283**, 74-81.

68.   Harrison, P. and Kelsall, R.W. (1996) Theoretical studies of subband carrier lifetimes in an optically pumped 3-level terahertz laser, *Superlattices and Microstructures* **20**, 1-4.

69.   Henini, M., Rodgers, P.J., Crump, P.A., Gallagher, B.L. and Hill, G. (1995) The Growth and Physics of Ultra-High-Mobiliity 2-dimensional Hole Gas on (311) GaAs Surface, *Crystal Growth* **150**, 451-454.

70.   Cole, B.E., Chamberlain, J.M., Henini, M., Cheng, T., Ardavan, A., Polisski, A., Hill, S.O., Singleton, J., Batty, W., Nakov, V. and Gobsch, G. (1996) Collective Motion in GaAs/(Al,Ga)As 2D Hole System: A two component Fermi liquid, *Proceedings of the 23rd International Conference on the Physics of Semiconductors*, World Scientific, Singapore, pp. 2475-2478.

71.   Cole, B.E., Hill, S.O., Imanaka, Y., Shimamoto, Y., Batty, W., Singleton, J., Chamberlain, J.M., Miura, N., Henini, M. and Cheng, T. (1996) Effective mass anisotropy and many body effects in 2D GaAs/(Ga,Al)As hole gases observed in very high magnetic fields: comparison of theory and experiment, *Surface Science* **361/362**, 464-467.

72.   Cole, B.E., Chamberlain, J.M., Henini, M., Cheng, T., Batty, W., Wittlin, A., Perenboom, J.A.A.J., Ardavan, A., Polisski, A. and Singleton, J. (1997) Cyclotron Resonance in ultra-low- hole-density narrow p-type GaAs/(Al,Ga)As quantum wells, *Physical Review B,* in Press.

73.   Kim, B.W. and Majerfeld, A. (1995) Electronic and intersubband properties of p-type GaAs/AlGaAs superlattices for infrared photodetectors, *J.Appl.Phys.* **77**, 4552-4563.

74.   Xu, Z., Wicks, G.W., Rella, C.W., Schwettman, H.A. and Fauchet, P.M. (1996) Temperature dependence of the intersubband hole relaxation time in p-type quantum wells, in K.Hess *et al.* (eds.), *Hot Carriers in Semiconductors*, Plenum Press, New York, pp. 65-68.

75.   Shayesteh, S.F., Dumelow, T., Parker, T.J., Mirjalili, G., Vorobjev, L.E., Donetsky, D.V. and Kastalsky, A. (1996) Far infrared spectra of reflectivity, transmission and hot-hole emission in p-doped GaAs/AlGaAs multiple quantum wells, *Semicond. Sci. Technol.* **11**, 323-330.

76.   Erickson, N.R. (1982) High Efficiency Frequency Tripler for 230GHz. *Proceedings of 12th. European Microwave Conference,* Helsinki, Finland, 13-17 Sept, 1982, 288-292.

77.   Kollberg, E. L., Stake, J., Dilner, L., *Phil. Trans. R. Soc. Lond.* **A354**, 2383-2398.

78.   Brauchler, F.T., East, J.R. and Haddad, G.I. (1994) Novel Varactor Diode Structures for improved Power Performance, in F.T. Ulaby and C. Kukkonen (eds.), *Proceedings of 5th International Symposium on Space Terahertz Technology*, pp. 460-473.

79.   Kollberg, E. and Rydberg, A. (1989) Quantum-well high-efficiency millimetre-wave frequency tripler, *Electronics Letters* **25**, 348-349.

80.   Tanguy O., Lippens, D., Burston, J., Pernot, J.C., Beaudin, G., Nagle, J. and

Vinter, B. (1994) Frequency conversion to 368 GHz using Resonant Tunnel Diodes, in F.T. Ulaby and C. Kukkonen (eds.), *Proceedings of 5th International Symposium on Space Terahertz Technology*, pp. 524-530.

81. Chen, K.J., Maezawa, K., and Yamamoto,M. (1995) Novel current voltage characteristic of an n-InP based RTD-HEMT *Appl. Phys.Lett.* **67**, 3608-3611.

82. Kollberg, E. and Rydberg, A. (1989) Quantum-barrier-varactor diodes for high-efficiency millimetre wave-wave multipliers, *Electronics Letters* **25**, 1696-1698.

83. Hong-Xia, L., Liu, L.B., Sjogren, L.B., Domier, C.W., Luhmann, N.C., Sivco, D.L. and Cho, A.J. (1993) Monolithic Quasi-opical Frequency Tripler Array with 5-W Output Power at 99GHz., *IEEE Electron Device Letters* **EDL-14**, 329-331.

84. Krishnamurthi, K., Harrison, R., Rogers, C., Ovey, J., Nilsen, S. and Missous, M. (1994) Stacked Heterostructure Barrier Varactors on InP for Millimeter Wave Triplers, *Proceedings of the 24th European Microwave Conference*, Cannes, France, Sept 1994, 758-763.

85. Liu, H.C., Li, J.M., Brown, E.R., McIntosh, K.A., Nichols, K.B. and Manfra, M. (1995) Quantum-Well Intersubband Heterodyne Infrared Detection up to 82 GHz, *Appl. Phys. Lett.* **67**, 1594-1596.

86. Brown, E.R., McIntosh, K.A., Nicholls, K.B. and Dennis, C.L. (1995) Photo-mixing up to 3.8 THz in Low-Temperature Grown GaAS, *Appl. Phys. Lett.* **66**, 285-287.

87. Pine, A.S., Suenram, R.D., Brown, E.R. and McIntosh, K.A. (1996) A Terahertz Photomixing Spectrometer, *J.Mol. Spectroscopy* **175**, 37-47.

88. Sigg, H., Kwakernak, M., Margotte, B., and Erni, D. (1995) Ultrafast far-infrared GaAs/AlGaAs photon drag detector in microwave transmission line topology, *Appl. Phys. Lett* **67**, 2827-2829.

89. Gornik, E., Boxleitner, W., Rosskopf, V., Hauser, M., Wirner, C. and Weimann, G. (1994),Smith-Purcell Effect in GaAs/AlGaAs Heterostructures, *Superlattices and Microstructures* **15**, 399-404.

90. Fetterman, H.A. (1997) Lightwave/Terahertz Interaction, *this Volume*.

91. Frankel, M.Y., Caruthers, T.F. and Kyono, C.S. (1995) Analysis of Ultrafast Photocarrier Transport in AlInAs-GaInAs Heterojunction Bipolar Transistor, *IEEE J. Quant. Electronics* **31**, 278-285.

92. Bava, G.P., Debernardi, P. and Osella, G. (1996) THz frequency conversion in injection locked semiconductor laser oscillators, *IEEE Proc. Optoelectronics* **143**, 41-47.

93. Yao, X.S. and Maleki, L. (1996) Converting Light into spectrally pure microwave oscillation, *Optics Letters* **21**, 483-485.

94. Bhattacharya, D., Bal, P.S., Fetterman, H.R. and Streit, D. (1995) Optical Mixing in Epitaxial Lift-off Pseudomorphic HEMTs, *Photonics Tech. Lett.* **7**, 1171-1173.

95. Wang, W., Chen, D. and Fetterman, H.R. (1995) Optical Heterodyne-Detection of 60 GHz Electrooptic Modulation from Polymer Waveguide Modulators, *Appl. Phys. Lett.* **67**, 1806-1808.

96. Auston, D. H., Cheung, K. P. and Smith, P. R. (1984) Picosecond Photoconducting Hertzian Dipoles, *Appl. Phys. Lett.* **45**, 284-286.

97.   Roskos, H.G., Pfeifer,T., Heiliger, H-M., Löffler, T. and Kurz, H. (1997) Electro-optic and photoconductive techniques for probing and imaging of THz electric signals, *this Volume*.

98.   Eaglesham, D.J., Pfeiffer, L.N., West, K.W. and Dykaar, D.R. (1991) Limited thickness epitaxy in GaAs molecular beam epitaxy near 200°C, *Appl. Phys. Lett.* **58**, 65-67.

99.   Dykaar, D.R., Greene, B.I., Federici, J.F., Levi, A.F.J., Pfeiffer, L.N. and Kopf, R.F. (1991) Log-periodic antennas for pulsed terahertz radiation, *Appl. Phys. Lett.* **59**, 262-264.

100.   Nahata, A., Wu, C., Yardley, J.T. and Auston, D.H. (1995) Generation of Terahertz Radiation from a Poled Polymer, in *Ultrafast Electronics and Optoelectronics*, Optical Society of America Technical Digest Series **13**, PD1-2 -PD1-4.

101.   Ironside, C. (1997) Multi-Gigahertz Optoelectronic Devices, *this Volume*.

102.   For a review, see: Kurz, H., Roskos, H.G., Dekorsky, T. and Köhler, K. (1996) Bloch Oscillations, *Phil. Trans. R. Soc. Lond.* **A354**, 2295-2310.

103.   Martini, R., Klose, G., Roskos, H.G., Kurz, H., Hey, R., Grahn, H.T. and Ploog, K. (1997) *Proceedings of the 23rd International Conference on the Physics of Semiconductors*, World Scientific, Singapore pp. 1771-1774.

104.   Mitin, V, Guzinskis, V, Starikov, E. and Shiktorov, P. (1994) Submillimeter Wave Generation and Noise in InP diodes, *J. Appl. Phys* **75**, 935-941.

105.   Maas, S. A. (1986) Microwave Mixers, *Artech House Inc, Massachusetts*.

106.   Bhartia, P. and Bahl, I. J. (1984) Millimeter Wave Engineering and Applications, *John Wiley and Sons Inc., New York*.

MICROWAVE MIXERS

N.J. CRONIN
School of Physics,
The University of Bath,
Claverton Down, Bath BA2 7AY, UK.

1. Basic Principles

The function of a mixer is to reduce the frequency of a signal by beating (or mixing) it with a second locally generated signal from a Local Oscillator (LO).

Mixing takes place in a non-linear device - any non-linearity will do in principle, e.g. consider a square law device, where the current (I) and voltage (V) are related by

$$I = \alpha V^2 \qquad \alpha \text{ is a constant}$$

Suppose that we apply a voltage given by:

$$V = V_s \cos(\omega_s t) + V_{LO} \cos(\omega_{LO} t)$$

$$\uparrow \qquad\qquad \uparrow$$

signal   local oscillator

The resulting current in the device is:

$$I = \alpha (V_s \cos(\omega_s t) + V_{LO} \cos(\omega_{LO} t))^2$$

expanding this expression gives:

$$I = \alpha \{ V_s^2 + V_{LO}^2 + V_s^2 \cos(2\omega_s t) + V_{LO}^2 \cos(2\omega_{LO} t)$$
$$+ 2V_s V_{LO} \cos[(\omega_s + \omega_{LO})t]$$
$$+ 2V_s V_{LO} \cos[(\omega_s - \omega_{LO})t] \} / 2$$

Note: The non-linearity of the device has produced currents at frequencies other than the two driving frequencies ω_s and ω_{LO}.

In particular, the frequency $(\omega_s - \omega_{LO})$ is called the intermediate frequency (IF). It is this component which is usually considered to be the output from the mixer.

A basic heterodyne downconverter using the mixer would be:

J.M. Chamberlain and R.E. Miles (eds.), New Directions in Terahertz Technology, 29–51.

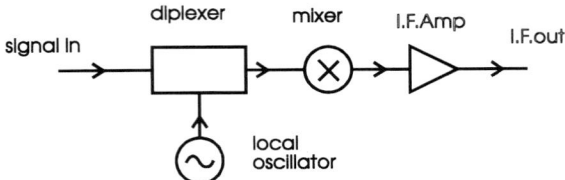

The diplexer is a device which overlays the signal and local oscillator before they enter the mixer. The IF amplifier amplifies only the IF rejecting all high frequency components such as $cos(2\omega_s t)$ etc as well as DC terms.

If the amplitude and phase of the LO are kept fixed, then the IF output contains all the amplitude/phase information in the signal.

2. Mixer Noise Temperature and Conversion Loss

The two parameters used to characterise the performance of a mixer are the *noise temperature*, T_N and the *conversion loss*, L_0.

If we consider a resistor at temperature T, the available noise power per unit bandwidth at its terminals is given by:

$$P_N = kT$$

Using this, any component generating white noise can be assigned a temperature by dividing the noise power per unit bandwidth by Boltzmans Constant.

Consider now a noisy two-port device connected to a matched source and load:

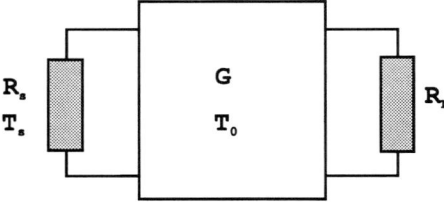

The noise per unit bandwidth delivered to the load R_L is the noise generated by R_s multiplied by the gain of the two port G plus the noise (kT_0) generated by the two port itself. Thus:

$$P_L = k(GT_s + T_0)$$

The noise temperature of the two-port is defined by reference to an equivalent *noiseless* two port:

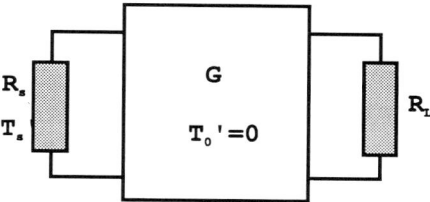

The power delivered to the load is now given by:

$$P_L = kGT_s'$$

If this is the same as previously:

$$GT_s' = GT_s + T_0$$

$$T_s' = T_s + \frac{T_0}{G}$$

The noise temperature of the two port is defined as the increase in T_s required, i.e.

$$T_N = \frac{T_0}{G}$$

[Note: T_0 may not be the physical temperature of the two port - it depends on the noise process involved].

3. Noise Temperature of an Attenuator

If the two port is simply an attenuator, then the total noise power into the load is given by $k[GT_s + T_{AO}]$. However an attenuator can be thought of as a network of resistors, which can be combined with R_s to give an equivalent resistance, still at temperature T_s (since there are no additional noise sources in an attenuator). Thus the power into the load is simply kT_s. Thus:

$$kT_s = k[GT_s + T_{A_0}]$$

i.e.

$$T_{A_0} = T_s(1 - G)$$

From above:

$$T_{A_N} = \frac{T_{A_0}}{G}$$

Thus:

$$T_{A_N} = T_s(\frac{1}{G} - 1)$$

Or using $L = \frac{1}{G}$ and $T_s = T$ in this case:

$$T_{A_N} = T(L - 1)$$

e.g. at 290K a 10dB attenuator has a noise temperature given by $290(10-1) = \underline{2600K}$.

Returning to the simple mixer downcoverter:

Let: f_{LO} =   Local Oscillator frequency
                f_{IF} =   Centre frequency of the IF amplifier passband
                Δf =   Bandwidth of the IF amplifier.

There are two input bands which will product beat frequencies with the LO which fall within the passband of the IF Amplifier. These are called the signal and image sidebands (which is the signal and which is the image depends on how the mixer is being used).

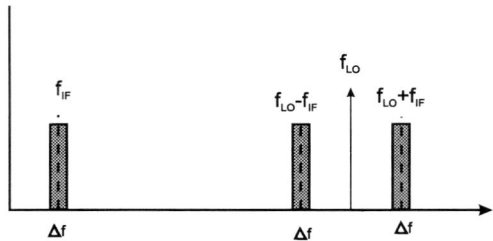

The fact that there are two input bands but only one output band complicates the definition of the noise temperature of this system - do we assign the noise generated in the mixer to one, or both, of the input sidebands?

Schematically:

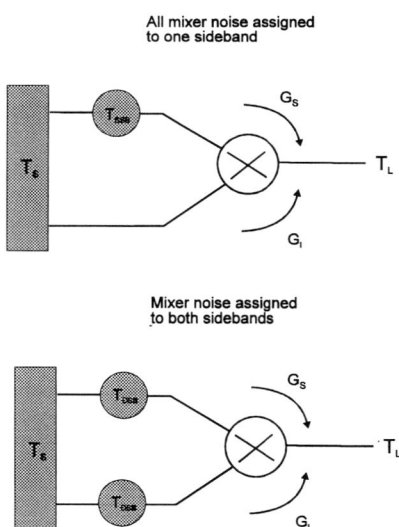

Clearly, since the actual noise generated by the mixer is the same in both cases, if $G_I = G_s$. then the double sideband mixer noise temperature is half the single sideband noise temperature.

In fact, most mixers exhibit loss rather than gain. This is called Conversion Loss and is given by:

$$L = \frac{1}{G}$$

If we again consider a simple system:

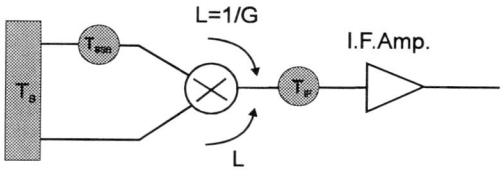

(using SSB noise temp.)

The noise power into the IF Amplifier is given by:

$$P_N = k[(2T_s + T_{SSB})G + T_{IF}]$$

Subtracting the noise due to the source alone and dividing by Boltzmans Constant and the "gain" of the mixer gives the single sideband noise temperature of the system:

$$T_{sys,SSB} = \frac{k[(2T_s + T_{SSB})G + T_{IF}] - 2kT_sG}{kG}$$

or, in terms of the conversion loss:

$$T_{sys,SSB} = T_{SSB} + L_0 T_{IF}$$

Components following the IF amplifier have little effect since their noise contribution is divided by the large gain of the IF amplifier. [Note, here I have included the loss of the diplexer in with that of the mixer - also, I have assume that there is no pre-amplifier between the source and the mixer - always the case at terahertz frequencies].

With these provisions, we see that the system noise temperature is dominated by three factors: Mixer Noise Temperature, Mixer Conversion Loss and IF Amplifier Noise Temperature

Question: Why is system noise temperature important?
Answer: Because it determines the minimum detectable power level in a given integration time.

In *radiometry*, the signals are themselves noise and can be characterised by a temperature. The minimum detectable temperature by a system having noise temperature T_{sys} and bandwidth Δv in an integration time τ is given by the Radiometer equation:

$$T_{min} = \frac{T_{sys}}{\sqrt{\Delta v \cdot \tau}}$$

i.e. if you double the system temperature, it takes four times as long to get the same result.

4. Diode Mixer Theory

The analysis of a diode mixer proceeds in three phases:

- The Voltage and Current waveforms produced in the diode by the Local Oscillator are determined using non-linear circuit analysis.
- Small signal analysis is then performed to obtain the mixer input and output impedances and the conversion loss.
- Down converted thermal and shot noise components produced in the diode are determined and the mixer noise temperature calculated.

Before proceeding, we need to develop a few ideas concerning transmission lines and impedances:

4.1 TRANSMISSION LINES

The parallel wire transmission line consists of two wires running parallel to each other with the spacing small compared to the wavelength.

If an oscillator is connected as above, a wave will run along the line to the right. If we now terminate the line in an impedance in general, there will be a reflected wave moving to the left:

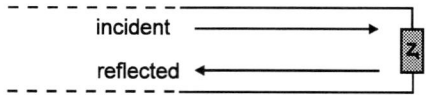

Every transmission line has a *Characteristic Impedance*, Z_o. If $Z_t = Z_0$, there is no reflected wave. The transmission line is then said to be matched.

All prevailing conditions repeat every wavelength along the line. Thus if a line is an integer number of wavelengths long the current, voltage and impedance are the same at both ends. Thus, at a fixed frequency, we can insert such a length of transmission line without any effect on a circuit.

As far as the termination is concerned, the situation is:

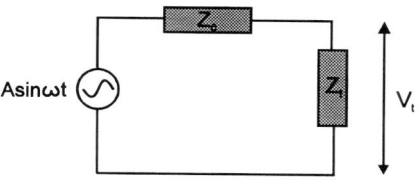

i.e. the transmission line looks like a source with output impedance Z_o.

Note: if $Z_t = Z_0$, then $V_t = \dfrac{A \sin \omega t}{2}$, thus, the voltage produced at the termination by a wave travelling to the right is half of that produced by the source. Thus if we know V_t and need an equivalent transmission line source, we must use:

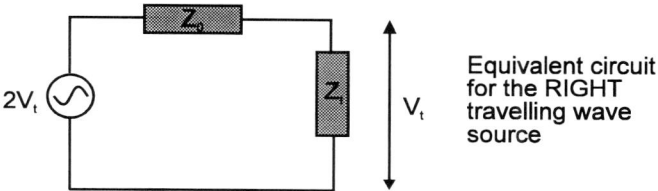

Equivalent circuit
for the RIGHT
travelling wave
source

4.2 EMBEDDING IMPEDANCE

The operation of the mixer is dependant upon the electromagnetic environment in which the diode finds itself. For our purpose, this is characterised by *embedding impedances.* These can be defined as follows:

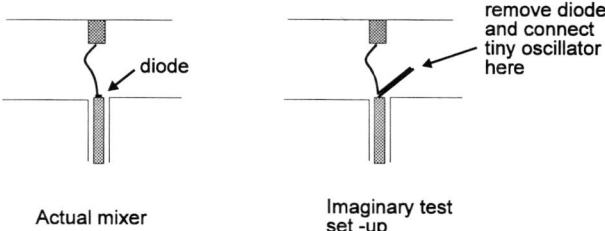

Actual mixer

Imaginary test
set -up

The tiny oscillator oscillates at the frequency of interest, .e.g. signal, L.O. etc. The oscillator causes currents to flow into the terminals where the diode was located. If the applied voltage is $Ve^{j\omega t}$ and the resulting current is $Ie^{j\omega t}$, then the embedding impedance at frequency ω is:

$$Z_e(\omega) = \frac{Ve^{j\omega t}}{Ie^{j\omega t}}$$

36

The value of $Z_e(\omega)$ depends upon the physical structure of the mixer, e.g. whisker length, backshort position etc.

Embedding impedances may be determined by measurements on scale models, from equivalent circuit models or computer modelling of the electromagnetic fields.

5. Large Signal Analysis

A large signal analysis is carried out to determine the voltage and current waveforms produced at the diode by the Local Oscillator. A Harmonic Balance Technique is used which treats the nonlinear element (the diode) in the frequency domain and the linear embedding circuit in the time domain.

5.1 LARGE SIGNAL EQUIVALENT CIRCUIT:

We need an equivalent circuit which represents the mixer at DC, the local oscillator frequency and all of its harmonics.

DC:

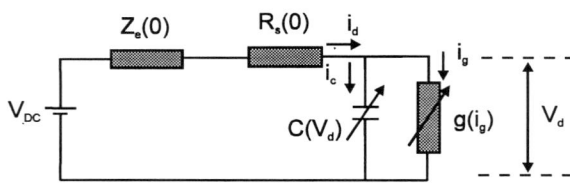

L.O.:

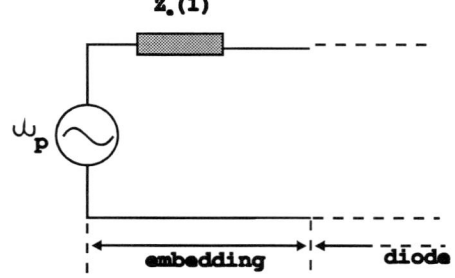

Harmonic $n\omega_p$

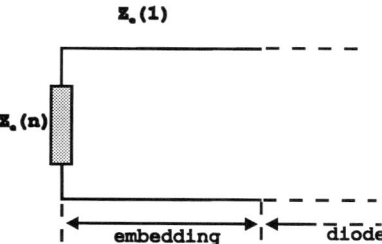

The first step is to express V_d and I_d as Fourier Series:

$$V_d(t) = \sum_{n=0}^{\infty} I_{d_n} e^{jn\omega_p t} \tag{1}$$

$$I_d(t) = \sum_{n=0}^{\infty} I_{d_n} e^{jn\omega_p t} \tag{2}$$

From the equivalent circuits we can now derive the following constraints on V_{d_n} and I_{d_n}:

$$\frac{V_{dc} - V_{d_o}}{I_{d_o}} = Z_e(o) + R_s(o) \qquad \text{D.C.} \tag{3}$$

$$\frac{V_{LO} - V_{di_1}}{I_{d_1}} = Z_e(1) + R_s(1) \qquad \text{LO} \tag{4}$$

$$\frac{-V_{d_n}}{I_{d_n}} = Z_e(n) + R_s(n) \qquad \text{Harmonics} \tag{5}$$

Once we have the correct solutions for $V_d(t)$ and $I(t)$ the Fourier components will satisfy these equations.

6. Multiple Reflections

In the method of Held and Kerr [1] we now introduce an imaginary transmission line between the "intrinsic" diode (diode minus R_s)- which is lumped in with the embedding network- and the embedding network.

This transmission line is an integer number of wavelengths long at the L.O. frequency. Let the characteristic impedance of this line by Z_0 (arbitrary).

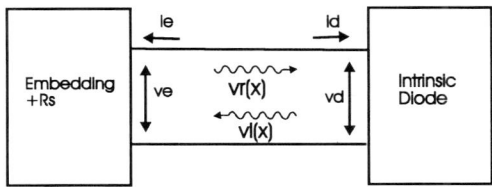

In operation, at any instant, there will be waves, at all the harmonic frequencies, travelling in the left and right directions. Because of the length of the line v_e and v_d should be equal at all the frequencies.

The total voltage and current on the transmission line are given by:

$$V(x) = V_r(x) + V_e(x) \tag{6}$$

$$i(x) = i_r(x) + i_e(x) = \frac{[v_r(x) - v_e(x)]}{z_o} \tag{7}$$

Here, x is a position co-ordinate along the line and the "minus" comes from general transmission line theory.

Because the line is an integer number of wavelengths long:

$$v(x = o) = v(x = l)$$

and
$$i(x = o) = i(x = l) \quad \text{(where } l = \text{the length of the line)}.$$

We begin the calculation by assuming that at t=0, the diode is removed and replaced by a matched load Z_o which generates no reflected wave.

At t=o, the load is removed and replaced by the diode - we assume that the voltage on the diode is initially as it was with the load:

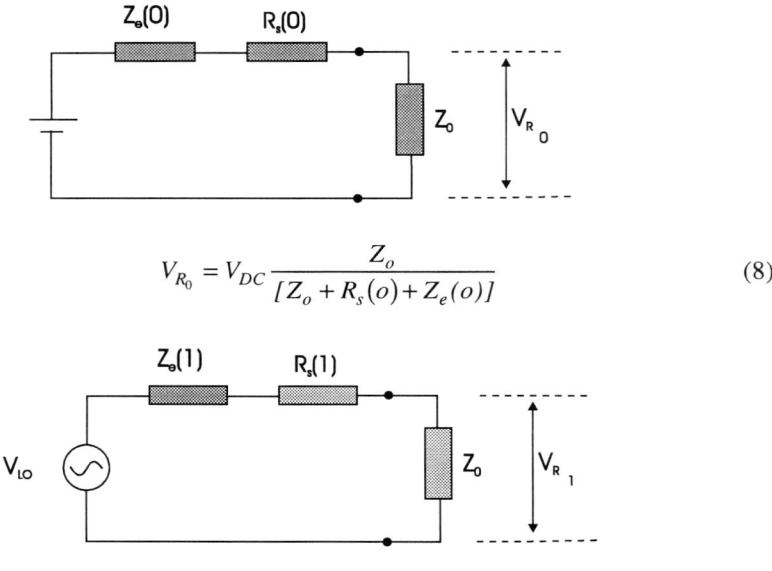

$$V_{R_0} = V_{DC} \frac{Z_o}{[Z_o + R_s(o) + Z_e(o)]} \tag{8}$$

$$V_{R_1} = V_{LO} \frac{Z_o}{[Z_o + R_s(1) + Z_e(1)]} \tag{9}$$

At t=o, these are the only voltages, therefore the total voltage across the intrinsic diode (where $w_p = LO$ or 'pump' frequency) is given by:

$$V_d(t) = V_{R_o} + V_{R_1} e^{jw_p t}$$

After connection, this voltage is applied across the diode by the transmission line, the equivalent circuit in the time domain is:

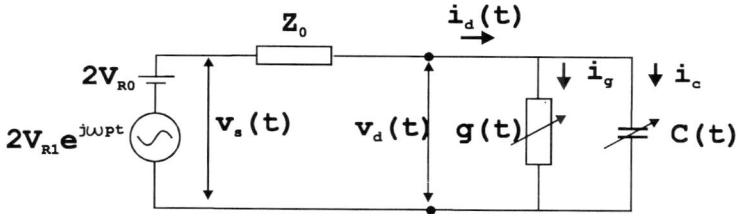

Note: The factor of two described earlier.

The state equation for this circuit is:

$$c(t)\frac{dv_d}{dt} = [\frac{V_s - V_d}{z_o} - i_g(t)]$$ (10)

Using the diode characteristics this equation can be integrated numerically giving a first estimate of the diode voltage and current waveforms which will now include higher harmonics of the LO frequency.

These voltages now give rise to new left travelling waves on the transmission line. Solving (6) and (7), we can obtain the solutions for such waves from the currents and voltages at the end of the line:

$$v_l(x = l) = \frac{v_d(t) - i_d(t)Z_o}{2}$$ (11)

This can now be Fourier analysed to yield an equivalent expression for each of the harmonics:

$$V_{l_n} = \frac{V_{d_n} - I_{d_n}Z_0}{2} \qquad for \ n = 0,1,2,\cdots$$ (12)

These waves now propagated to the LEFT where they encounter the corresponding embedding impedance in series with R_s. This results in reflections generating a new set of RIGHT travelling waves. The reflection coefficient ρ_n for each of the harmonics is given by:

$$\rho_n = \frac{Z_e(n) + R_s(n) - Z_0}{Z_e(n) + R_s(n) + Z_0}$$ (13)

$$[\quad \rho_n = \frac{Z_t - Z_0}{Z_t + Z_0}, \quad Z_t = \text{the terminating impedance} \quad]$$

The new equivalent circuit for the time domain solution at the diode end now becomes:

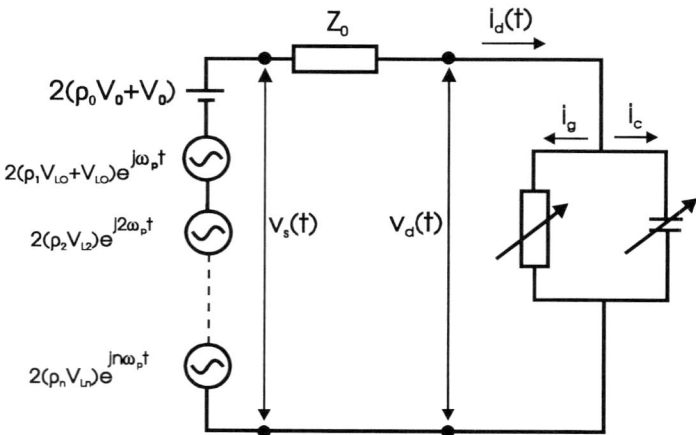

Equation (10) - the state equation can now be re-solved to give a new solution for $i_d(t)$ and $v_d(t)$.

7. Convergence

The above procedure is repeated until equation (5) is satisfied for all of the harmonic frequencies and equation (4) is satisfied for the LO frequency. The values of $V_d(t)$ and $I_d(t)$ obtained are then taken to represent the true solution.

8. Small Signal Analysis

From the large signal analysis we have determined the current and voltage waveforms in the diode i.e. we now know $I_d(t)$ and $V_d(t)$. From the known characteristics of the Schottky Barrier diode:

$$I_d = I_0 [exp(\frac{qV_d}{\eta kT}) - 1]$$

The conductance

$$g = \frac{dI_d}{dV_d}$$

$$= \frac{qI_0}{\eta kT} exp(\frac{qV}{\eta kT}) \tag{14}$$

$$\approx \frac{q}{\eta kT} I(V_d) \qquad for \quad V > \frac{\eta kT}{q}$$

Since V_d is a known function of time, we can substitute into (14) and determine $g(t)$. As we know that g must be periodic with the period of the LO pump we can express $g(t)$ as a complex Fourier series:

$$g(t) = \sum_{n=-\infty}^{\infty} G_n \, exp(\, jn\omega_p t)$$ (15)

where (since g(t) is real) $G_{-n} = G_n{}^*$

Similarly, from the known diode properties and the large signal analysis we can express the time varying diode capacitance as:

$$C(t) = \sum_{-\infty}^{\infty} C_n \, exp(\, jn\omega_p t)$$ (16)

where $C_{-n} = C_n{}^*$

As far as the signal is concerned the pumped diode behaves as a linear component with time varying conductance and capacitance given by equations (15) and (16).

8.1 SMALL SIGNAL MIXING FREQUENCIES

When a signal is applied to a pumped diode, currents and voltages are generated at many frequencies, as illustrated below:

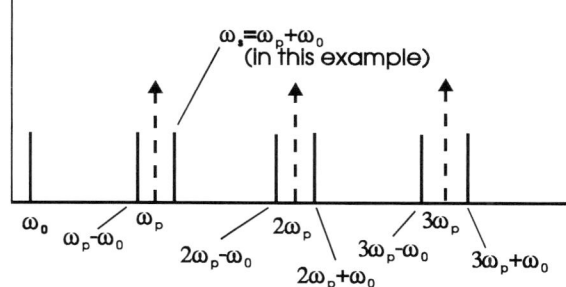

The frequencies indicated by solid lines are called the small signal mixing frequencies. The pumped diode couples them all together.
We can group these together as follows:

I.F.	Upper sidebands	lower sidebands
ω_0	$\omega_p + \omega_0$	$\omega_p - \omega_0$
	$2\omega_p + \omega_0$	$2\omega_p - \omega_0$
	$3\omega_p + \omega_0$	$3\omega_p - \omega_0$
	•	•
	•	•
	$n\omega_p + \omega_0$	$n\omega_p - \omega_0$

In phasor notation, the upper sideband currents and voltages can be written as:

$$I_n \, exp \, j(n\omega_p + \omega_0) \quad n = 1 \cdots \infty$$ (17)
$$V_n \, exp \, j(n\omega_p + \omega_0) \quad n = 1 \cdots \infty$$ (18)

The lower sideband frequencies are of the form:

$$I_n \, exp \, j(n\omega_p - \omega_0) \quad n = 1 \cdots \infty \tag{19}$$

$$V_n \, exp \, j(n\omega_p - \omega_0) \quad n = 1 \cdots \infty \tag{20}$$

By examining these phasors in details we can make a useful simplification. For example, consider equation (20). All voltages are actually *real*, therefore when we write $V_n \, exp \, j(n\omega_p - \omega_0)$, we actually mean:

$$\Re\{V_n \, exp \, j(n\omega_p - \omega_0)\}$$

V_n is a complex amplitude, therefore let:

$$V_n = \tilde{V}_n \, exp(j\varphi)$$

so that:

$$\Re\{V_n \, exp \, j(n\omega_p - \omega_0)\} = \Re\{\tilde{V}_n \, exp(j\phi) exp \, j(n\omega_p - \omega_0)\}$$

$$= \tilde{V}_n \, cos(n\omega_p - \omega_0 + \phi)$$

Using $\cos(-\theta) = \cos(\theta)$ we can re-write this as:

$$\tilde{V}_n \, cos(\omega_0 - n\omega_p - \phi)$$

This function can now be written as:

$$\Re\{\tilde{V}_n \, exp(-\phi) exp \, j(\omega_0 - n\omega_p)\}$$

or in phasor notation:

$$V_n^* \, exp \, j(\omega_0 - n\omega_p)$$

If we now define the integers specifying the lower sidebands to be negative then equation (7) becomes:

$$V_n^* \, exp \, j(\omega_0 + n\omega_p) \quad n = -1 \cdots -\infty$$

The IF frequency can be written as $V_0 \, exp(j\omega_0)$, hence all of the sidebands now fall into the series:

$$V_n \, exp \, j(\omega_0 + n\omega_p) \quad n = -\infty \cdots -1,0,1,\cdots +\infty$$

Where, for negative n, V_n is the complex conjugate of the actual value. Similarly the diode current can be written:

$$I_n \, exp \, j(\omega_0 + n\omega_p) \quad n = -\infty \cdots -1,0,1,\cdots +\infty$$

with I_n complex conjugated for negative integers.

9. The Conversion Admittance Matrix

The pumped diode can be thought of as a multi-port circuit with one pair of terminals for each of the mixing frequencies:

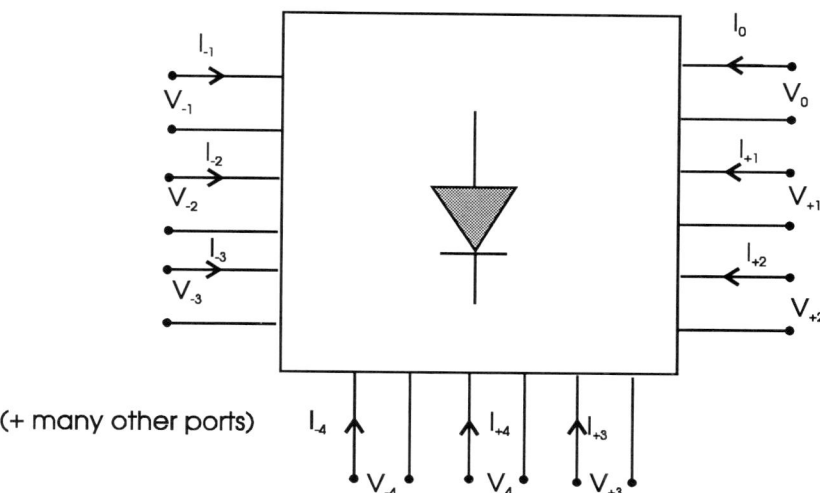

(+ many other ports)

Note: In practice, we only need to consider a finite number of frequencies. Let this number be N.

The currents and voltages in the ports of the intrinsic diode can be represented as a matrix equation:

$$
\begin{bmatrix}
I_N \\
I_{N-1} \\
\vdots \\
I_1 \\
I_0 \\
I^*_{-1} \\
\vdots \\
I^*_{-N+1} \\
I^*_{-N}
\end{bmatrix}
=
\begin{bmatrix}
 & & & Y_{N0} & & & & \\
 & & & \vdots & & & & \\
 & & & \vdots & & & & \\
 & & Y_{11} & Y_{10} & Y_{1-1} & & & \\
Y_{0N} & \cdots \quad \cdots & Y_{01} & Y_{00} & Y_{0-1} & \cdots \quad \cdots & Y_{0-N} \\
 & & Y_{-11} & Y_{-10} & Y_{-1-1} & & & \\
 & & & \vdots & & & & \\
 & & & \vdots & & & & \\
 & & & Y_{-N0} & & & &
\end{bmatrix}
\begin{bmatrix}
V_N \\
V_{N-1} \\
\vdots \\
V_1 \\
V_0 \\
V^*_{-1} \\
\vdots \\
V^*_{-N+1} \\
V^*_{-N}
\end{bmatrix}
$$

So, if for example we were to short port N, the current which would flow as the result of voltages applied to the other ports would be given by:

$$
I_N = Y_{NN}V_N + Y_{N,N-1}V_{N-1} + \cdots Y_{N0}V_0 + \ldots Y_{N,-N+1}V^*_{-N+1} + Y_{N,-N}V^*_{-N}
$$

44

The Matrix Y is called *the conversion admittance matrix.* We can show that the components of Y are given by:

$$Y_{mn} = G_{m-n} + j(\omega_o + m\omega_p)c_{m-n}$$

Where G_{m-n} and C_{m-n} are the *Fourier components* of the conductance and capacitance waveforms.

10. The Augmented Network

The Y-matrix of the intrinsic diode is now *augmented* by including its spreading resistance, R_s and the embedding network in parallel with the ports

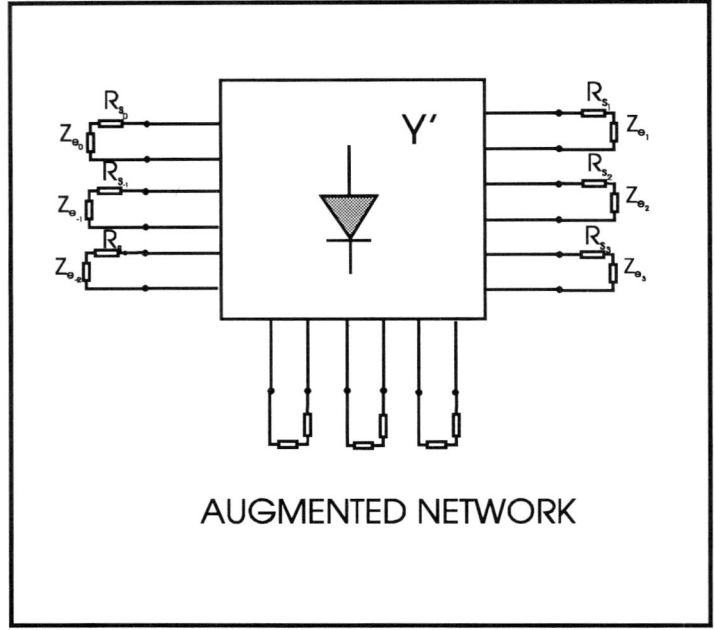

AUGMENTED NETWORK

Note: the ports of the augmented network have remained as for the intrinsic diode.

e.g. consider port 1:

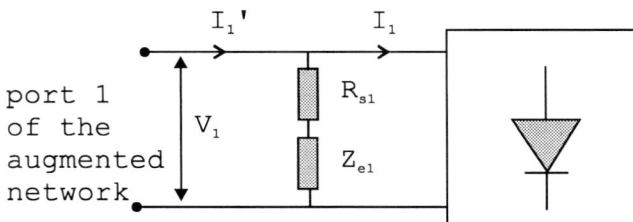

- R_{s1} and Z_{e1} are linear and do not add to the frequency conversion process. Therefore applying voltage V_1 to port 1 of the augmented network results in the same currents in all the other ports as generated by the intrinsic diode. Thus all the off diagonal elements of Y' are the same as Y.
- From the definition of Y:

$$I_1 = Y_{11}V_1$$

The current through R_{s_1} and Z_{e_1} is given by:

$$\frac{V_1}{R_{s_1} + Z_{e1}}$$

Thus:

$$I_1' = Y_{11}V_1 + \frac{V_1}{R_{s_1} + Z_{e_1}}$$

i.e.

$$I_1' = [Y_{11} + \frac{1}{R_{s_1} + Z_{e_1}}]V_1$$

hence:

$$Y_{11}' = Y_{11} + \frac{1}{R_{s_1} + Z_{e_1}}$$

In general, for the diagonal terms are:

$$Y_{nn}' = [Y_{nn} + [R_{s_n} + Z_{e_n}]^{-1}$$

11. Determination of Conversion Loss

For the augmented network we have the matrix equation:

$$\hat{I}' = Y'\hat{V} \tag{21}$$

where: $\quad \hat{I}' = (I_N', I_{N-1}', \dots I_1', I_o', I_{-1}' *\dots I_{-N}' *)^T$

and $\quad \hat{V} = (V_N, V_{N-1}, \dots V_1, V_0, V_{-1} *\dots V_{-N} *)^T$

Inverting equation (21) gives:

$$\hat{V} = Z'\hat{I}' \tag{22}$$

where $Z' = [Y']^{-1}$ is the *augmented conversion impedance matrix*. From Z' we can determine the conversion loss.

If we assume that the signal input to the mixer is at a frequency ω_1 ($= \omega_p + \omega_o$) and the IF is extracted at frequency ω_o these are the only two ports of interests - the circuit is therefore effectively a two port:

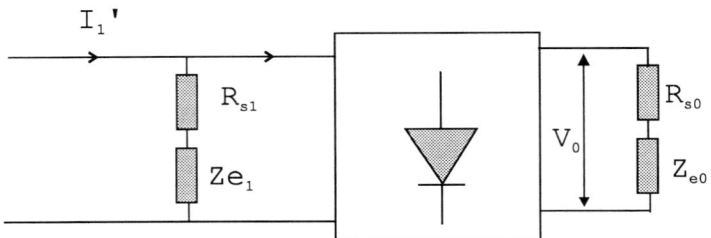

We excite the mixer by injecting a current into port 1 of the augmented network. From equation (22) - assuming there are no inputs at other ports:

$$V_0 = \hat{Z}_{10} I_1'$$ (23)

The voltage which appears across Z_{eo} - the actual IF load impedance is given by:

$$V_{IF} = V_o \frac{Z_{e_0}}{Z_{e_0} + R_{s_0}}$$

The current flowing through Z_{e_0} is given by:

$$I_{IF} = \frac{V_o}{Z_{e_0} + R_{s_0}}$$

The time average power delivered to Z_{e_0} is given by:

$$P_{del} = \frac{1}{2} Re[V_{IF} I_{IF}{}^*]$$

Substituting:

$$P_{del} = \frac{1}{2} \Re[V_o \frac{Z_{e_0}}{(Z_{e_0} + R_{S_0})} \cdot \frac{V_o{}^*}{(Z_{e_0} + R_{S_0})^*}]$$

$$= \frac{1}{2} \frac{|V_0|^2}{|Z_{e_0} + R_{s_0}|^2} \Re(Z_{e_0})$$

Using equation (23) this becomes:

$$P_{del} = \frac{1}{2} \frac{|\hat{Z}_{10}|^2 |I_1|^2}{|Z_{e_0} + R_{s_0}|^2} \Re(Z_{e_0})$$ (24)

On the input side (Port 1) we can use Thevenins theorem to show that:

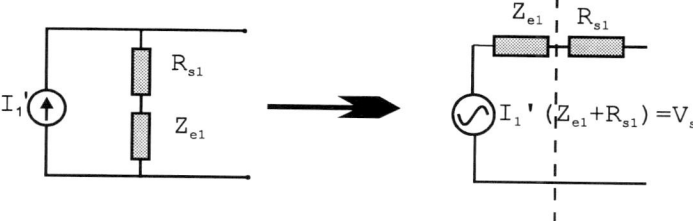

The power *available* from the actual signal source is that which it would deliver to a matched load:

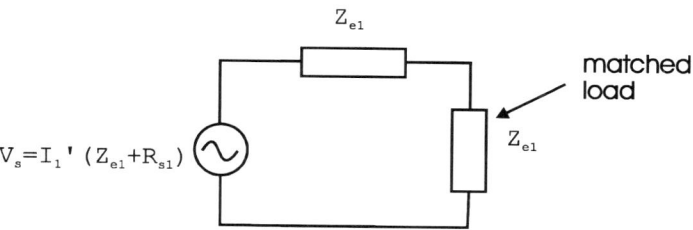

The voltage across the matched load is:

$$V_{ml} = \frac{1}{2} V_s = \frac{1}{2} I'_1 (Z_{e_1} + R_{s_1})$$

The current is given by:

$$I_{ml} = \frac{V_s}{2Z_{e_1}} = \frac{I'_1 (Z_{e_1} + R_{s_1})}{2Z_{e_1}}$$

Therefore power delivered to the matched load is:

$$P_{AV} = \frac{1}{2} \Re[V_{ml} \cdot I^*_{ml}]$$

So that:

$$P_{AV} = \frac{1}{2} \Re \left[\frac{1}{2} I'_1 (Z_{e_1} + R_{s_1}) \frac{I_1^* (Z_{e_1} + R_{s_1})^*}{2Z_{e_1}^*} \right]$$

$$= \frac{1}{8} |I_1|^2 \frac{|Z_{e_1} + R_{s_1}|^2}{\Re[Z_{e_1}]}$$

(25)

The conversion loss is now defined to be the power available from the source divided by the power delivered to the IF load. Using (24) and (25) above, we now have the result:

$$L = \frac{|Z_{e_1} + R_{s_1}|^2 |Z_{e_0} + R_{s_0}|^2}{4|\hat{Z}_{10}|^2 \Re[Z_{e_0}]\Re[Z_{e_1}]}$$

48

12. Mixer Noise Analysis

Sources of noise in a mixer are:

- Thermal noise in R_s
- Shot noise in current through the diode junction
- Lattice scattering noise
- Hot electron noise
- Thermal noise from resistive elements in the embedding network.

Usually only the first two are considered to be significant. The *noise equivalent circuit* in this case is:

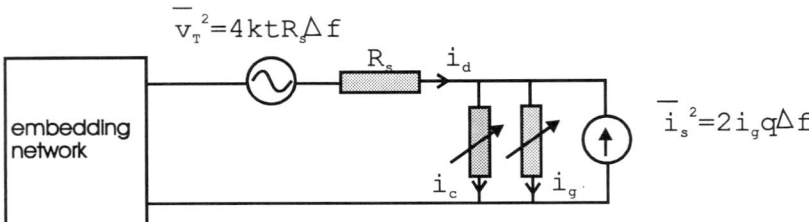

These noise components appear at all of the 'ports' of the augmented network. Some of this noise power is converted to the IF frequency where it contributes to the noise temperature of the mixer. A knowledge of the conversion impedance matrix $\hat{Z}'$ enables the noise temperature of the mixer to be calculated:

$$T_{SSB} = \frac{\left\langle \left|\delta V_N\right|^2 \right\rangle \left|Z_{e_1} + R_{s_1}\right|^2}{4k\Delta f \left|\hat{Z}_{01}\right|^2 \Re[Z_{e_1}]}$$

where:

$$\left\langle \left|\delta V_N\right|^2 \right\rangle = \hat{Z}'_0 (C_s + C_t)\hat{Z}'_0{}^{*T}$$

$\hat{Z}'_0$ is the centre row of the Augmented Impedance Conversion Matrix. C_s and C_t are matrices representing the correlation properties of the noise at the mixer output terminals. C_s and C_t have been evaluated as:

$$C_{s_{mn}} = 2qI_{m-n}\Delta f$$

where I_{m-n} is the (m-n)th Fourier component of the diode conductance current (available from the large signal analysis).

$$C_{t_{mn}} = 0 \qquad for \quad m \neq n$$

$$C_{t_{mn}} = \frac{4kTR_{s_m} \Delta f}{\left| Z_{e_m} + R_{s_m} \right|^2} \qquad for \quad m = n \neq 0$$

$$C_{t_{mn}} = \frac{4kTR_{s_0} \Delta f}{\left| Z_{e_0} - R_{s_0} \right|^2} \qquad for \quad m = n = 0$$

13. Frequency Multipliers

- Use the non-linearity of the SB diode to generate harmonics of a pump source to provide solid-state LO sources above about 100 Ghz.
- Harmonic generation is possible through either the non-linear conductance or capacitance of the diode. However, modem multipliers almost always use the non-linear capacitance of the diode as the main harmonic generation mechanism.
- Diodes used in multipliers differ from those used in mixers as they are optimised to give the greatest capacitance variation possible. Such diodes are referred to as *varactor diodes*.
- Varactor diodes are operated under *reverse bias* to give the best possible capacitance variation and limit the current thereby increasing the power handling capability of the device.

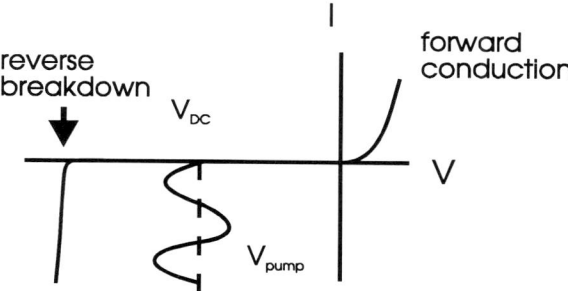

In the reverse bias region, the depletion layer thickness varies with bias giving the capacitance variation. The epi-layer thickness must be great enough to ccommodate this variation - therefore varactor diodes have thicker epi-layes than mixer diodes.

- It can be shown that a multiplier using a non-linear capacitance can have a conversion efficiency ($P_{harmonic}/P_{pump}$) of at most $1/n^2$ where n is the harmonic number. Varactors have no such limitations - efficiencies of 100% are theoretically possible (but of course not practically achievable).

- To achieve the maximum efficiency at the third harmonic, for example, it may be necessary to allow significant current flow at the second harmonic into a low loss resonator. This is an example of an *idler* circuit.
- The Analysis of a frequency multiplier is - in principle - the same as the large signal analysis of a mixer.

13.1. FREQUENCY TRIPLER - SCHEMATIC

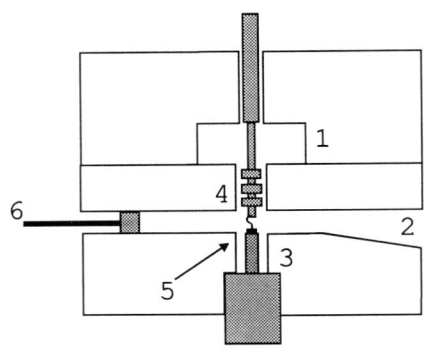

1.   pump waveguide
2.   3rd harmonic output waveguide
3.   Coaxial resonator, the 2nd harmonic idler
4.   Filter to prevent 3rd harmonic loss to the pump waveguide
5.   The diode
6.   Moveable backshort (also in pump waveguide - not shown)

13.2. MULTIPLIER ANALYSIS

The frequencies of interest in the multiplier are the pump frequency ω_p and its harmonics $n\omega_p$. Once again, we assume that we know the embedding impedances at all of these frequencies (up to a reasonable number of harmonics) and the diode conductance and capacitance characteristics enables us to carry out the harmonic balance calculation to determine diode voltage, current, capacitance and conductance waveforms:

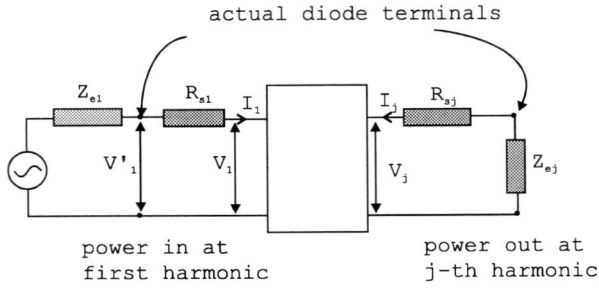

V_i, I_i, V_j, and I_j, are available from the harmonic balance calculation.

The pump power "absorbed" by the diode is given by:

$$P_{abs} = \frac{1}{2}\Re[V'_1 I_1{}^*]$$

$$= \frac{1}{2}\Re[\{V_1 + I_1 R_{s_1}\}I^*{}_1]$$

The power "delivered" to the load impedance Z_{ej} is given by:

$$P_{out} = \frac{1}{2}\Re[I_j{}^* V_j]$$

$$= \frac{1}{2}\Re[I^*{}_j I_j Z_{e_j}]$$

$$= \frac{1}{2}|I_j|^2 \Re[Z_{e_j}]$$

The efficiency is therefore:

$$\varepsilon = \frac{P_{out}}{P_{abs}} = \frac{|I_j|^2 \Re[Z_{e_j}]}{\Re[\{V_1 + I_1 R_{s_1}\}I^*{}_1]}$$

14. Reference

1. Siegel P.H., Kerr A.R. and Huarg W., (1984) Topics in the optimisation of millimetre wave mixers, (NASA technical paper 2287).

FUNDAMENTALS OF RECEIVERS FOR TERAHERTZ SYSTEMS

G.BEAUDIN AND P.J.ENCRENAZ
Observatoire de Paris, DEMIRM,URA CNRS 336 - ENS PARIS
61 avenue de l'Observatoire - 75014 Paris - France

1. Introduction

The submillimeter wavelength spectral band, covering the frequency range 0.3 THz ($\lambda = 1mm$) to 3 THz ($\lambda = 0.1$ mm), represents one of the the least explored yet information rich segments of the electromagnetic spectrum. This frequency span encompasses all of the critical spectral emissions from the key molecules involved in atmospheric chemistry on Earth (and on the planets and comets). These include those molecular transitions which have been identified as crucial to our understanding and monitoring of the global ozone depletion problem. The submillimeter-wave regime also contains spectral line emissions which can further our understanding of interstellar chemistry, new star formation and galactic structures. Due to high atmospheric opacity (Figure 1) both astrochemical and stratospheric observations in the submillimeter-wave spectral bands must be made from high altitude aircraft : Kuiper Airborne Observatory (KAO/NASA) and SOFIA ; balloons : PIROG (SSC), Programme National d'Astronomie Submillimétrique (PRONAOS/CNES) or satellites :

There are three funded space missions which will carry submillimeter wave radiometers : Submillimeter-Wave Astronomy Satellite (SWAS/NASA), ODIN (Swedish Space Centre, SSC) combining Astronomy and Aeronomy objectives, and Earth Observing System Microwave Limb Sounder (EOS-MLS/NASA) . Four other missions are in phase A/B study : the Submillimeter Observation of PRocesses in the Atmosphere Noteworthy for Ozone (MASTER- SOPRANO/ESA) for Aeronomy, ROSETTA for planetary and cometary observations, SAMBA-COBRAS for cosmology and the Far-Infrared Space Telescope (FIRST/ESA) for Astronomy .

SWAS has two radiometers at frequencies of 490 and 550 GHz ; ODIN has three radiometers, at 120, 480 and 550 GHz ; EOS-MLS is currently configured with radiometers at 210, 440 and 640 GHz and potential channels at 1.2 and 2.5 THz.
MASTER - SOPRANO is a project having two instruments on at 200, 325 and 350 GHz; the second at 500, 630 and 950 GHz frequency bands ; a complementary instrument called PIRAMHYD will cover 1.25 and 2.5 THz. FIRST is designed to have broad spectral coverage beginning at 500 GHz and going up to 1.2 THz.

J.M. Chamberlain and R.E. Miles (eds.), New Directions in Terahertz Technology, 53–62.

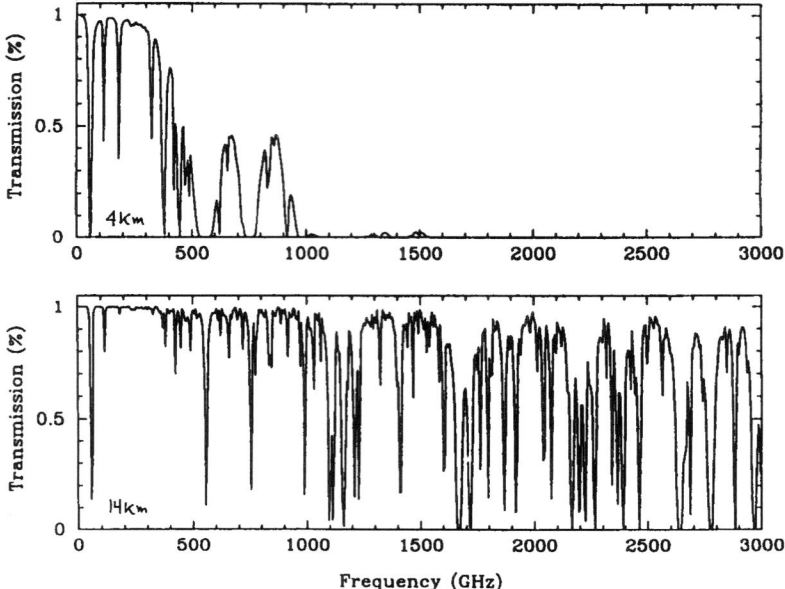

Figure 1. Atmospheric transmission in the submillimetre and far-infrared from a very good high-altitude ground-based site (Mauna Kea at 4.1 km with 1 mm of precipitable water vapour) and from the altitude of an airborne observatory (e.g. KAO at 14 km altitude). The blocked regions are mostly caused by molecular absorption.

2. Detection Techniques for Submillimeter Waves

There are two basic ways to analyse electromagnetic radiation at submillimetre and far-infrared wavelengths, either by (super-) heterodyne (coherent detection) or by direct (incoherent) detection techniques. This part of the spectrum lies between what traditionally can be regarded as the radio and infrared domains and the two techniques reflect this fact :

2.1. DIRECT AND HETERODYNE DETECTION SYSTEMS

In direct detection (Figure 2) the detectors respond to the signal photons themselves ; in heterodyne detection (Figure 3) the signal is converted to a lower convenient intermediate frequency (IF) by "mixing" with a generated stable monochromatic local oscillator (LO) signal before signal processing.

The fundamental difference between the two types of detection is the retention or destruction of the phase in the detected signal. The quantum mechanical uncertainty principle shows that heterodyne detection can never be more sensitive than direct detection, at least in principle. Fixing the phase causes a measurement uncertainty of order one photon in a heterodyne conversion process ; this is equivalent to imprecision introduced by a noise source. The lack of phase sensitivity in incoherent detection enables the direct detection of individual photons.

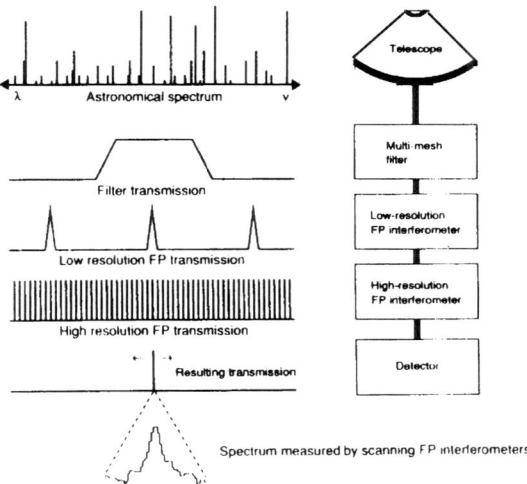

Figure 2. Principle of direct detection system. The spectral bandwidth is reduced by a succession of filters and interferometers. Approximate wavelength selection is made by means of a bandpass filter isolating one Fabry-Pérot interferometer transmission fringe. By adding one additional interferometer at higher resolution (larger mesh spacing) a final higher resolution is achieved. The transmitted, very narrow bandpass radiation, can be scanned in wavelength by tuning the two interferometers enabling the detectors to sample different parts of the spectrum.

Since instruments based on direct detection respond to signal photons alone, spectroscopy must be done by separating individual frequency components in the incoming signal (Figure 3) before detection. With the exception of Fourier transform instruments, incoherent systems measure one frequency channel per detector, requiring some scanning of the predetection filter to obtain a spectrum. A typical heterodyne receiver consists of two separate parts : a heterodyne mixer (the "front- end), which shifts a high frequency band of frequencies from one center frequency to a lower one without altering the spectral information within the band, and a separate ("backend") spectrometer which obtains the spectrum of the lower frequency band (Figure 4). Since the spectroscopy is performed at low frequencies, simple filters with modest resolution can be used. The backend spectrometers of heterodyne systems analyze the entire instantaneous receiver bandwidth, which in practical cases will cover the spectral line and baseline. The frequency multiplex advantage of the heterodyne backend can be offset by the simpler spectral or spatial multiplexing of incoherent array elements.

The choice of heterodyne or direct detection for a given application at submillimetre and far-infrared wavelengths is not always obvious, as it is in this range of the spectrum that the two methods both cross in sensitivity and become technologically possible. Tradeoffs, involving for instance observing frequency,

56

spectral and spatial resolution and coverage, required sensitivity and detector availability, will determine whether Fourier transform, grating, Fabry-Pérot, or heterodyne instruments will be best suited (see Figure 4).

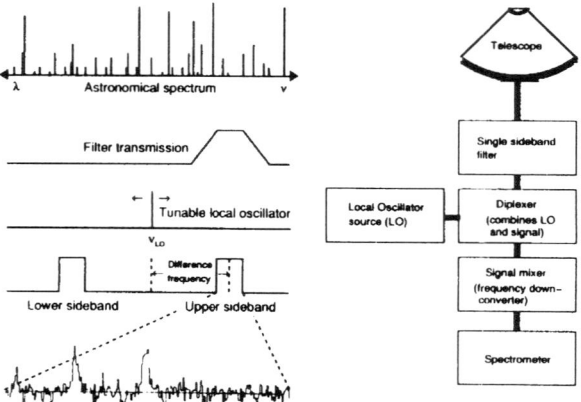

Figure 3. Principle of heterodyne detection system. Downconversion of the received signal to an easier to process frequency is achieved by adding a tuneable line signal (local oscillator) to the received one and extract the different frequency. A filter can be used to select any of the two received sidebands. Signal merging is performed in a diplexer before feeding the signal to the mixer. After further amplification the downconverted signal can be analysed at the chosen frequency resolution in the backend spectrometer. This device samples simultaneously a very large number of frequency channels.

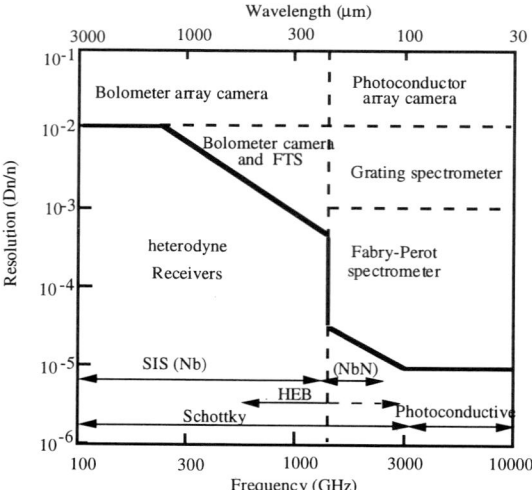

Figure 4. There are two basic ways to analyze electromagnetic radiation at submillimetre and far-infrared wavelengths, either by (super) heterodyne (coherent detection) or by direct (incoherent) detection techniques. This part of the spectrum lies between what tradionally can be regarded as the radio and infrared domains and the two techniques reflect this fact.

3. Submillimeter Heterodyne Technologies

All of the missions requiring high sensitivity and high spectral resolution use the heterodyne detection technique (Figure. 5). Such receivers generally consist of a low loss signal coupling structure (waveguide feed horn of planar antenna), a local source of RF power (local oscillator, L.O.) at a frequency very close to that the observed signal, a frequency diplexer which efficiently couples the RF signal and LO into the low noise down converting (mixer) element (Schottky barrier diode or SIS superconducting tunnel junction), a low noise intermediate frequency (IF) usualy in the microwave band, and finally a high resolution spectrometer to separate out the spectral lines (filter banks, digital autocorrelators, surface acoustic-waves filters, acousto-optic spectrometers), are used.

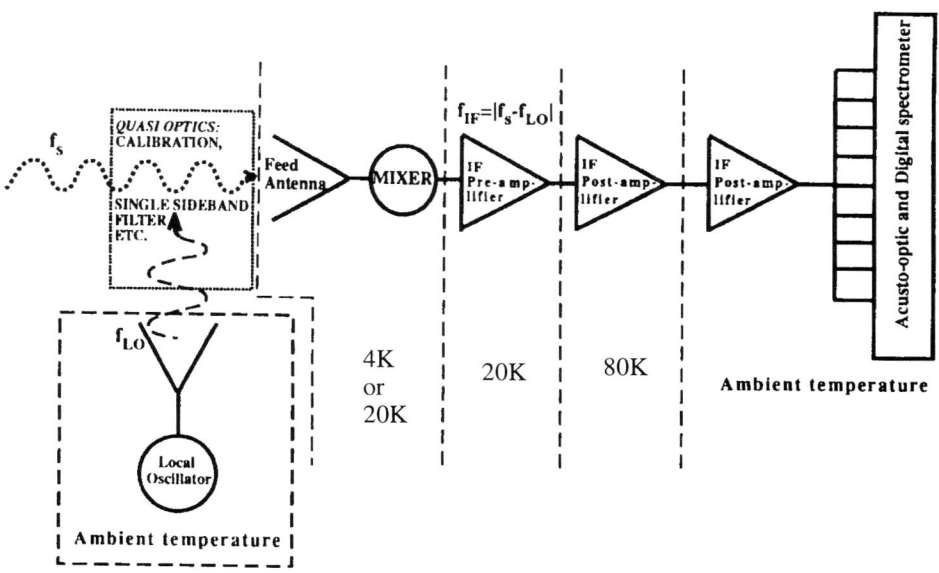

Figure 5. Schematic outline of a heterodyne receiver using common quasi-optics and spectrometers. Each channel has SIS mixers operating at 4 K, or Schottky mixers operating between 20 K and room temperature.

3.1. SUBMILLIMETER HETERODYNE MIXERS

Both Schottky diode and superconducting tunnel junction (Superconducting-Insulator-Superconducting = SIS) mixers could be used on the submillimeter heterodyne receivers. Research into the atmosphere of the Earth (or other planets) does not need so high sensitivity, and is more often accomplished using Schottky diode mixers ; astrophysical research needs the highest sensitivity, currently employing SIS mixers. Waveguides with horns and Quasi-optical mixer technologies are both employed up to 2.5THz. Arrays have been developed for astronomical focal imagery.

3.1.1. Submillimeter Schottky-Diode Developments

For more than two decades the best uncooled heterodyne radiometers for use in the 100 GHz - 3 THz frequency range have been composed of waveguide or open structure mixers with whisker-contacted metal-semiconductor Schottky-barrier honeycomb diodes.

In order to reduce the assembly cost and improve the reliability and reproducibility of heterodyne receivers for space missions throughout the millimeter and submillimeter wavelength bands, two major changes must be incorporated into current radiometer design. First, the whisker-contact honeycomb diode must be replaced by a more reliable, easier to handle, integrated structure similar to the beam-lead diodes now routinely used below 100 GHz. Second, for applications up to (or above) 600 GHz, the diode must be integrated with the remaining, physically larger, mixer circuitry to increase flexibility and simplify assembly. An added benefit of this latter approach is the potential of going one step further and replacing the last remaining mechanically fabricated component, the waveguide mount, with an all-planar photolithographic structure scalable to frequencies well beyond one THz (JPL, SHP mixer at 650GHz). A major goal is to advance the state-of-the-art in millimeter-wave quasi-planar-diode technology to the point at which it can be used readily at frequencies as high as 2.5 Thz. The Schottky diode must be cooled to about 20-30 K for optimum performances, but works even at room temperature.

3.1.2. SIS Tunnel junctions developments

In the push to obtain ever higher sensitivity, shorter observation times and the use of smaller collecting surfaces, the submillimeter-wave astrophysics community has devoted much of their resources towards the developement of radiometer front-ends based on the refractory superconductor niobium nitride. At present, the most prevalent form of high frequency superconducting heterodyne receiver is the small area superconductor-insulator-superconductor Nb(SIS) tunnel junction which offers the potential of near quantum-limited sensitivity throughout the millimeter-wave bands up to 700 GHz (Figure 6) and possibly at frequencies as high as 1.2-1.4 THz with normal metal tuning stubs circuits (Al). 2-3THz could be achievable in the near future by using SIS with NbN superconducting junctions or by using Hot Electron Bolometer heterodyne mixers. The SIS mixers must be physically cooled to temperatures well below the superconductor transition temperature, i.e. to 4 K for $Nb/Al_xO_y/Nb$ elements.

However, the requirement for a liquid helium ambient environment poses a significant limitation for remote, long lifetime space operation.

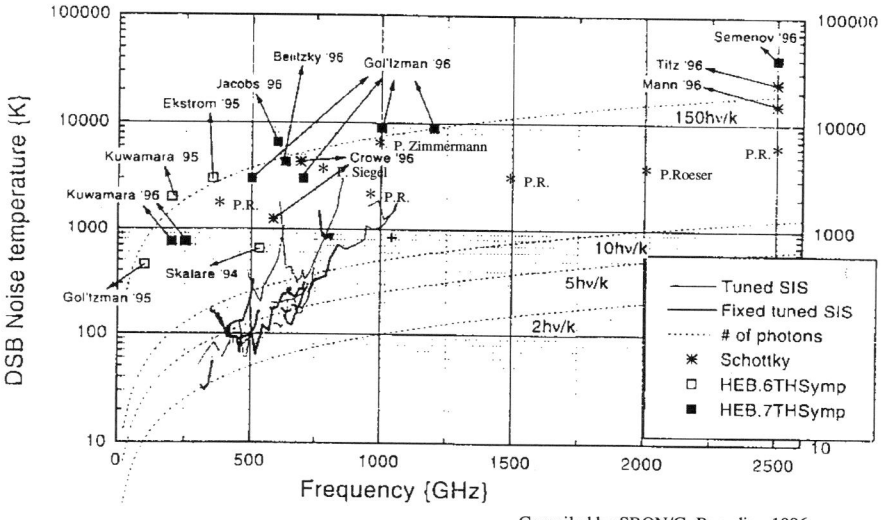

Compiled by SRON/G. Beaudin, 1996

Figure 6. Heterodyne SIS receiver noise temperatures. The currently best experimentally obtained receiver noise temperature (T_{rec}) vs frequency for SIS mixers. It is estimated that T_{rec} for the 1 THz cooled Schottky receiver will be approximately 1000 K.

3.1.3. *Local Oscillator Generation Technologies*

The L.O. power needed for Schottky diodes or SIS junctions is currently obtained by Gunn oscillators cascaded with frequency multipliers using whiskered varactor diodes. This technology is able to provide enough power up to THz for Schottky diode and 1.5 THz for SIS mixers. New planar components are under development HEMT oscillators, Quantum well oscillators are able to provide enough power up to 300-400 GHz ; long Josephson junctions and flux-flow oscillators arrays are also capable of driving the SIS mixers up to 600-700ghz.

CO$_2$ lasers pumping submillimeter masers are used as LO sources in the range 300 GHz - 3 THz, but they have too high electrical power consumption for space applications : laser mixer technology is now under investigation.

4. Submillimeter Technologies in the FIRST Project

The Far InfraRed and Submillimetre Space Telescope (FIRST) is one of the four "Cornerstone" projects in the ESA Long-Term Programme for Space Science, "Horizon 2000". This mission is devoted to high throughput spectroscopy and photometry in the submillimetre and far-infrared wavelength range. FIRST is foreseen as having a 3m diameter radiatively cooled Cassegrain telescope equipped with a payload consisting of a multichannel, very high spectral resolution heterodyne receiver (Table 1), an imaging medium, spectrometer and photometer, covering the 85-900 µm wavelength band (Table 2). FIRST will open up this virtually unexplored part of the spectrum which cannot be observed from the ground, and is only partially accessible from airborne platforms.

With its high throughput, low thermal background, extensive wavelength coverage, and high spatial and spectral resolution, FIRST will offer superb sensitivity for both photometry and spectroscopy. Its multiband instruments will give unprecedented information on the physics, chemistry and dynamics of interstellar, circumstellar, planetary and cometary gas and dust, resulting in a quantum step forward in the study of the cold universe. It will be the first multi-purpose submillimetre and far-infrared space observatory available to the world-wide astronomical community.

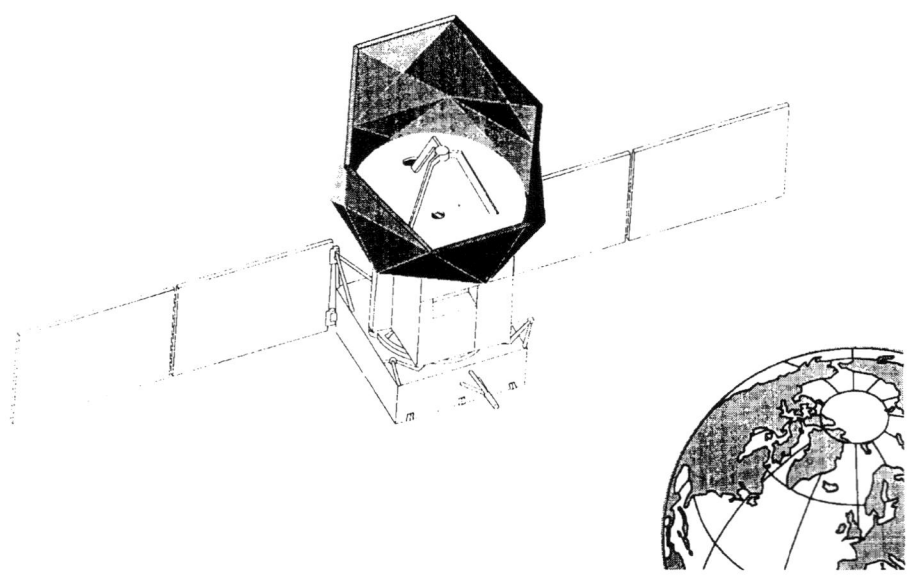

Figure 7. The FIRST spacecraft in orbit (artist view)

Table 1. FIRST, high resolution MultiFrequency Heterodyne (MFH) receiver frequency bands

Receiver band #	1 & 2	3 & 4	5 & 6	7 & 8	9
Mixer type	SIS	SIS	SIS	SIS	SIS
ν (GHz)	490 - 650	600 - 750	700 - 850	800 - 950	950-1200
Beam FWHM	48 - 36	39 - 31	34 - 28	29 - 25	25-20

Table 2. FIRST direct detection channels

Wavelength range	85-210 μm	85-210 μm	210-300 μm	85-210 μm	210-280 μm	280-600 μm	600-900 μm
Filter bands	5	5	1	2	1	1	1
mode	hi-res.	med-res.	med-res.	photom.	photom.	photom.	photom.
Detector	photo-conductor	photo-conductor	short wavelength bolometer	photo-conductor	short wavelength bolometer	short + long wavelength bolometer	long wavelength bolometer

5. Conclusion

For most observations of the atmospheres of the Earth and of other planets and comets, sensitivity is not nearly as critical an issue as it is for stellar astrophysics. A large number of key molecular transitions can be observed with the sensitivity available from current room-temperature or passively cooled semiconductor-diode radiometers. For the most part, the emphasis for millimeter and submillimeter-wave Earth remote-sensing applications has been on pushing to higher frequencies (up to 3 THz), increasing the instantaneous bandwidth, improving device reliability and reducing radiometer complexity and cost.

For astrophysics, very sensitive and very high spectral resolution heterodyne spectroscopy are required. SIS mixers in the 500 to 1200 GHz range and flexible backend spectrometers with more than 4 GHz instantaneous bandwidth will be used on the FIRST project. Significant progress is being made on both mixer performance and junction reliability as well as cryocooler technology (Sterling closed cycle cryogenerator and JT stage, Pulsed Gas Tube, etc.). There is now no doubt that SIS heterodyne submillimetric receivers will fly in space sometime in the very near future ...

Acknowledgements : to the FIRST heterodyne payload team, with *ESTEC-ESA, (The Netherlands),* as well as *Erico Armandillo (ESTEC), Raymond Blundel (Smithsonian, Harvard), Thomas Crowe (U. Va.), Brian Ellison (RAL), Neal Erickson (U. Mass), Margaret Frerking (JPL), Jean Marc Goutoule(MMS-F), William Mc Grath (JPL), A.I.Harris(MPIfeP), Karl Jacobs (KOSMA), Liam Kelly (NMRC), Nigel Keen (THz Consultants,ex MPIfR), Eric Kollberg (Chalmers), Jacob Kooï (JPL), Christine Letrou (INT), Didier Lippens (IEMN), Chris Mann (RAL), Anti Raïsänen (H.U.T.), Gabriel Reibeiz (U. Michigan), Peter Röser (DLR.), Herman van de Stadt (SRON), Peter Siegel (JPL), Jonas Smuidzinas (Caltech), Guy Thomas (CNES), Serge Toutain (ENSTBr), Bill Wilson (JPL), Nick Wyborn (SRON), Peter Zimmermann (RPG) and numerous other colleagues* for their contributions to the background of this paper. We thanks also CNES, ESA and NASA for funding the submillimeter developments.

62

References

This paper has been made from the listed articles on the following but not exhaustive list of references

- F.T. Ulaby et al. (1984) *Microwave Remote Sensing*, AR Tech House Inc..
- Kenneth J. Button (ed) (1985) *Infrared and Millimeter Waves* **1-13**, Academic Press Inc. (London) Ltd..
- *Radio Astronomy*, Kraus 2e edition (1986), Cygnus Quasar Books, Ohio (USA).
- *Coherent detection at millimeter wavelengths and their applications*, Nova science publications New York, Centre de physique des Houches, France, march 1990.
- *Proceedings of the 29th Liege International Colloquium* (July 1990), Institut d'Astrophysique, Liege, Belgium, ESA 8-10 rue Mario Nikis, 75738 Paris Cedex 15 - France.
- *International Symposium of Space Terahertz Technology*: JPL-NASA, Pasadena California (Feb 1991), University of Michigan (Mar 1992), UCLA (Mar-Apr 1993), U. Michigan (May 1994), Caltech (Mar 1995), University of Virginia Charlottesville (Mar 1996).
- G. Beaudin, D. Scouarnec, G. Thomas (Nov 1991) *Radiométrie en ondes millimétriques au sol et dans l'espace*, IEEE workshop millimétrique, Carry le Rouet, France.
- G. Beaudin, M. Gheudin, G. Thomas, P. Encrenaz (14-15 Jan, 1992) *Etat de l'art en détection cohérente dans la domaine millimétrique et submillimétrique*, 2_mes Journ_es d'Etudes Micro-ondes et Espace, Toulouse.
- *ODIN, a swedish small satellite project for astronomical and atmospheric research* (1992-96), SSC reports.
- P. Siegel et al. (May 1993) *Earth Observing System Microwave Limb Sounder (EOS-MLS)*, SPIE conf Proc 1874 - IRMMW engineering.
- G. Pilbratt et al. (6 Sept, 1993) *FIRST, far infrared and submm space telescope*, ESA phase A report, SCI 93.
- JPL-Caltech et al. (Dec 1995/Jan, May 1996) *HET HIFI meeting*, reports SRON.
- *MASTER - SOPRANO submillimetre observation of processes in the atmosphere noteworthly for ozone*, ESA proposal (Oct 1993, Jun 1994).
- *ESA WPP Workshop on millimeter wave technology and applications* (Dec 95).

MILLIMETRE WAVE AND TERAHERTZ WAVEGUIDES AND MEASUREMENTS

ROGER D. POLLARD
Department of Electronic and Electrical Engineering
The University of Leeds
Leeds LS2 9JT
U.K.

1. Introduction

One of the major challenges encountered in working at Terahertz frequencies is that of making measurements. This contribution is intended to review the problems encountered as transmission lines are used at these higher frequencies, to describe the options available for making measurements and to discuss techniques which can be employed at Terahertz frequencies to make reflection, transmission and spectrum measurements.

2. Review of Transmission Line Fundamentals

The properties of a microwave transmission line are usually obtained from a solution of the equations (the "Telegrapher's Equations") which result from the analysis of an elemental section of a line considered to be constructed from lumped elements. The components comprising the basic section are as shown in Figure 1, where L, R, C and G are the inductance, series resistance, capacitance and shunt conductance per unit length. For many realisable types of line with air dielectric the shunt conductance can be very closely approximated to zero but series resistance is always present owing to skin effect losses in the metallic conductors. If it is assumed that the current and voltage are sinusoidal (i.e. v and i are of the form $e^{j\omega t}$), then the solutions of the Telegrapher's Equations are of the form

$$v(z) = K_0 e^{-\gamma z} + K_1 e^{+\gamma z} \tag{1}$$

where

$$\gamma = \sqrt{(R + j\omega L)(G + j\omega C)} \tag{2}$$

63

J.M. Chamberlain and R.E. Miles (eds.), New Directions in Terahertz Technology, 63–78.
© 1997 *Kluwer Academic Publishers. Printed in the Netherlands.*

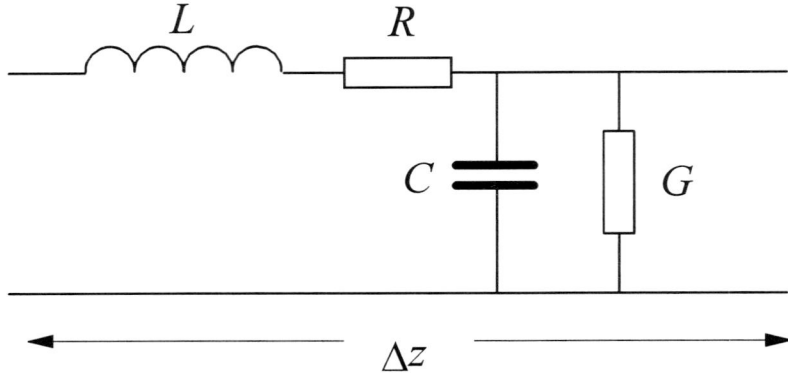

Figure 1. Elementary Section of Transmission Line

is the propagation constant and

$$Z_0 = \sqrt{\frac{R + j\omega L}{G + j\omega C}}$$ (3)

is the characteristic impedance. For the case of a lossless line ($R = G = 0$) then

$$Z_0 = \sqrt{\frac{L}{C}}$$ (4)

However, no practical implementation of a transmission line is truly lossless - the dominant effect being the conductivity (σ) of the metallic conductors. It is there necessary to retain

$$\gamma = \alpha + j\beta$$
$$= \sqrt{j\omega\mu(\sigma + j\omega\varepsilon)}$$ (5)

where

$$\alpha = \omega\sqrt{\frac{\mu\varepsilon}{2}\left(\sqrt{1 + \frac{\sigma^2}{\omega^2\varepsilon^2}} - 1\right)} \qquad \text{Np / m}$$

and

$$\beta = \omega\sqrt{\frac{\mu\varepsilon}{2}\left(\sqrt{1 + \frac{\sigma^2}{\omega^2\varepsilon^2}} + 1\right)} \qquad \text{rad / m}$$

The conductor loss mechanisms are the d.c. resistive loss

$$R_{dc} = \frac{\rho\ell}{A}$$

and that due to skin effect. The latter has both resistive and inductive components given from

$$R_s = \frac{1}{\sigma\delta} = \sqrt{\frac{\pi f \mu}{\sigma}} \quad \text{and} \quad L_i = R_s / \omega$$

thus

$$Z_s = R_s + j\omega L_i$$

Where the line is filled with dielectric, it is necessary to take account of dielectric loss (usually expressed as tan δ), leakage loss can be significant where transmission lines are laid out on semiconductor and radiation loss can be a problem with open structures. [1]

It is important to recall that the definition of characteristic impedance is that impedance which would be measured at the terminals of an infinite transmission line. A signal transmitted along such a line would never be reflected. To provide a reflectionless environment with a finite length of line requires that it be terminated in its characteristic impedance. In any other situation there will be both incident and reflected waves travelling along the line and the net voltage and current at any point is determined by the magnitude of these waves.

$$V = E_i + E_r \quad \text{and} \quad I = \frac{E_i - E_r}{Z} \tag{6}$$

Since voltages and current cannot easily be measured at high frequencies, a more useful relationship is the reflection coefficient

$$\Gamma = \frac{E_r}{E_i} = \frac{Z_L - Z_0}{Z_L + Z_0} \tag{7}$$

which is a complex quantity with magnitude ρ and angle θ and is a measure of the quality of the impedance match between the load and the characteristic impedance of the line. Alternatively, this may be expressed as the (Voltage) Standing Wave Ratio (VSWR or SWR)

$$\text{VSWR } \sigma = \frac{|E_i| + |E_r|}{|E_i| - |E_r|} \tag{8}$$

Note that the reflection coefficient is zero when matched and one when short- or open-circuited, while VSWR is one when matched and infinite for short or open.

2.1 SCATTERING PARAMETERS

The characterisation of networks at microwave frequencies and above is usually considered in terms of scattering coefficients. There are a number of reasons for this practice including the fact that it is almost impossible to measure voltages and currents, and the working environment is that of a transmission line with specified characteristic impedance. Also, active devices will often be unstable when operated into the highly mismatched terminations required to measure voltage- or current-reference parameters. Consider the situation in Figure 2; the device under test is embedded in a transmission line of characteristic impedance Z_0 and is characterised in terms of the travelling waves a and b. The scattering parameters are then defined as

$$b_1 = S_{11}a_1 + S_{12}a_2$$

$$b_2 = S_{21} a_1 + S_{22} a_2 \qquad (9)$$

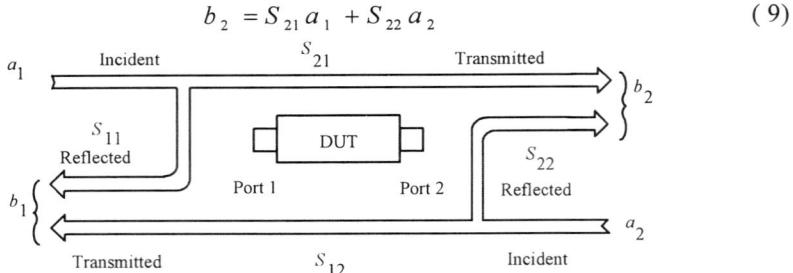

Figure 2. Scattering Parameters

with zero values of S_{ii} (the reflection coefficient at port i) being the matched condition. It is important to appreciate that the S-parameters of a given network will change if they are referred to a different characteristic impedance. a and b are functions of the current, voltage and a normalising factor. They are the scattering waves and are normalised such that $|a|^2$ represents the incident power and $|b|^2$ represents the reflected power. [2]

3. Performance of Conventional Transmission Lines

3.1 COAXIAL LINE

Figure 3 defines the general structure of a coaxial transmission line and Table 1 gives the important parameters of those types of lines which have accepted definitions.

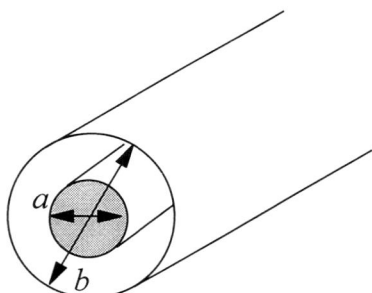

Figure 3. Structure of Coaxial Transmission Line.

TABLE 1. Limits for Operation of Coaxial Lines

OD (b) mm	ID (a) mm	max. frequency GHz	Loss at 1 GHz dB/m
14	6.08	8.5	0.1
7	3.04	18	0.2
3.5	1.52	34	0.4
2.82	1.22	40	0.5
2.4	1.04	50	0.6
1.85	0.80	65	0.8
1.0	0.43	120	1.4

The values given are for air dielectric lines of high quality made of copper. The limits of operation are determined by the onset of waveguide modes, tolerances and manufacturability as well as loss. If mode-free operation at 170 GHz was required, a coaxial line would require and outer diameter of 0.5 mm and have a loss of at least 33 dB per metre assuming perfect surface finish.

3.2 RECTANGULAR WAVEGUIDE

Figure 4 defines the dimensions of a rectangular waveguide and the specifications of waveguides for use at the high end of the microwave band and the millimeter wave bands are shown in Table 2.

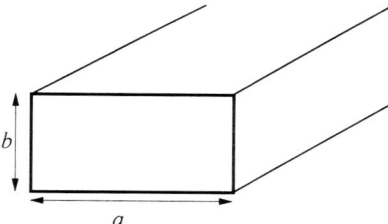

Figure 4. Rectangular Waveguide

TABLE 2. Specifications for Rectangular Waveguides

Band	Frequency GHz	Waveguide EIA	ID (a x b) mm	Tolerance μm	Cutoff GHz	Loss dB/m
K	18-26.5	WR-42	10.668 x 4.318	±5.1	14.047	0.35
Ka	26.5-40	WR-28	7.112 x 3.556	±3.81	21.081	0.5
Q	33-50	WR-22	5.690 x 2.845	±2.54	26.342	0.7
U	40-60	WR-19	4.775 x 2.388	±2.54	31.357	0.9
V	50-75	WR-15	3.759 x 1.880	±2.54	39.863	1.3
E	60-90	WR-12	3.099 x 1.549	±2.54	48.350	1.7
W	75-110	WR-10	2.540 x 1.270	±2.54	59.010	2.3
F	90-140	WR-8	2.032 x 1.016	±1.27	73.840	3.3
D	110-170	WR-6	1.651 x 0.8255	±1.27	90.840	4.6
G	140-220	WR-5	1.295 x 0.6477	±1.27	115.750	6.5
Y	170-260	WR-4	1.092 x 0.5461	±1.27	137.520	8.5
J	220-325	WR-3	0.8636 x 0.4318	±1.27	173.280	11.6

The loss is specified for silver waveguide at the high end of each band. It should be noted that the losses increase rapidly with frequency. There a very considerable fabrication difficulties with the smaller sizes of guide and it is impossible to scale the tolerances properly with the dimensions of the structure. There are considerable problems with the construction of components inside waveguide at millimeter wave frequencies and beyond. The surface resistance is typically 25% or more worse than the theoretical values even with the best available calculations of the effects of surface roughness. The power handling capability decreases with increasing frequency since it is determined by the breakdown fields across the guide.

3.3 DIELECTRIC WAVEGUIDE

Considerable success has be achieved with rectangular dielectric waveguide (Figure 5) applied at millimeter wave frequencies. The guide can be made from a wide range of materials and is relatively easy to fabricate. The resulting waveguide has manageable dimensions (10 mm x 5 mm of material with a relative dielectric constant of 2 will support 80 to 150 GHz with a loss of less than 1 dB per metre). The field distribution for the simplest mode is illustrated in Figure 6. It should be notes that a significant proportion of the energy is carried by fields which are outside the guide.

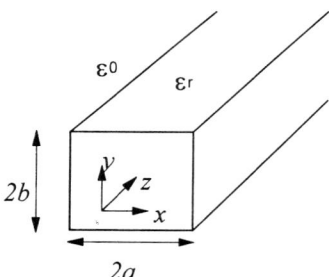

Figure 5. Rectangular Dielectric Waveguide

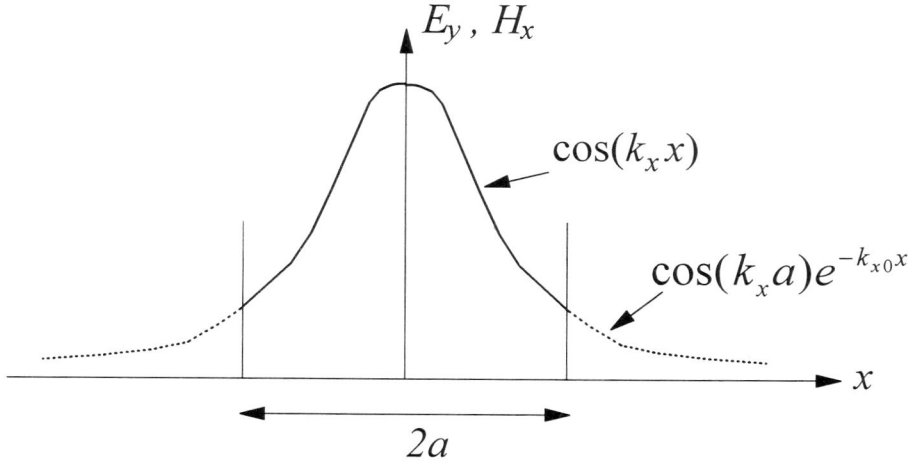

Figure 6. Field Distribution is Rectangular Dielectric Waveguide

The analysis of non-rectangular structures is considerably more difficult, but the same principles apply. It is immediately recognisable that circular dielectric waveguide is better known as optical fibre and that all the technology used in fibre optics can be applied at terahertz frequencies.

3.4 PRINTED TRANSMISSION LINES

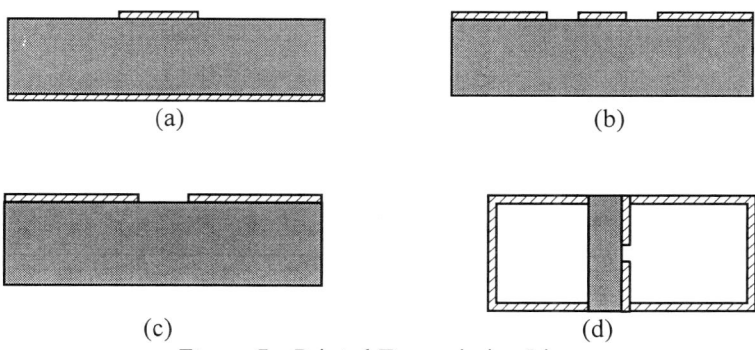

Figure 7. Printed Transmission Lines
(a) Microstrip; (b) Coplanar Waveguide; (c) Slot Line; (d) Finline

A wide range of transmission lines which can be fabricated by printed circuit techniques are in use at microwave frequencies. Four of the most common structures are shown in Figure 7. Although these types of structure a readily fabricated using photolithographic techniques, where the substrate is semiconductor the loss can be unacceptably high as the frequencies approach the terahertz regime. Considerable success has been achieved, however, employing membrane technology for the fabrication of such lines. One important advantage of printed transmission line structures is the ease with which circuits can be integrated with antennas.

4. Quasi-Optical Systems

Dielectric guides (fibres) work very well for wavelengths at the visible and near-visible. Free-space has extremely low loss and it should be as well-suited for millimeter and submillimetre applications as for light. However, conventional ray optics employing beams, mirrors, lenses and the like assume that component sizes are thousands of wavelengths and transmission by plane-wave beams. Attempts to design a system for operation even at submillimetre waves using conventional geometric optics will have poor performance as a results of effects such as diffraction, poor collimation, focussing and coupling to detectors. A highly compact arrangement can be constructed using Gaussian Beam Mode where the beam incident on the mirrors or lenses has a Gaussian distribution of intensity across the cross-section.[3] As shown in Figure 8, the Gaussian form is maintained as the beam propagates. The beam is one of an infinite series of free-space modes which have power localised in the cross-section with distribution that is unchanged as the beam propagates. The fundamental mode is Gaussian with amplitude

$$|u| \propto \exp\left(\frac{-r^2}{w^2}\right) \tag{10}$$

where r is the distance from the axis of propagation and w a parameter giving $1/e$ at the half-width points. The minimum value occurs at the "beam-waist" w_0 which is a quasi-focus. The fundamental mode as no cutoff frequency; the higher-order modes have modulated cross-section, but the Gaussian envelope varies in the same way as the fundamental.

The beam modes are solution to the wave-equation

$$\nabla^2 \psi + k^2 \psi = 0 \tag{11}$$

where $k = 2\pi/\lambda$. If the excitation is sinusoidal and propagation is in the z-direction, then

$$\psi = u(x,y,z)\exp(-jkz)\exp(j\omega t) \tag{12}$$

and assuming the paraxial limit

$$\frac{\partial^2 u}{\partial z^2} \to 0$$

then the equation becomes

$$\frac{\partial^2 u}{\partial x^2} + \frac{\partial^2 u}{\partial y^2} - 2jk\frac{\partial u}{\partial z} = 0 \tag{13}$$

The fundamental mode solution is

$$u = \frac{w_0}{w}\exp\left(\frac{-r^2}{w^2}\right)\exp\left[-j(kz-\phi)\right]\exp\left(\frac{-jkr^2}{2R}\right) \tag{14}$$

where

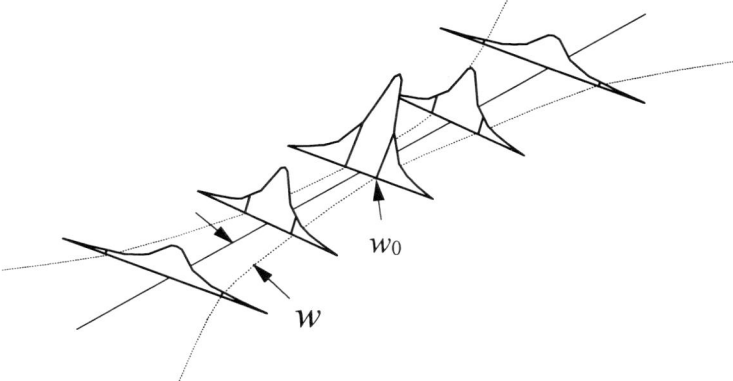

Figure 8. Gaussian Beam Amplitude Profile

$$w^2 = w_0^2 \left[1 + \left(\frac{\lambda z}{\pi w_0^2} \right)^2 \right] \qquad R = z \left[1 + \left(\frac{\pi w_0^2}{\lambda z} \right)^2 \right]$$

$$\phi = \arctan\left(\frac{\lambda z}{\pi w_0^2} \right) \qquad r^2 = x^2 + y^2$$

for large values of $|z|$, $\theta = \dfrac{\lambda}{\pi w_0}$

These parameters describe a spherical wavefront with radius of curvature R, Gaussian distribution half-width w and a phase angle ϕ modifying that expected for plane-wave propagation along the z-axis. The behaviour is shown in Figure 9.

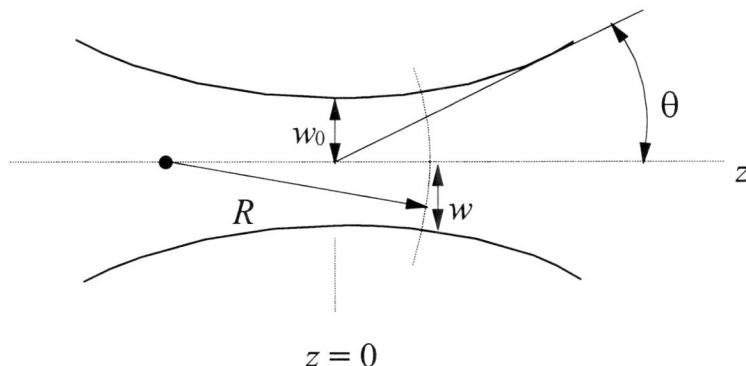

Figure 9. Parameters of Gaussian Beam

It can be seen that the Gaussian beam modes suffer dispersion, similar to the modes in waveguide. However, conventional lenses and mirrors will affect the beams in a manner similar to geometric optics and systems can be designed using simple lens formulae but relating the positions of the beam waists. It is readily demonstrated that for a wavelength of 1 mm (300 GHz) and a feed aperture ratio of 10, the beam waist is approximately 6.4 mm. We could the use up to 200 mm between 40 mm lenses and mirrors. This allows adequate space to accommodate signal processing components such as beam splitters and the power loss due to diffraction is calculated to be no more than .005 dB.

5. Network Characterisation at Sub-Millimetre Wavelengths

The need for accurate measurements at the top end of the microwave frequency range and beyond is not disputed, but much that is taken for granted at lower frequencies cannot be achieved beyond 100 GHz or so. Although sources to provide test signals are available, adequate power levels, suppression of spurious signals and good match remain a challenge to designers and really broadband sources are uncommon. Good detectors based on schottky diodes are, however, readily available offering high sensitivity and wide bandwidth with frequency response well into the terahertz region. Network and signal measurements also require downconverters - good mixers, both fundamental and harmonic, have been developed using schottky diodes although local oscillators remain the weakness.

The greatest problem in making high quality measurements remains the difficulties encountered in interconnecting components and devices and the design and fabrication of calibration components. The frequency limit of precision coaxial connectors is at about 120 GHz and even here the required mechanical tolerances make the components expensive and delicate. Similarly, waveguide flanges (routinely available up to 325 GHz) are small and require skill to achieve repeatable connections. It would appear that dielectric waveguide or free-space (quasi-optical) techniques offer the best method of interconnection at terahertz frequencies, although there remain issues with launchers and connections between these media and the environment in which the devices to be measured are located.

5.1 REFLECTOMETERS AND NETWORK ANALYZERS

The basis of the network analyzer is a reflectometer - essentially a component which can separate the incident and reflected waves on a transmission line. Conventionally, this is accomplished with a dual-directional coupler; the detection system determines whether it is possible to determine on the magnitude of the reflection coefficient (scalar) or both magnitude and phase (vector). Figure 10 shows the block diagram of the conventional network analyzer, capable of bi-directional reflection and transmission measurements.

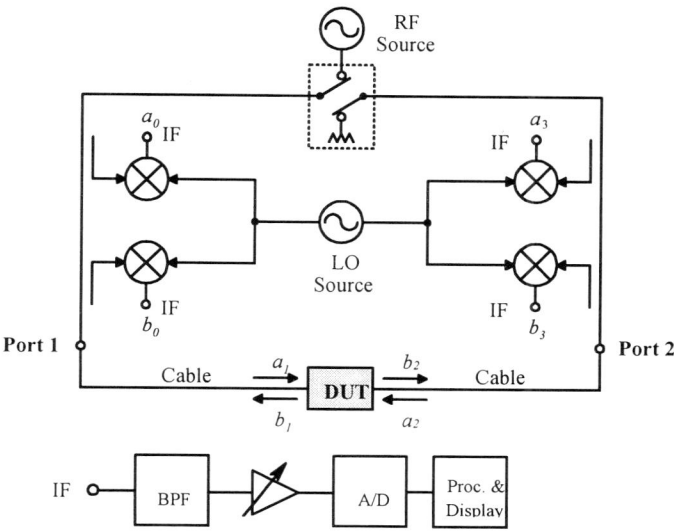

Figure 10. Block Diagram of a Network Analyzer.

Provided that signal separation components are available, it is possible to extend the frequency range of this block diagram by using frequency multipliers to provide the test signal and harmonic mixers to permit the use of a tunable local oscillator at a frequency in the microwave band. The basic scheme is shown in Figure 11. It is important to note that both the source and local oscillator must be derived from an accurate frequency synthesizer or the local oscillator must be phase locked to the source to ensure that the IF is at precisely the correct frequency for input to the receiver. The frequency resolution of the source and the multiplication factor determine the stimulus frequencies that can be provided. This basic arrangement has been implemented at frequencies beyond 300 GHz.

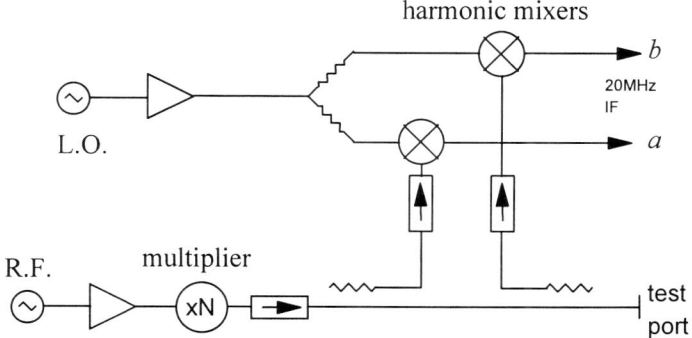

Figure 11. Vector Network Analyzer Frequency Extension

At higher frequencies, however, the availability of components makes this arrangement increasingly difficult to implement. An alternative vector reflectometer which is more amenable to implementation at terahertz frequencies is the six-port (Figure 12).[4]

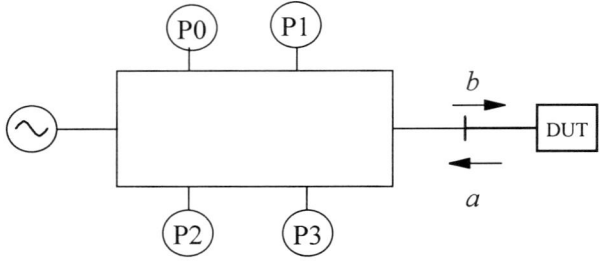

Figure 12. Six-port reflectometer - basic principle.

A vector reflectometer required four data items (magnitude and phase of incident and reflected signals) to determine the complex reflection coefficient. This information can be inferred from four *scalar* measurements - in principle, P0, P1, P2 and P3 can be power sensors. There are a number of successful six-port arrangements in use, but a particularly economical one is the multi-state reflectometer (Figure 13). In this arrangement, the use of a highly repeatable switched reflection permits sufficient power measurements to made to satisfy the equations and thereby solve for the unknown reflection coefficient.

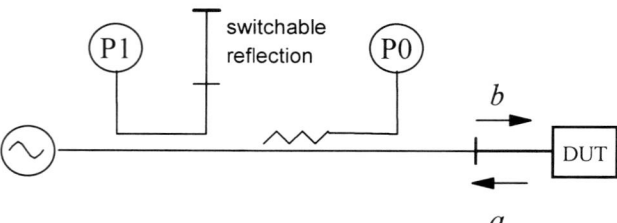

Figure 13. Multi-state reflectometer

It is possible to implement this arrangement in quasi-optical form, replacing the direction couplers with mirrors and beam splitters. The schematic is shown in Figure 14. The source and DUT ports are connected to the system by means of corrugated horns and the Gaussian beams are polarised as a consequence of using wire grids. The variable reflector is a grid which is rotated and moved along a track by means of precision motor drives. This system has a bandwidth which is determined by the spacing of the grid wires (upper limit, currently beyond 1.2 THz) and the diameters of the mirrors and grids (lower limit). The most significant limitations are the availability of sources and means to launch the beams.

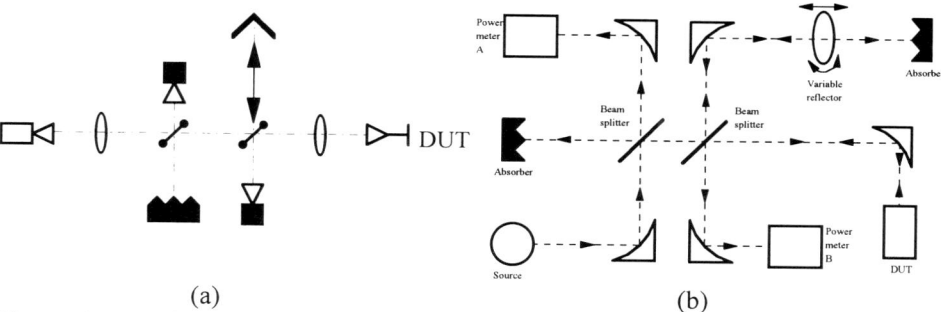

(a)   (b)

Figure 14. Quasi-optical reflectometer: (a) basic principle, (b) practical implementation.

It is well known that the systematic errors due to the imperfections of the instrumentation make error correction essential for scattering parameter measurements at high frequency. The error correction techniques are well established and the key to accurate measurements lies in the provision of precision artifacts which can be used for calibration. The basic principle assumes that the measurement system is linear and that all the errors may be included in a network which lies between an ideal measurement system and the test ports. It is thus only necessary to measure a number of known devices to generate sufficient equations to solve for the parameters of the error network and permit calculation of the parameters of the unknown device on subsequent measurements. The key to calibration is then the provision of defined devices which can be used in the calibration step.

The basic principle can be described in terms of the system equations - the parameters are as shown in Figure 10. The scattering parameters of the device under test (DUT) are

$$\begin{bmatrix} a_1 \\ a_2 \end{bmatrix} = \mathbf{S}_A \begin{bmatrix} b_1 \\ b_2 \end{bmatrix} \tag{15}$$

whilst those measured are

$$\begin{bmatrix} b_0 \\ b_3 \end{bmatrix} = \mathbf{S}_M \begin{bmatrix} a_0 \\ a_3 \end{bmatrix} \tag{16}$$

We can write the parameters of the error adaptor as

$$\begin{bmatrix} b_0 \\ b_3 \\ a_0 \\ a_3 \end{bmatrix} = \mathbf{T} \begin{bmatrix} a_1 \\ a_2 \\ b_1 \\ b_2 \end{bmatrix} \tag{17}$$

where

$$\mathbf{T} \equiv \begin{bmatrix} \mathbf{T}_1 & \mathbf{T}_2 \\ \mathbf{T}_3 & \mathbf{T}_4 \end{bmatrix} = \begin{bmatrix} t_{00} & t_{03} & t_{01} & t_{02} \\ t_{30} & t_{33} & t_{31} & t_{32} \\ t_{10} & t_{13} & t_{11} & t_{12} \\ t_{20} & t_{23} & t_{21} & t_{22} \end{bmatrix}$$

Thus a set of measurements with known values of $\mathbf{S_A}$ allow the determination of $\mathbf{T}$ from a set of simultaneous linear equations

$$\mathbf{S_M} = \left(\mathbf{T_1 S_A} + \mathbf{T_2}\right)\left(\mathbf{T_3 S_A} + \mathbf{T_4}\right)^{-1} \tag{18}$$

and the determination of $\mathbf{S_A}$ once $\mathbf{T}$ is known from

$$\mathbf{S_A} = \left(\mathbf{T_1} - \mathbf{S_M T_3}\right)^{-1}\left(\mathbf{S_M T_4} - \mathbf{T_2}\right) \tag{19}$$

A variety of schemes have been developed for the solution of equation (18) and the choice of method depends on the transmission line medium and the availability or possibility of fabrication of suitable known artifacts. Most of the methods are suitable for implementation at millimetre wavelengths and beyond; it must be emphasised that in all cases a reference impedance (length of transmission line or reflectionless termination) and a port-to-port connection are required. The most widely used method employs a Short circuit, Open circuit, Load and Through (SOLT) and is one of the simplest to formulate and solve. However, the provision of so many devices and the need to make at least seven connections has led to the development of other techniques. For application at the highest frequencies, variations on the TRL technique have proved to be the most popular. The method requires only the Through connection of the two ports, the connection of a length of Line defining the reference impedance between them and the connection of a Reflection (which may be unknown) at each port. This requires the use of only one component (the Line) which is in any sense a standard and, although the solution of the equations is somewhat laborious, it is usually possible to fabricate the required components.

5.2 SPECTRUM ANALYZER FREQUENCY EXTENSION

The measurement of signals at ever-increasing frequencies is made easier by the use of a spectrum analyzer which can display all the frequency components present in a single sweep. However, the frequency range of a receiver is limited by the tuning range of the local oscillator. By employing harmonic mixing, it is possible to extend the frequency range which can be displayed limited only by the capabilities of the mixer. The basic arrangement is shown in Figure 15.

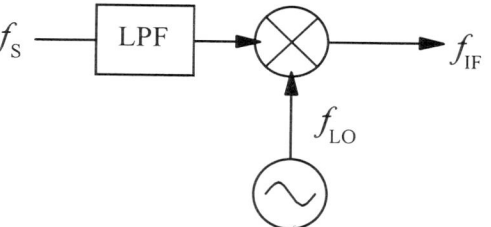

Figure 15. Signal Detection by harmonic mixing

If the local oscillator amplitude is high, then the frequencies are related by

$$f_S = N.f_{LO} \pm f_{IF} \qquad (20)$$

and a wide range of signal frequencies can be detected at a fixed IF for a small range of local oscillator tuning. An example of the tuning curves derived from equation 20 are shown in Figure 16, at higher frequencies the line-pairs are steeper and closer together.

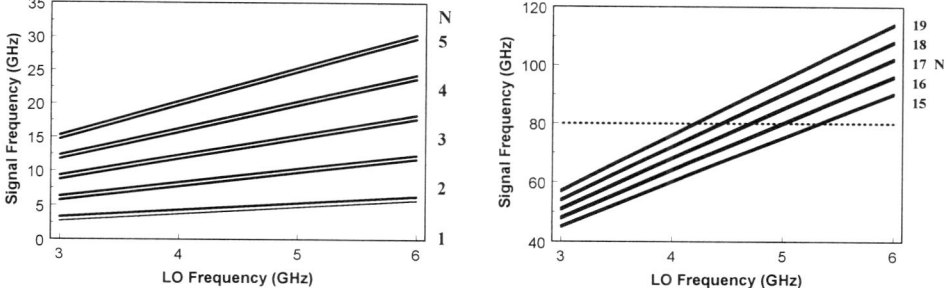

Figure 16. Harmonic Mixing Tuning Curves

It can be seen that, unless precautions are taken, there will be ambiguity in the response seen at the IF from a single input frequency. This ambiguity is normally resolved at lower frequencies by the use of a preselector - a tunable bandpass filter which is swept synchronised with the local oscillator. At higher frequencies the situation may be resolved using image identification techniques which rely on identification of the value of n in equation 19. The process makes very precise changes to the local oscillator frequency which will cause the response to shift in frequency by a calculated amount. Thus if the local oscillator is moved up and down by $2 \times f_{IF}/N$ then a response at the calculated place with higher LO means that the $N-$ mixing mode has been identified, a response at the lower LO means that the $N+$ mixing mode is identified and no response means that the value of N is incorrect. With a microprocessor-controlled instrument and a digitally synthesised local oscillator, this process can be carried out entirely automatically and can be made transparent to the user.

6. Conclusions

Conventional transmission line technologies and components are difficult to manufacture and have high loss at terahertz frequencies. From consideration of their properties, it would appear that dielectric waveguide (optical fibres) and quasi-optical techniques can provide an environment in which the transmission line techniques developed at microwave frequencies can be used in the sub-millimetre range and beyond. Both measurement and calibration have been demonstrated and employ extensions of the techniques which have been used successfully at lower frequencies.

The most important aspects which enable good quality measurements to be made are, however, still very weak at frequencies above about 100 GHz. Firstly, the lack of any connector technology makes it especially difficult to achieve precise, repeatable, low loss, good match connections which are necessary to make measurements. Secondly, the problems associated with fabrication of passive components which are suitable for use as calibration standards, particularly high quality transmission lines or their equivalent which exhibit known and uniform characteristic impedance.

7. References

1. Wadell, B.C. (1992) *Transmission Line Design Handbook,* Artech House.
2. Kurokawa, K. (1965) Power Waves and the Scattering Matrix, *IEEE Transactions on Microwave Theory and Techniques,* **MTT-13**, 194-202.
3. Martin, D.H. and Lesurf, J. (1978) Submillimetre-Wave Optics, *Infrared Physics,* **18**, 405-412.
4 . Engen, G.F. (1977) The Six-Port Reflectometer: An Alternative Network Analyzer, *IEEE Transactions on Microwave Theory and Techniques*, **MTT-25**, 1075-1079.

Theme 2

Device Update

UNIPOLAR AND BIPOLAR RESONANT TUNNELING COMPONENTS

C. VAN HOOF, J. GENOE, S. BREBELS, PH. PIETERS, E. BEYNE,
G. BORGHS
IMEC
Kapeldreef 75, B-3001 Leuven, Belgium

Abstract

Resonant tunneling based components have potential application in ultra-high speed analog and digital applications. These components can also act as high-speed light-emitters which makes them particularly attractive for chip-to-chip optical communication. In addition, novel hybrid means of integrating such III-V components may open new areas of application. In this article we briefly overview existing resonant tunneling transistor schemes. The resonant tunneling light-emitting transistor and diode are discussed in more detail. Integration of microwave elements in multi-chip-module technology and the integration with resonant tunneling components is discussed.

1. Introduction - Resonant Tunneling Transistors

Every transistor in which resonant tunneling occurs could be called a resonant tunneling transistor (RTT) [1,2]. If we only consider three terminal devices in which carriers tunnel resonantly from the emitter to the collector through a double barrier structure and in which a quantum well (QW) base layer is added in order to obtain control, we can classify transistors in four groups according to their carrier nature (unipolar and bipolar) and according to the position of the base QW, i.e. equal to or different from the tunneling QW. Figure 1 and Table I summarize the resulting classes.

TABLE 1. Four classes of resonant tunneling transistors

	Unipolar devices	Bipolar devices
Base QW = tunneling QW	Bound-State Resonant Tunneling Transistor (BSRTT)	Bipolar Quantum well Resonant Tunneling Transistor (BiQuaRTT)
Base QW ≠ tunneling QW	Resonant tunneling Hot Electron Transistor (RT-RHET)	Resonant Tunneling Light Emitting Transistor (RTLET)

J.M. Chamberlain and R.E. Miles (eds.), New Directions in Terahertz Technology, 81–95.

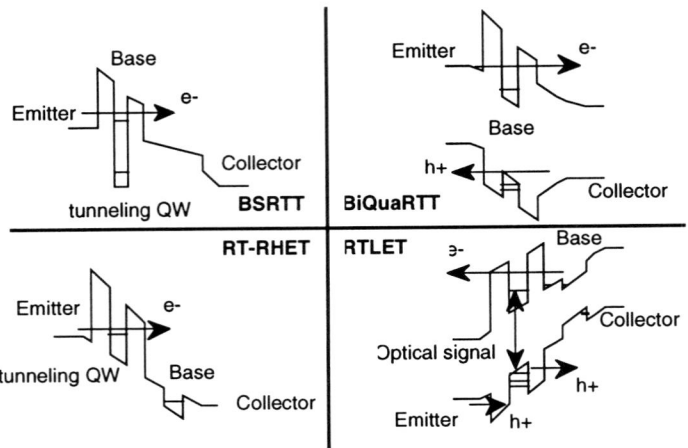

Figure 1. Existing classes of resonant tunneling transistors.

1.1. UNIPOLAR DEVICES WITH ONE QUANTUM WELL

The Bound-State Resonant Tunneling Transistor (BSRTT) (Figure 1) has been proposed by Schulman *et al.* [3] and Haddad *et al.* [4]. The principle of operation is as follows. The first energy level is strongly populated by confined electrons to obtain a good base conductivity. Changing the bias applied to this layer will change the band structure and in this way also the position of the second energy level. The second energy level is used for resonant tunneling from the emitter to the collector.

The base quantum well is made of a semiconductor with a lower bandgap than the emitter layer so that the first energy level is far below the conduction band of the emitter. Tunneling can only occur using the second energy level.

The second barrier consists of 2 parts: the first part is a narrow but high barrier to have a high tunneling rate through the second energy level combined with a good confinement at any bias. The second part is an additional barrier to prevent tunneling escape from the first energy level even under a high base-collector bias.

1.2. UNIPOLAR DEVICE WITH SEPARATE BASE QUANTUM WELL

The Room-Temperature Resonant-tunneling Hot-Electron Transistor (RT-RHET) evolved from a transistor with a wide base and no confinement [5, 6] to a device with a 10 nm base and hence confinement [7] . Initially, the wide base was used to obtain a low base resistance (Figure 1). The major part of the electrons that tunnel through the double barrier structure remains hot until they are above the base-collector separation barrier. The electrons that relax contribute to the base current. To obtain a good room temperature operation, the base width was reduced from 60 nm to 10 nm [7,8] .

1.3. BIPOLAR DEVICES WITH ONE QUANTUM WELL

The bipolar alternative of the RTT has been proposed first by Ricco and Solomon [9] and fabricated by Seabaugh and Reed [10-12] . It was realized using an n-type emitter and collector and a p-type double barrier tunneling structure. A superlattice is added in both the emitter and the collector to restrict the tunneling in the quantum well to the second electron energy level. The base contact is realized by a p-implantation, providing the required isolation. This device requires a high doping concentration in the base, which deteriorates the device characteristics.The multiple-quantum-well base equivalent of this structure has also been realized [14] and more recently, clear negative differential resistance characteristics have been observed [15] .

1.4. BIPOLAR DEVICES WITH SEPARATE BASE QUANTUM WELL - THE RTLET

Due to the reduced density of states, electrons in a 2DEG need to reside on a higher energy (Pauli principle) which makes a 2DEG base unable to screen an electric field completely [17, 18] . Though this generally deteriorates transistor characteristics, it is used in the RTLET design to its advantage. A base quantum well with low charge density acts as a transparent base which allows to maintain the emitter-collector tunneling characteristics, including the NDR feature and oscillation region. At the same time, the quantum-well base layer can be used to inject minority carriers into the tunneling structure [19] . When using an npn structure the resonance features will be more pronounced due to majority-carrier electron tunneling, but the base contact technology is lacking. A pnp structure [19] leads to lower maximum speed but non-spiking ohmic contact technology can be used. In the following section, the resonant tunneling light-emitting transistor is discussed.

2. Resonant Tunneling Light Emitting Transistors and Diodes

2.1. RESONANT TUNNELING LIGHT-EMITTING TRANSISTORS

In the design of the RTLET [19] a separation layer is added between the base quantum well and the tunneling structure as shown in Figure 2.This layer on top of the base quantum well has several crucial tasks. First, it makes the base contact etch more controllable. A GaAs layer is a better etch-stop than an InGaAs layer. It also allows a small consumption during the deposition of the base contact. Such consumption (3 to 4 nm) effectively occurs in both the PdGe regrowth and the MBE regrowth. Furthermore, it acts as a barrier for tunneling into surface traps. Finally, it creates an electron accumulation layer under the double barrier structure which participates in the base transport process.

The base layer is delta-doped. This is also important under the base contact since it creates a pn-junction as an isolation between base and collector. The delta doping is at a small distance from the base quantum well to maintain the high mobility in the base layer. We can distinguish two areas in the current-voltage characteristics (figure 3). As long as the base doping layer is not completely depleted ($V_{EC} < 3.3$ V), bipolar amplification occurs. The transistor acts as a bipolar transistor with a tunneling structure between emitter and base. At higher emitter voltages, we find no bipolar amplification, but the injected base current is still observed and appears as a surplus in the emitter current. There are no electrons in the base doping layer but only in the quantum well. The quantum-well base layer has a small effect on the field across the tunneling structure but still allows electron transport and injection in the tunneling structure. The NDR position is independent of the base voltage. The steplike behavior in the collector current is present for all base bias voltages indicating that the structure is always in oscillation in the NDR region. A higher electron injection decreases the PVR due to increased scattering.

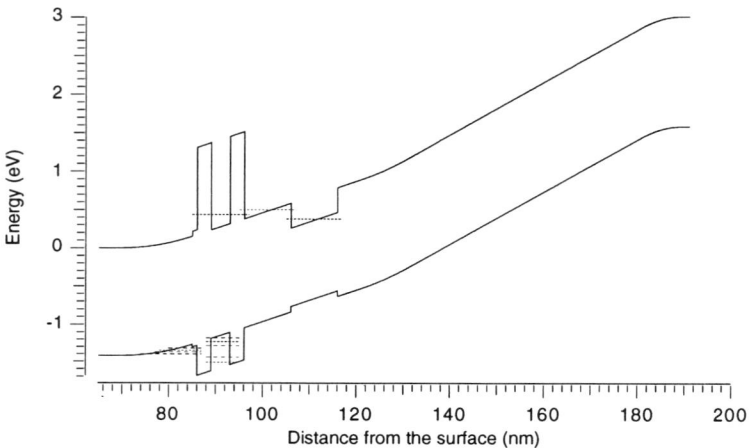

Figure 2. Band structure of the RTLET under forward bias operation showing line-up of the base quantum well with the tunneling quantum well.

Concerning the LED emission, the quantum-well electroluminescence at 1.59 eV dominates the GaAs luminescence (1.43 eV) (Figure 4), which indicates that the major part of the recombination takes place in the tunneling structure [20]

External quantum efficiency is low but can be improved as will be shown in the next section. Also, speed aspects of the device are covered in section 2.2.

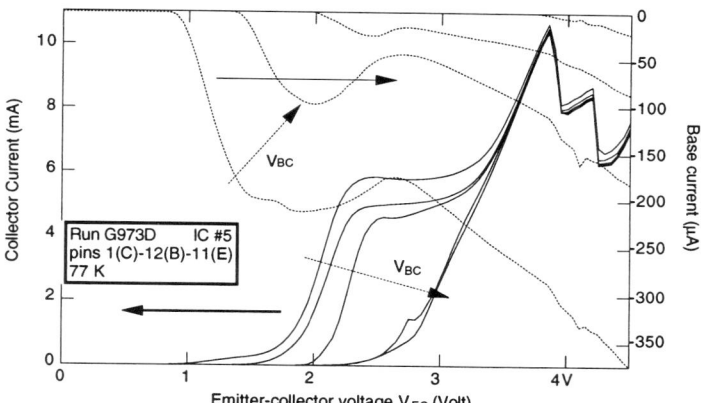

Figure 3. The collector current and the base current as a function of the emitter-collector voltage at 77 K. Below VEC=3.3 V bipolar amplification occurs. Above that, the base injection current adds to the emitter current and the peak-to-valley ratio decreases.

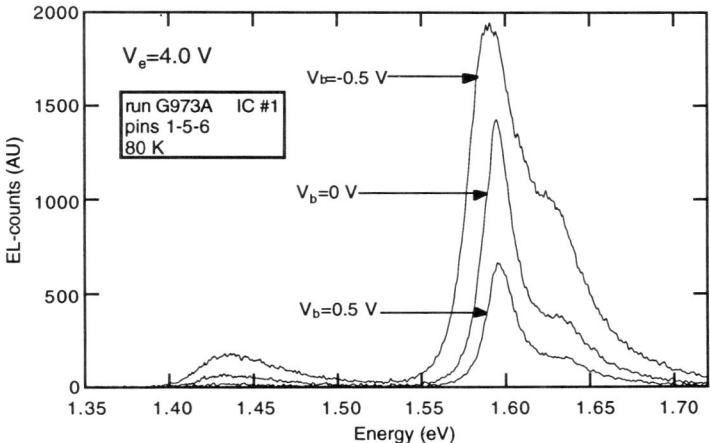

Figure 4. Emission spectra of the RTLET biased in the oscillation region The base voltage varies from 0.5 V to -0.5 V. The tunneling QW is the main light-emitter.

2.2. RESONANT TUNNELING LIGHT-EMITTING DIODES

A resonant tunneling light-emitting diode (RTLED) [20-22] consists of an undoped double-barrier resonant tunneling structure inside the intrinsic region of a p-i-n junction.

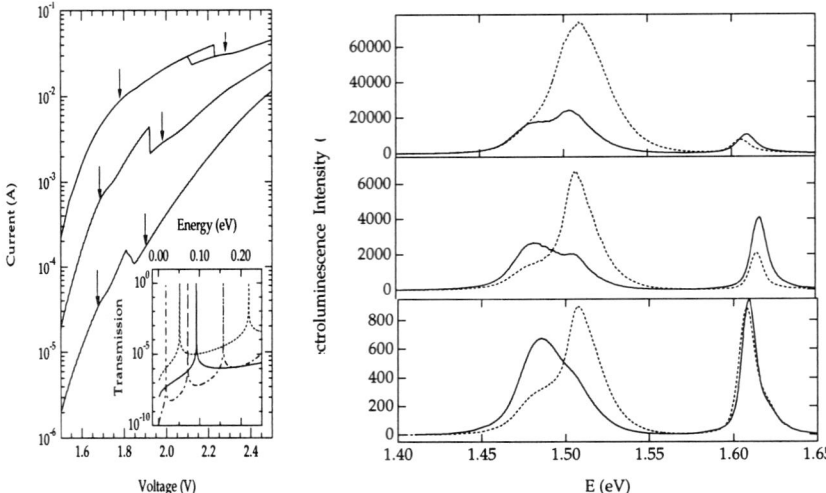

Figure 5. (a) I(V) characteristics of RTLEDs with 3 nm, 4 nm and 5 nm barrier (top, middle and bottom). The large resonances are electron resonances and the small resonances indicate light-hole resonances. The insert shows the electron and hole transmission coefficients. (b) LED spectra of the RTLEDs with 3 nm, 4 nm and 5 nm barrier (top, middle and bottom) biased at and beyond the electron resonance. Charge redistribution between QW and accumulation layers takes place.

Figure 5 shows the I(V) characteristics of three RTLEDs at cryogenic temperature. Both electron and hole resonances occur, each causing charge redistribution between the quantum well and the electron and hole accumulation layer on either side of the tunneling structure. This is clear from the electroluminescence spectra. Electroluminescence from three distinct and spectrally separated regions occurs: the quantum well and the electron and hole accumulation layers. Between the electron resonance and the off-resonance, the electron population in the quantum well is drastically reduced, leading to electron and hole charge redistribution in order to maintain global bandbending. This is pronounced for all RTLEDs: the electron charge in the electron accumulation layer increases drastically to compensate for the loss of charge in the quantum well, as can be seen from the increased n-GaAs emission at 1.51 eV. The emission from the p-GaAs hole accumulation layer decreases between resonance and off-resonance, in agreement with the reduced electron current.

Room temperature operation leads to decreased current peak-to-valley ratio, but the inherent speed of the device remains unaffected. In addition, the external quantum efficiency at room temperature amounts to 0.25 % as follows from figure 6 where the power conversion is plotted versus drive current.

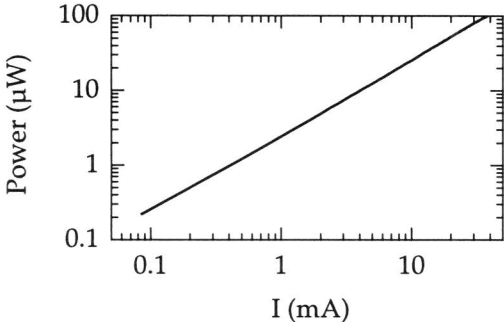

Figure 6. Room-temperature optical power conversion efficiency of the RTLED at room temperature. An external quantum efficiency of 0.25 % is obtained.

In contrast to conventional light-emitters, the optical response is not limited by radiative recombination but rather by tunneling escape. The *room temperature* optical response of the RTLED to a rectangular voltage pulse is shown in Figure 7. Characteristic rise and decay time constants are 180 ps demonstrating a 3 dB frequency of above 2 GHz.

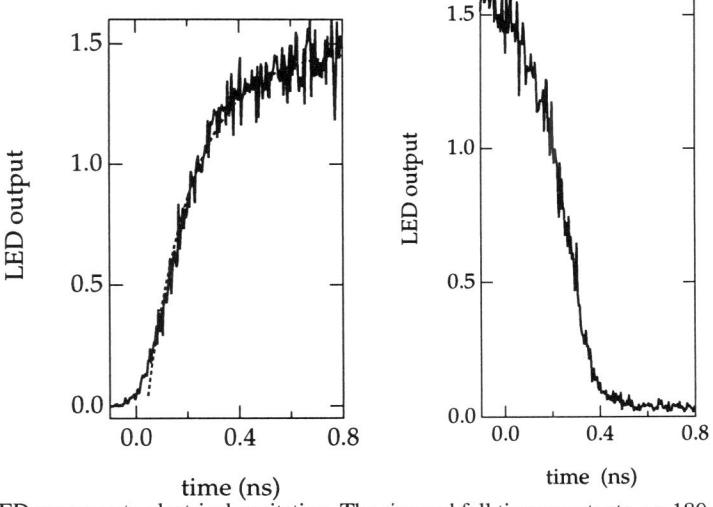

Figure 7. LED response to electrical excitation. The rise and fall time constants are 180 ps each. The external quantum efficiency of the RTLED is 0.25 %.

Due to their high-speed combined with fair efficiency, these structures can be useful for chip-to-chip optical communication where there is no need for coherent radiation nor narrow linewidth.

88

3. Integration of microwave elements in MCM-D
3.1. INTRODUCTION

An efficient way of integrating resonant tunnel diodes (or other GaAs devices) together with their biasing circuits, matching networks and antennae on the same substrate is using thin film multi-chip module technology. The MCM-D (multi-chip module deposition) interconnection technology uses multiple layers of metals and dielectric films [23] and was thus far primarily used to interconnect high speed digital integrated circuits. But as this thin film technology has excellent dimension control, it may also be used to interconnect microwave integrated circuits (MICs). Active devices may be interconnected to the substrate using wire bonding, flip-chip [24] and epitaxial lift-off [25] and the passives may be integrated in the substrate itself.

A typical cross-section of a microwave MCM-D structure is depicted in figure 8. The substrate is generally thin film graded alumina or low loss cordierite [26]. The advantage of the cordierite is the low dielectric constant ($\varepsilon_r = 4.9$) which enables a size enhancement of the high frequency components. The lower metallisation is a 2 μm thick sputtered Ti/Cu/Ti layer. Here, a 30 nm thick Ti layer is used as an adhesion layer. The top metallisation is a 5 μm thick electroplated Cu layer. As dielectric, a 10 μm thick photosensitve BCB, Cyclotene™, from DOW [27,28] is applied. This spin coatable dielectric material has a permittivity of 2.7, a low loss tangent and low moisture absorption. With this build-up, a wide range interconnections and components can be realised. Some typical passive microwave components integrated in the MCM-D substrate will be discussed next.

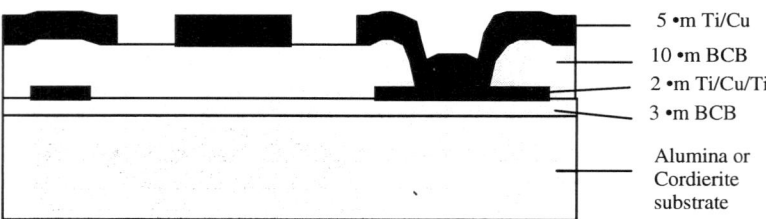

Figure 8. A typical microwave MCM-D build-up.

3.2. INTEGRATED INDUCTORS

Depending on the required inductance value, quality factor and the available area, different types of inductors (e.g. spiral and line inductors) can be integrated in the MCM-D substrate [29]. The spiral inductors can be made circular as coplanar cut-outs in the ground plane. An example of such an inductor can be seen in figure 9. The inside of the coil is connected to the outside with an underpass on the bottom metallization. With this kind of inductors, inductance values from a few nH upto tens of nH and quality factors upto almost 30 may be obtained.

Line inductors are useful if low inductance values are required or if the inductor has to interconnect parts over larger distances. These line inductors are simply coplanar lines with a large spacing to the ground plane. In this way the high impedance transmission line has a dominating inductive part and inductance values of typically 1.02 nH/mm are obtained. The maximum quality factor is about 20.

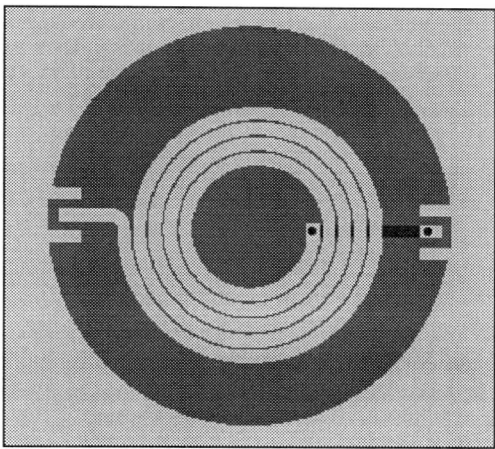

Figure 9. Photograph of a typical MCM-D spiral inductor, with dimensions : Dout = 1200 μm, N = 4.5, W = 50 μm, S = 20 μm.

3.3. INTEGRATED CAPACITORS

Metal-insulator-metal (MIM) capacitors can easily be realised in the multi-layer MCM-D technology : the two parallel metal plates of the capacitor are made in the two metal layers. The size of the plates, together with the dielectric in-between determines the capacitance. For small capacitance values, the BCB dielectric can be used (e.g. for 4 μm of BCB we obtain about 7 pF/mm^2), while for larger ones, anodised Ta is very useful (values upto 1 nF/mm^2 are possible).

A more unusual type of integrated capacitor is the radial stub capacitor. This type of capacitor is used as a capacitor to ground, in shunt with the signal line. Its fan-like shape is basically a microstrip structure, narrow at the beginning and broad at the end. At low frequencies the component act as a simple capacitor to the ground plane underneath, but when the frequency raises it starts to act as a microstrip open stub. The specific length and dimension realises the desired capacitor behaviour for a given frequency range.

3.4. INTEGRATED RESISTORS

In passive microwave circuits, the need for integrated resistors is obvious. They are e.g. used in matched loads, attenuators, couplers and baluns. These resistors can be made by means of inserting a thin NiCr resistive layer in the layered MCM-D structure. Such a NiCr layer of 90 nm thickness gives a resistance of about 23 Ω/o.

3.5. INTEGRATED MICROWAVE FILTERS

Besides the use in biasing circuits for RTDs or other active devices, the integrated passive components can be used to realise a given filter function. Since this is fully described in [29], we will briefly discuss here the design procedure and an example.

Passive microwave filters integrated in MCM-D may be realised as an appropriate combination of inductors and capacitors. A clear advantage of this integration is the fact that the filters can be incorporated in the interconnection between different parts of the MCM, in this way realising a functional interconnection. The design of these filters starts from the given filter specifications, from which we calculate the ideal values of the LC-elements of the filter ladder network. These ideal components are translated to practical MCM-D structures using a fitting procedure with the HP MDS software tool. Figures 10,11 and 12 show the different steps in the realisation of a 3th order chebyshev band pass filter (with a pass band from 5 to 10 GHz). The comparison between measurements and simulations of figure 6 indicates a good agreement. This shows that microwave filters may be integrated efficiently in the MCM-D interconnection substrate.

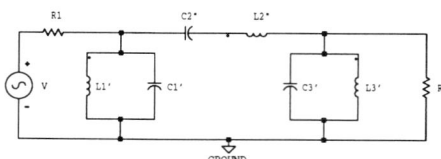

Figure 10. Lumped elements equivalent ladder network of a 3th order chebyshev band pass filter.

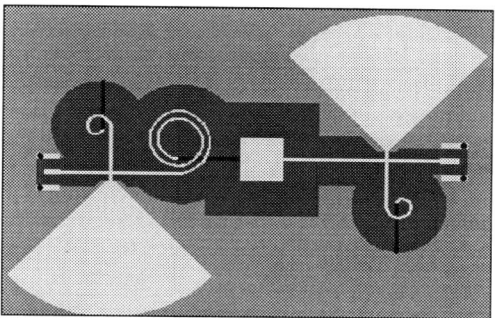

Figure 11. Photograph of a 3th order chebyshev band pass filter integrated in the MCM-D substrate.

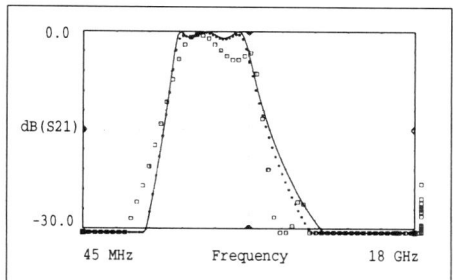

Figure12. Comparison of S21 transmission measurements (o), MDS model simulations (•) and ideal LC network simulations (◯) of a MCM-D 3th order chebyshev band pass filter.

3.6. INTEGRATION OF ACTIVE COMPONENTS BY EPITAXIAL LIFT-OFF

The integration of the active components into the circuit can be done by realising the active devices onto the MCM-D substrate before the deposition of the dielectric and metal layers for the passive circuit. These active components can then be connected using the multilayer structure of the MCM-D as can been seen in figure 13. This method uses only lithographic steps for the realisation of the contacts, hence reducing the parasitics of the interconnection. However, it uses a high dielectric substrate which is defavourable for the realisation of integrated antennae.

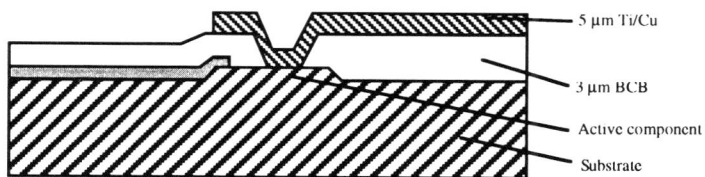

Figure 13. MMIC with multilayer interconnection circuit.

In a hybrid approach, the MCM-D and the active devices are realised separately and connected to each other in a next step e.g. by using wire bonding. This makes an optimal processing of the separate circuits possible, but has high interconnection parasitics.

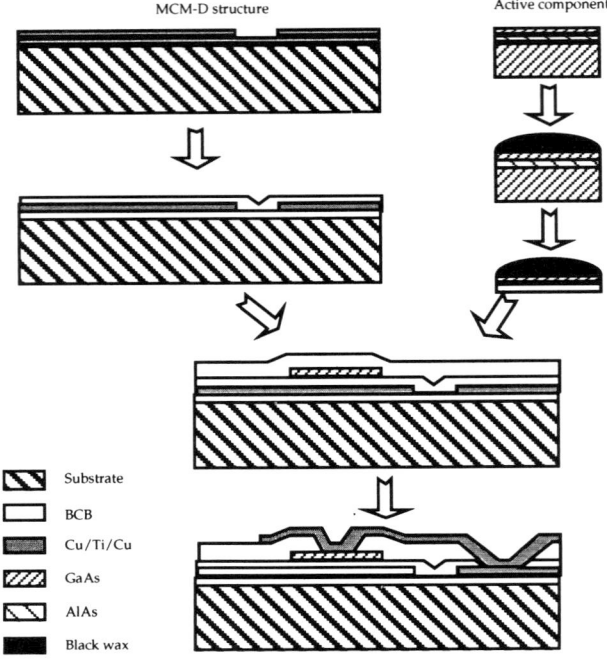

Figure 14. Epitaxial lift-off: method.

A third method is epitaxial lift-off [25,32,33]. Epitaxial lift-off is a pseudo-monolithic integration method : the circuits are processed separately first and are connected afterwards by photolithography. This method combines the advantages of both the monolithic and the hybrid bonding, i.e. a large processing flexibility with low interconnection parasitics. Figure 14 depicts the epitaxial lift-off method: The MCM-D structure and the GaAs active devices are first processed separately. The extreme etch selectivity in a $HF:H_2O$ solution of a AlAs layer sandwiched between the GaAs substrate and the GaAs device layers makes the release of the thin (between 200 nm and 5 μm) device layers possible. These released layers can then be transplanted and adhered to an other substrate using the Van der Waals forces. The device layers are protected while etching and supported during the transplantation by a black wax layer. A thin (e.g. 3 μm) BCB layer is deposited on top of the MCM-D to improve the adhesion of the transplanted devices with the new substrate. The connection of these extremely thin active devices can now proceed using deposition and photolithographic techniques as in a monolithic integration.

Figure 15. Transplanted GaAs light emitting thyristor on glass.

An example of the successful transplantation of a GaAs thyristor on glass is shown in figure 15. The devices are shown from the glass side : the 200 nm thick thyristors are more or less optical transparent. The performance of the devices remains unaffected.

3.7. APPLICATION: SHORT RANGE RADAR IN MCM-D

The schematic of a FMCW radar system for measuring the distance and velocity of a reflecting object can be seen in figure 16. Figure 17 shows the practical implementation of the circuit by using a Resonant Tunnelling Diode as self-oscillating mixer.

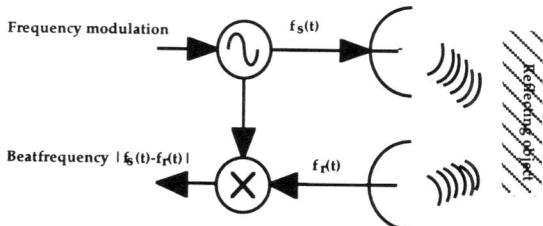

Figure 16. Schematic of the short range FMCW radar.

The RTD generates an high frequency signal which is directly fed to the microstrip patch antenna. The reflected signal is also received by the antenna and will be mixed in the RTD with the oscillation signal. The resulting low frequency beat signal can be measured in the bias circuit. Frequency sweeping of the oscillation signal can be done by adjusting the bias voltage over the RTD.

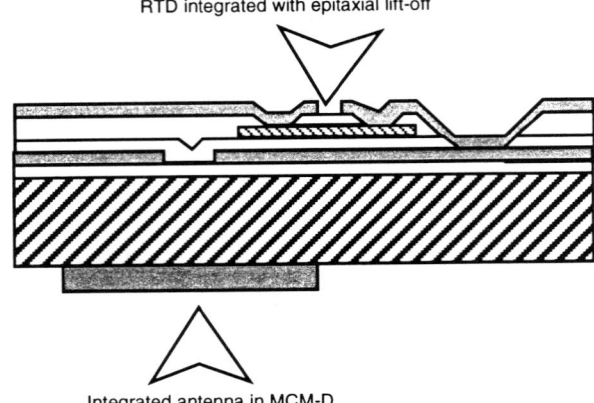

Figure 17. Implementation of the short range radar by using the RTD as a self-oscillating mixer.

4. Conclusions

In this overview article, resonant tunneling transistors have been described. Special focus was on light-emitting devices because of possible application in chip-to-chip communication. Switching speeds of the light-emitters was shown to be in the Gighahertz range. An overview of integration of active and passive components using MCM-D technology was provided including several example circuits.

References

1. Tsu, R. and Esaki, L. (1973) Tunneling in a finite superlattice, *Appl. Phys. Lett.* **22**, 562-564.
2. Capasso,F. , Sen, S., Cho,A. and Sivco, D. (1988) Multiple negative transconductance and differential conductance in a bipolar transistor by sequential quenching of resonant tunneling, *Appl. Phys. Lett.* **53**, 1056-1058.
3. Schulman, J. and Waldner, M. (1988) Analysis of second level resonant tunneling diodes and transistors, *J. Appl. Phys.* **63**, 2859-2861.
4. Haddad,G.I., Mains,R.K., Reddy,U. K. and East, J. R. (1989) A proposed narrow-band-gap base transistor structure, *Superlattices and Microstructures* **5**, 437-441.
5. Mori, T., Ohnishi, H., Imamura, K., Muto, S. and Yokoyama, N. (1986) Resonant Tunneling hot-electron transistor with current gain of 5, *Appl. Phys. Lett.* **49**, 1779-1780.
6. Yokoyama, N., Imamura, K., Ohnishi, H., Mori, T., Muto,S. and Shibatomi, A. (1988) Resonant-Tunneling Hot Electron Transistor (RHET), *Solid-State Electronics* **31**, 577-582.
7. Seabaugh, A. C., Kao, Y.-C., Randall, J., Frensley, W. and Khatibzadeh, A. (1991) Room Temperature Hot Electron Transistors with InAs-Notched Resonant-Tunneling-Diode Injector, *Jpn. J. of Appl. Phys.* **30**, 921-925.
8. Activities of the Research and Development Association for future Electron Devices, *FED Journal* 44-60.

9. Ricco, B. and Solomon, (1984) P. M,. *IBM Tech. Dig. Bull.* **27**, 3053.

10. Seabaugh, A. C., Reed, M. A., Frensley, W. R., Randall, J. N. and Matyi, R. J. (1988) Realization of Pseudomorphic and Superlattice Bipolar Resonant Tunneling Transistors, Proceedings of *IEDM* , Washingthon DC, 900-902.

11. Reed, M. A., Frensley, W. R., Matyi, R. J., Randall, J. N. and Seabaugh, A. C. (1989) Realization of a three-terminal resonant tunneling device: The bipolar quantum resonant tunneling transistor, *Appl. Phys. Lett.* **54**, 1034-1036.

12. Seabaugh, A. C., Frensley, W., Randall, J., Reed, M. A., Farrington, D. L. and Matyi, R. J. (1989) *IEEE Trans. Electron Dev.* **36**, 2328.

13. Capasso, F. (1990) Is the Resonant Tunneling Transistor a Reality?, *Physics Today* September 1990, 132.

14. Waho, T., Maezawa, K., and Mizutani,T. (1991) Resonant Tunneling in a Novel Coupled Quantum well base Transistors, *Jpn. J. of Appl. Phys.* **30**, L2018-L2020.

15. Waho, T., Maezawa,K., and Mizutani, T. (1993) Clear Negative Characteristics Observed in Coupled Quantum well base Resonant Tunneling Transistors, *IEEE Electron Device Lett.* **14**, 202-204.

16. Iogansen, L. V., Malov, V. V. and Xu, J. M. (1993) Time dependent theory of a double quantum well resonant interband tunnel transistor, *Semicond. Sci. Technol.* **8**, 568-574.

17. Luryi, S. (1988) Quantum capacitance devices, *Appl. Phys. Lett.* **52**, 501-503.

18. Luryi, S. (1989) Coherent versus incoherent resonant tunneling and implications for fast devices, *Superlattices and Microstructures* **5**, 375-382.

19. Genoe, C. Van Hoof, K. Fobelets, R. Mertens and G. Borghs, (1992) pnp resonant tunneling light emitting transistor, *Appl. Phys. Lett.* **61**, 1051-1053.

20. Van Hoof, C., Genoe, J., Bertram, D., Grahn, H.T., Borghs, G. (1995) Bipolar charge redistribution in resonant tunneling light-emitting diodes, *Phys. Rev.* **B51**, 13491-13498.

21. Van Hoof, C., Genoe, J., Portal, J.C., Borghs, G. (1995) Spatially indirect transitions due to coupling between a hole accumulation layer and a quantum well in resonant tunneling diodes, *Phys. Rev.* **B51**, 14745-14748.

22. Van Hoof, C., Genoe, J., Portal, J.C., Borghs, G. (1995) Charge accumulation in the two-dimensional electron gas emitter of a resonant tunneling diode, *Phys. Rev.* **B52**, 1516-1519.

23. Doane, D.A., Franzon, P.D. (1993) *Multichip Module Technologies and alternatives : The Basics*, Van Nostrand Reinhold, NY.

24. Richter, H. et al. (1994) Flip Chip Attach of GaAs-Devices and Application to mm-Wave Transmission Systems, *Proc. of Microsystems Technologies, VDE-Verlag GmbH.*, 535-543.

25. Young, P. et al., (1993) *A 10-GHz amplifier using an Epitaxial Lift-Off pseudomorphic HEMT device*, IEEE-MGW letters, vol. 3, 107.

26. Pieters, Ph., Beyne, E., Roggen, J., Heidinger, R., Nazaré, S., Schüßler, A., Gnappi, G., Montenero, A. (1996) Microwave Structures on a new Cordierite substrate, *Proc. of ISHM ICET International Conference on Electronics Technologies, focusing on microwave materials, processes, devices and applications, Brighton, UK.*

27. Beyne, E., Van Hoof, R., Achen, A. (1995) The use of BCB and photo-BCB dielectrics in MCM-D for high speed digital and microwave applications, *4th International Conference & exhibition on MultiChip Modules, Denver, Colorado, April 19-21, 1995.*

28. Chinoy, P.B., Tajadod, J. (1993) Processing and Microwave Characterisation of Multilevel Interconnects Using Benzocyclobutene Dielectric, *IEEE-CHMT*, **16**, no. 7.

29. Pieters, Ph., Brebels, S., Beyne, E. (1996) Integrated microwave filters in MCM-D, *Proc. of IEEE MCMC'96 Multi-Chip Module Conference, Santa Cruz, CA, 6-7 February, 1996.*

30. Pozar, D. M. (1992) Microstrip Antennas, *Proc. of IEEE*, **80**, no. 1.

31. Rebeiz, G. M. (1992) Millimeter-Wave and Terahertz Integrated Circuit Antennas, *Proc. of IEEE*, **80**, no. 11.

32. Yablonovitch, E., Gmitter, T., Harbison, J. P., Bhat, R. (1987) Extreme selectivity in the lift-off of epitaxial GaAs films, *Appl. Phys. Lett.*, **51** (26).

33. Van Hoof, C., De Raedt, W., Van Rossum, M., Borghs, G. (1989) MESFET lift-off from GaAs substrate to glass host, *Electron. Lett.*, **25**.

SUPERCONDUCTING MIXERS FOR SUBMILLIMETRE WAVELENGTHS

E. L. KOLLBERG
Chalmers University of Technology
S- 412 96 Göteborg, SWEDEN

1. Introduction

For radio astronomy and remote sensing applications at frequencies of the order THz there is a strong need for receivers with much higher sensitivity than is available at present. Today, most receivers for frequencies near and above 1 THz have to rely on Schottky-diode mixers, with rather poor sensitivity [1,2]. Low noise SIS mixers based on superconductors have excellent performance and have replaced Schottky-diode mixers for frequencies up to about 650 GHz, corresponding to the energy gap of niobium [1,2,4,5]. Since niobium tri-layer technology is by far the most successful SIS-mixer technology and since the RF loss will be significant above the energy gap of niobium [6], it may be very difficult to realise SIS mixers with a noise temperature limited to a few times the quantum limit ($T_{mixer} \approx hf/k$) above about 700 GHz.

Superconductor Hot-Electron Bolometer (HEB) mixers utilising thin superconducting films in the resistive state have recently emerged as a serious alternative to the traditional mixers used in THz receivers [7-9]. In one version, phonon cooled devices are used. In another, one is utilising diffusion-cooling of hot electrons [10,11]. Which version which will come out as the better is unclear at the moment (May 1996). These recently proposed devices have a very simple structure and can be realised with a technology which is relatively uncomplicated compared to SIS and Schottky diode fabrication. Based on the frequency insensitive nature of the electromagnetic interaction and the non linearity, it should be possible to use these devices up to several THz, where they can be integrated with planar antennas.

The SIS and HEB mixers will be discussed below and since the HEB mixer is of a much more recent type, it will be described in more detail.

2. The SIS mixer

The SIS mixer uses properties of superconductivity. The acronym SIS stands for Superconductor-Insulator-Superconductor. It could also be called the "quasi particle mixer". SIS mixers are presently the most sensitive alternative for low noise receivers in the frequency range from about 75 GHz to about 1 THz. Since it is essentially a

J.M. Chamberlain and R.E. Miles (eds.), New Directions in Terahertz Technology, 97–117.

resistive mixer operating in the quantum regime, we will start by introducing some basic concepts valid.

2.1. THE CLASSICAL RESISTIVE MIXER

A mixer is a frequency down-converter, i. e. power at a signal frequency ω_s is converted to a much lower frequency. A mixer can be realised using a non-linear impedance element and the most common type for millimetre and waves is still the Schottky diode. The non linear impedance of a Schottky diode can be modulated at frequencies up to several THz.

When a local oscillator voltage $V_{LO} \cdot cos(\omega_{LO}t)$ is applied to a device with a non-linear impedance (i.e. where the impedance varies with applied current or voltage) a periodic variation of its impedance is obtained. This can be described by a Fourier series

$$G = G_0 + G_1 \cdot cos(\omega_{LO}t) + G_2 \cdot cos(2\omega_{LO}t) + \ldots \ldots \quad (1)$$

Adding a small signal voltage $\delta v \cdot cos(\omega_s t)$ (where the amplitude is too small to significantly effect the conductance waveform), a small signal current is obtained viz.

$$\delta i = \delta v \cdot G_0 \cdot cos(\omega_{LO}t) + \frac{1}{2} \cdot \delta v \cdot G_1 \cdot \left[cos\left((\omega_{LO} + \omega_s)t\right) + cos\left((\omega_{LO} - \omega_s)t\right) \right] +$$

$$+ \frac{1}{2} \cdot \delta v \cdot G_2 \cdot \left[cos\left((2\omega_{LO} + \omega_s)t\right) + cos\left((2\omega_{LO} - \omega_s)t\right) \right] + \ldots \quad (2)$$

Fourier components are created at $n\omega_{LO} \pm \omega_s$, or $m\omega_{LO} \pm \omega_{IF}$, where $\omega_{IF} = |\omega_{LO} - \omega_s|$ is the so called intermediate frequency (IF). Notice that IF current is created for $\omega_s > \omega_{LO}$ (upper sideband) as well as for $\omega_s < \omega_{LO}$ (lower sideband)[*]. Also notice that the signal δv will cause current frequency components at the image frequency $2\omega_{LO} - \omega_s$.

The more non-linear the device impedance is, the better conversion efficiency can be obtained. The SIS mixers device is extreme in this respect. Indeed the SIS mixers are working in the quantum limit (see below), which, for certain choices of embedding impedance network, will allow obtaining conversion gain larger than one [12,13].

2.2. THE SIS DEVICE

In a superconductor below the superconducting transition temperature, electrons form, so called Cooper pairs. When they do so, the energy of the electrons near the Fermi energy is lowered by a certain amount Δ. Hence to break up a Cooper-pair, an energy of

---

[*] When the mixer noise is measured by the Y-factor or "hot-cold body" technique, broad band noise sources at two different temperatures are used. This means that a so called "double sideband" (DSB) noise temperature is measured. For a mixer with the same conversion efficiency in the two sidebands, the the single sideband mixer noise temperature is twice that of the DSB. Compare section 2.3 below.

2Δ is required. This can described in terms of a bandgap with energy 2Δ, as shown in Figure 1.

In the quasiparticle or SIS mixer, a tunnelling phenomenon is used. Two superconductors are separated by a thin (≈ 20 Å) layer of insulator. The tunnelling can be viewed in the following way. Under bias, Cooper pairs on one side of the insulator break up into two electrons (quasiparticles) that individually tunnel through the insulator and may recombine on the other side. This is illustrated in Figure 1.

It is interesting to note that the density of states near the bandedges becomes "infinite". This is one important reason why there is such a sharp increase in the current when the device is biased to a voltage $V=2\Delta/e$ (e is the charge of the electron). The IV characteristic is shown in Figure 2. Notice that the voltage scale is in mV, and that *1 meV corresponds to 240 GHz*. Evidently the IV is strongly nonlinear within a fraction of a mV. This is the reason that this mixer is operating in the quantum regime. Compare also Figure 3, where the IV of an SIS device is compared with a Schottky diode. The Schottky diode is obviously not very non-linear within a voltage interval of 1 mV, and is therefore operating fully as a classical mixer for frequencies up to several Thz.

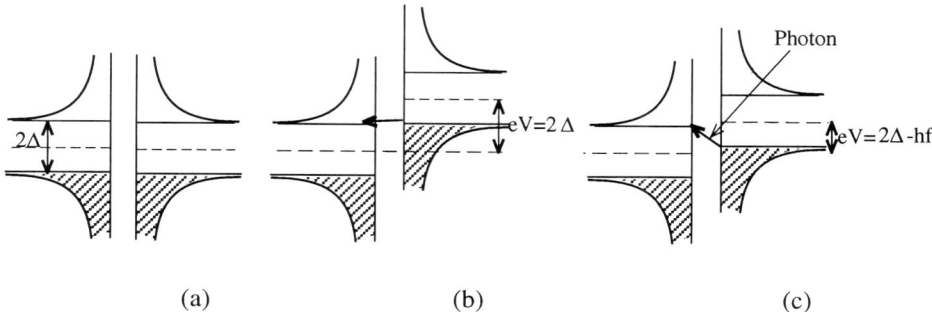

Figure 1. The SIS device under bias: a) No bias; b) enough bias so tjat electrons to the left can tunnel into empty states on the right; c) tunnelling assisted by a photon with energy hf.

In Figure 4 is shown a typical design of an SIS device using Nb tri-layer technology [25]. The device structure is $Nb/Al_2O_3/Nb$, where the ≈ 20 Å thick Al_2O_3 serves as the insulator in the SIS device. Although the structure looks complicated, it is not very difficult to fabricate. These devices are fabricated and used by many laboratories throughout the world.

For frequencies above about 700 GHz one is trying to develop devices based on NbN, which has a higher bandgap (≈ 1.2 THz). So far these attempts have not been very successful.

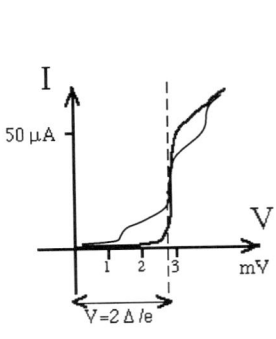

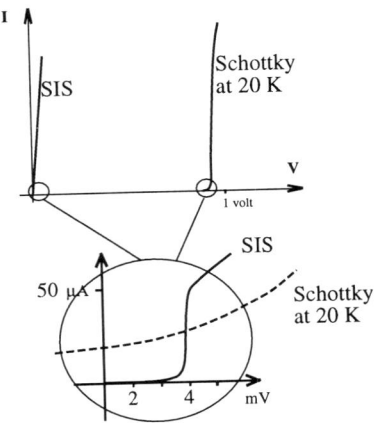

Figure 2. Typical IV characteristic of an SIS element. The thin line indicates the shape of the pumped IV-characteristic, where $f_{LO} \cdot h \gg 2\Delta$, while the fat line is for the unpumped device. The pump frequency corresponds approximately to 325 GHz.

Figure 3. Comparing the IV of an SIS element with that of a Schottky diode. Notice the enormous difference in non-linearity.

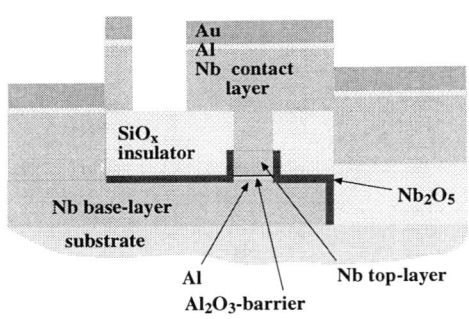

Figure 4. A typical layout of a tri-layer Nb/Al$_2$O$_3$/Nb type SIS device.

2.3. RECEIVER NOISE TEMPERATURE

The mixer is the heart of a low noise THz receiver. However, to determine the receiver sensitivity or equivalent noise temperature, it is necessary to account for attenuation in the quasioptics in front of the antenna, and the noise from the IF amplifier following the mixer. This is illustrated in Figure 5.

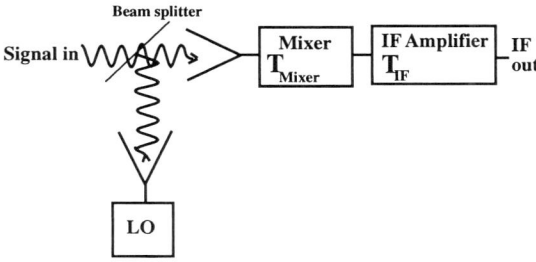

Figure 5. Schematic diagram of a mixer receiver.

The noise temperatures T_{mixer} and T_{IF} are defined at the input of the mixer and the IF amplifier respectively. If the conversion loss for the upper and the lower sidebands are equal, the receiver noise temperatures for a single sideband (SSB) and double sideband (DSB) receiver are respectively

$$T_{receiver,DSB} = T_{background} + 1/2 \times \left(T_{mxr,SSB} + L \cdot T_{IF}\right) \quad \text{(Double Sideband)} \quad (3)$$

$$T_{receiver,SSB} = 2 \cdot T_{background} + T_{mxr,SSB} + L \cdot T_{IF} \quad \text{(Single Sideband)} \quad (4)$$

In the double sideband case we have a useful signal in both sidebands, e. g. as in a radiometer observing broad band noise sources. Figure 6 shows a schematic of a receiver (omitting the input quasioptics).

2.4. NOISE IN SIS MIXERS

The basic source of noise in the SIS mixer is shot-noise, i. e. noise which is caused by electrons having a finite charge when passing through the device. The rms current fluctuations due to shot-noise can be expressed as

$$\left\langle \delta i^2 \right\rangle = 2eI \cdot \Delta f \quad (5)$$

where e is the charge of the electron, I is the current, and Δf is the bandwidth of the receiver. If the internal impedance of the device is equal to dV/dI, we have the equivalent output noise power

$$P_{noise} = \frac{\left\langle \delta i^2 \right\rangle}{4 \cdot dI / dV} = \frac{eI \cdot \Delta f}{2 \cdot dI / dV} \quad (6)$$

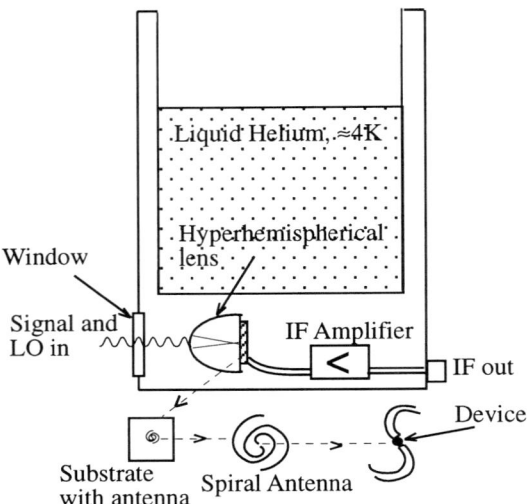

Figure 6. Schematic description of typical dewar with SIS or Hot-electron mixer. The LO and the signal enter together through the dewar window. The radiation is focused on to the antenna using a hyperhemispherical lens, down converted to the IF in the non-linear device and finally amplified by the IF amplifier. In this Figure a spiral antenna is illustrated; many other planar antenna structures are available.

Identifying P_{noise} with $kT_{eq} \cdot \Delta f$, we obtain

$$T_{eq} = \frac{e}{k} \cdot \frac{1}{2} \cdot \frac{1}{dI/dV} \tag{7}$$

Notice that equation (7) describes the output noise of a device biased with a DC current I. In Figure 3 the I-V characteristics of a SIS device and a Schottky diode are compared. From experimental data, equation (7) yields an *output noise temperature* of about 200K for a typical room temperature Schottky diode and about 100 K for a cooled diode. For a typical SIS device the corresponding temperature is less than 10 K. Moreover, the conversion loss of a Schottky mixer in the submillimetre wave frequency range is more than 10 dB, while for the SIS mixer it may be as small as 3 dB. This should lead to a noise temperature for a good cooled Schottky mixer of about 1000 K (SSB) and for a SIS mixer 20 K. In reality the theory is much more complicated than indicated here, mainly because the current is strongly modulated by the LO [12,13]. However, this estimate, where an average current is used, indicates that the SIS mixer should offer an order of magnitude better noise temperature than the Schottky mixer. This has been clearly demonstrated in practice.

2.5. MIXER CIRCUITS

The equivalent circuit of the SIS device is a non-linear resistance in parallel with a capacitance. This capacitance is basically the parallel capacitance (C) between the two superconductors separated by the insulating 20Å thick layer. The capacitance C is quite large and creates a circuit problem particular at higher frequencies. Often one defines the $\omega R_n C$-product where R_n is the normal resistance of the junction, measured above the step in the I-V curve. Typically the $\omega R_n C$-product is between 5 and 10 for good junctions.

There is another current caused by the IV non-linearity, but this is 90° out of phase with the ordinary quasi particle current. It can, however, in most practical cases be neglected. The main reason is that the parallel plate capacitance dominates the impedence of the device.

There are several circuits topologies that can be used for SIS mixers. Most common is to let the top conductor extend over a ground plane to form a microstrip line of a length which allows it to create an inductance in parallel with the junction. In this way the junction capacitance can be perfectly compensated at the *required* frequency. In another more elegant method two junctions are used. They are connected via a microstrip transmission line on the same substrate, such that the impedance of junction B at the terminals of the junction A is inductive, perfectly compensating the capacitance of junction A [14].

3. The Hot-Electron Bolometer Mixer

3.1. THE DEVICE

Bolometers are simple square law detectors and can in principle be used in heterodyne mixers. There is no instantaneous response at the RF, so there will be no harmonics of the LO or any signal power transformed to the image frequency as is the case for the SIS and Schottky mixer. The principal disadvantage of a bolometric mixer is the response time τ_o of the bolometer, limiting the maximum IF to $1/2\pi\tau_0$ In Figure 7 the basic elements of the bolometer are shown. The absorbing part of the bolometer (the electrons in the hot-electron bolometer) has a certain heat capacity C, and there is a thermal conductance from the absorber to the thermal bath. The response time is then $\tau_o = C/G$, i. e. the larger the heat capacity and the smaller the thermal conductance the longer is the response time.

104

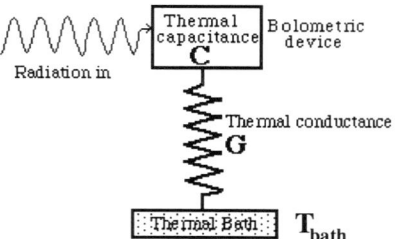

Figure 7. Basic elements of a bolometer.

A *superconducting* Hot Electron Bolometer consists of one or several superconducting thin film strips in parallel, deposited on a substrate, of for example, silicon, single crystalline quartz or sapphire. The strips are cooled to the superconducting state and then heated by DC and microwave power to temperatures near the superconducting to normal transition-temperature, where the superconductor will gradually become normal (Figure 8). We call this temperature interval the *resitive transition region*, while for temperatures above this region we talk about the *normal state*.

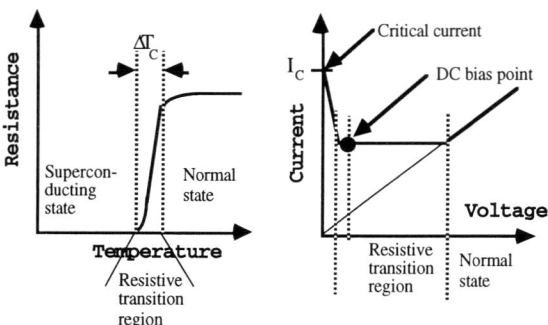

Figure 8. The three states of the bolometer: the superconducting, resistive and the normal state.

The resistance of the device in the resistive transition region, may be explained with the help of several possible physical phenomena, such as the formation of normal domains, phase slip centres, and moving magnetic vortices [15,16,17]. Let us adopt a simplistic model[1]: assume one resistive region (it may be in reality several that we combine into one). This situation is illustrated in Figure 9, where the electron temperature in the "resistive transition regions" is in the interval ΔT_C.

---

[1]Note that the validity of the circuit-based model which we will present in the next section does not depend on the details of the microscopic model, such as the one suggested here.

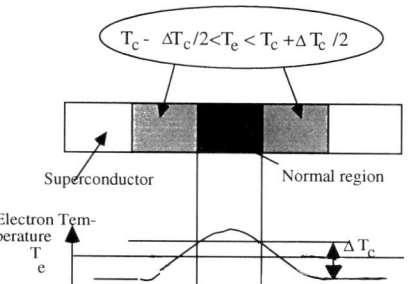

Figure 9. Different possible regions in the bolometer strip.

If the signal and LO frequency are high enough so that $f_{LO}, f_s > 2\Delta/h$, $(2\Delta$ is the energy gap; notice that $2\Delta \to 0$ when the electron temperature $T_e \to T_c$) the full length of the superconducting strip will be seen as a normal conductor strip by the RF. In fact the absorption of RF power by the normal electrons in the HEB is essentially frequency independent. Therefore, these mixers will operate at frequencies well into the THz range, i.e. to much higher frequencies than the gap frequency of the superconductor.

If the intermediate frequency $f_{IF} < 2\Delta/h$, an IV voltage will only develop across the normal regions while the superconducting parts are simply short-circuits. Since $f_{IF} < 1/2\pi\tau_0$, the IF impedance is essentially equal to the DC small signal impedance, i.e. dV/dI at the bias point.

There are two types of bolometric devices, "phonon cooled" and "diffusion cooled". Since the maximum IF-frequency is determined by the electron temperature relaxation time τ_0, i. e. $f_{IF} < 1/(2\pi\tau_0)$, a major issue is to find ways of making the time constant τ_0, short enough.

Typical dimensions of the *phonon cooled* devices are shown in Figure 10.a. Figure 11 shows the relaxation of the electron excess temperature. If the strips are made sufficiently thin, the time constant for phonons in the superconductor to escape to the substrate, τ_{ph-s}, may become shorter than the time constant for phonon electron interaction in the superconductor, τ_{ph-e}. If the strips are made narrow, then the back flow of phonons from substrate to superconductor will be reduced, and thus τ_{s-ph} will be long. At the same time [18], for Nb where $C_e \gg C_p$ at all temperatures below 10K, τ_{ph-e} is much shorter than the electron phonon relaxation, τ_{e-ph} (from the detailed balance equation $C_e/\tau_{e-ph} \approx C_p/\tau_{ph-e}$, where C_e and C_p are the specific heats of the electrons and phonons respectively). Consequently it is possible to heat the electrons *above* the temperature of the lattice and τ_0 will be dominated by τ_{e-ph}, which is not the case for ordinary bolometers where the electrons and the lattice are heated to the same temperature.

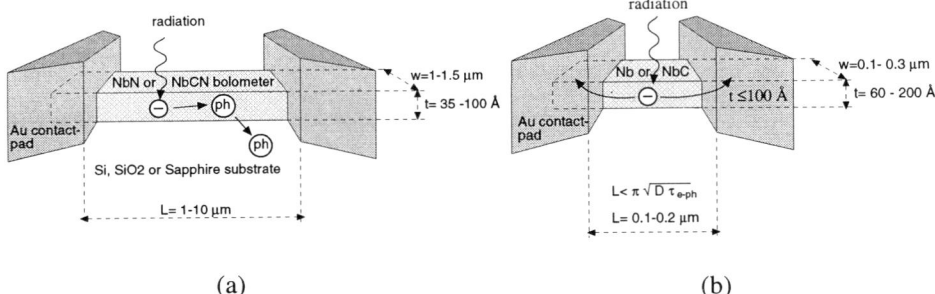

(a)   (b)

Figure 10. The two types of bolometric devices, phonon cooled (a) and diffusion cooled (b).

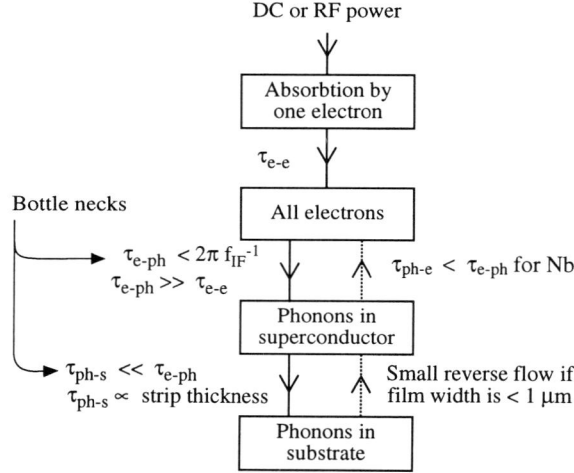

Figure 11.. Time constants for the hot electron bolometer.

In practice, to avoid lattice heating effects in the superconductor, the bolometer should be $\leq 50\text{Å}$ thick and ≤ 1 µm wide. For Nb the resulting bandwidth corresponding to $(2\pi\tau_{e\text{-}ph})^{-1}$ is of the order 100 MHz [7,9], and for NbN it is predicted to be of several GHz [8]. With higher T_c, a higher operating temperature can be used, yielding shorter $\tau_{e\text{-}ph}$. NbN mixers have so far demonstrated >3 GHz bandwidth [19,20].

The *diffusion* cooled device was first suggested by Prober [10,11]. Cooling of hot electrons in this device is caused by this rapid diffusion out of a submicron length strip of superconductor into normal metal contacts (Figure 10b). For the diffusion mechanism to dominate over the electron-phonon cooling, it is necessary to make the micro bridge extremely short. The appropriate length L can be estimated from [10]

$$L \approx 2\sqrt{D \cdot \tau_{ee}}$$

where D is the diffusion constant and τ_{ee} is the electron-electron energy relaxation time. Basically when an electron in the middle of the bridge absorbs energy from an RF photon, it will remain excited for the time τ_{ee} during which it can diffuse the distance L/2 to the normal metal contact which will serve as a heat sink. The pads must be normal metal since Adreev reflection [21] at the interface between the thin hot superconductor (with very small energy gap 2Δ) and the cold superconductor (with full size energy gap 2Δ) will otherwise trap the electron in the bridge. For a 10 nm thick Nb film, L≈0.2 μm. In other materials, such as NbN, it is much shorter, and hence is not likely to perform well in a diffusion cooled mode.

3.2. HEB MIXER THEORY

When operating as a mixer, the device is absorbing LO (P_{LO}) and signal (P_s) power as well as power from the DC-bias supply ($P_{DC}=V_o \cdot I_o$). When the power increases, the electron temperature increases and the resistance of the device increases as $\Delta R=(dR_o/dP)\Delta P$. The RF power $((V_s \, cos(\omega_s t)+V_{LO} \, cos(\omega_{LO}t))^2/R_0$ dissipates in the device. While the device impedance cannot follow the RF, it can follow the beat frequency between the LO and the signal and hence an IF modulation of the bolometer resistance is obtained.

Figure 12 shows an equivalent circuit of the mixer, where the device is biased by a constant DC current. Consequently the modulation at the IF of the resistance will cause an IF voltage to appear across the device, causing current through the load resistance R_L. The IF current superimposed on the DC bias current through the mixer device R_0 creates a "feed-back". The total power dissipated in the device is then:

$$P_o + \Delta P(t) = P_{DC} + \Delta P_{DC}(t) + P_{LO} + P_s + 2\sqrt{P_{LO}P_s} \times cos(\omega_{IF}t) \tag{8}$$

Notice that $\Delta P(t)$ consists of two contributions, one from the RF which is proportional to $(P_s \cdot P_{LO})^{1/2}$ and one from the IF $\Delta P_{DC}(t)=[R_0+dR) \cdot (I_0+dI)^2 - R_0 I_0^2$. It may be assumed that (at least approximately) the DC and RF *power dependencies* of the device resistance are equal. At the bias point of the device we have $V=V_o$ and $I=I_o$. Defining the device DC-resistance as $R_o=V_o/I_o$ one obtains [22] the conversion gain

$$G = \frac{P_{IF}}{P_s} = 2C_o^2 \frac{P_{LO} \, P_{DC}}{(R_L + R_O)^2} \cdot \frac{R_L}{R_O} \left(1 - C_o \frac{P_{DC}}{R_O} \cdot \frac{R_L - R_O}{R_L + R_O}\right)^{-2} \tag{9}$$

where $C_o=dR_o/dP$, R_L is the IF load resistance, P_{IF}, P_s, P_{LO} and P_{DC}, are the IF powers dissipated in the device at the LO and DC.

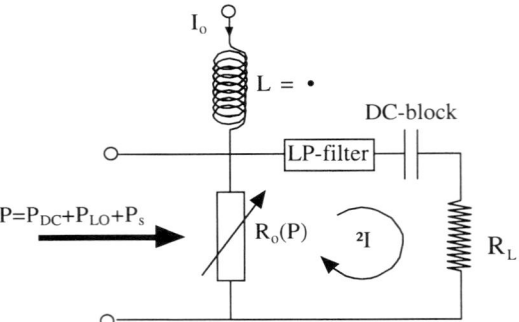

Figure 12. Equivalent circuit of bolometer with load.

Assuming that the total power, i. e. the sum of the DC power and RF power determines the value for R_0, (which is believed to be a fair assumption for the phonon cooled mixer device, and approximately true for the diffusion cooled device), we can evaluate the conversion gain as

$$G = 2\left(1 - \frac{I_0^2}{I_{oo}^2}\right) \frac{\left(C_0 I_0^2\right)^2}{\left(1 - C_0 I_0^2\right)^2} \frac{R_0 \cdot R_L}{\left(R_L + \left(\frac{dV}{dI}\right)_{DC}\right)^2}$$

(10)

Under the bias with the LO applied, we have a DC current I_0, while for the same resistance $(=V/I)$, we have I_{oo} when no LO is applied. The differential resistance $(dV/dI)_{DC}$ is taken at the bias point with LO applied. From this equation it is seen that for $R_L = (dV/dI)_{DC}$ a maximum conversion gain is obtained.

The differential resistance can be predicted if $C_0 I_0^2$ is known (see Appendix). We see that if $C_0 \cdot I_{oo}^2 > 1$, there is a negative resistance available for the unpumped IV. Using Eq. (10) we have calculated the conversion gain vs. $(I_0/I_{oo})^2$ $= P_{DC}/(P_{DC}+P_{LO})$, (see Figure 13).

The commonly assumed fundamental limit of -6 dB gain for hot-electron mixers [22] is obviously not valid if a negative differential resistance of the unpumped IV-curve is available. From Figure 13 it is seen that larger conversion gain is available for $R_L/R_0 > 1$ than for $R_L/R_0 < 1$.

As mentioned above the load resistance for maximum gain is equal to the differential resistance of the I-V-curve at the bias point of the pumped mixer. It is also possible to find the optimum LO power (or $(I_0/I_{oo})^2$) for a given R_L/R_0 or for R_L/R_0 equal to the differential resistance of the I-V-curve at the bias point.

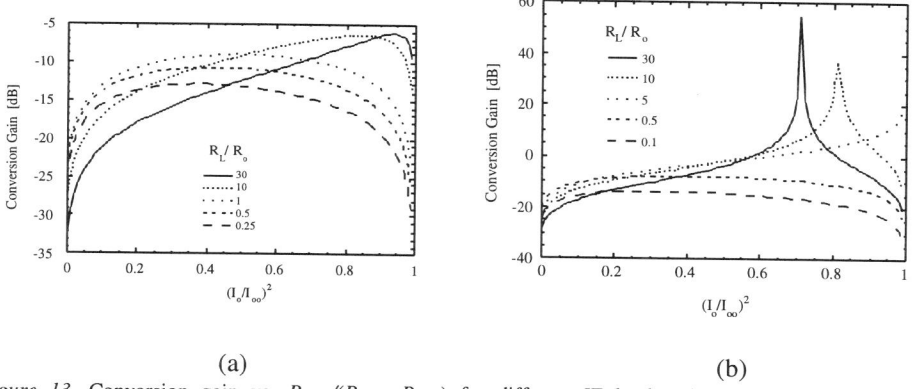

(a)   (b)

Figure 13. Conversion gain vs. $P_{DC}/(P_{DC}+P_{LO})$ for different IF load resistances. The differential resistance for the unpumped curve. (a): $dV/dI=\infty$ or $C_0 \cdot I_{oo}^2=1$; (b):$dV/dI=-5 \cdot R_o$, or $C_0 \cdot I_{oo}^2=1.5$.

3.3. CONVERSION GAIN VS IF

The coefficient $C_o=dR/dP$ can be expressed as $C_o=[dR/d\theta \cdot][d\theta/dP]$. Since the factor $d\theta/dP$ has a frequency dependence determined by the factor $1/(1+j\omega\tau_\theta)$, and assuming that the term $dR/d\theta$ is simply a constant, we may conclude that

$$C_o(\omega) = \frac{C_o(\omega = 0)}{1 + j\omega\tau_\theta}$$

(11)

Using this expression for C_o in the equations for G above, the IF dependence of the conversion gain of the mixer can be expressed as:

$$G = G(\omega = 0) \cdot \frac{1}{1 + (\omega\tau_{mix})^2} \quad where \quad \tau_{mix} = \frac{\tau_\theta}{1 - C_o I_o^2 \cdot \dfrac{R_L - R_o}{R_L + R_o}}$$

(12)

We have introduced τ_{mix}, which determines the -3 dB bandwidth. It is possible to make the bandwidth wider by making the factor $C_o I_o^2 \cdot (R_o - R_L)/(R_o + R_L)$ large $(R_o > R_L)$. In practise, one would look for a compromise between IF bandwidth and conversion gain. We can also conclude that it is undesirable to attempt to utilise a large conversion gain, a feature which the HEB mixer shares with other negative resistance devices.

The frequency dependence of the IF impedance can be derived in the same manner by using expression (A3).

3.4. MIXER NOISE IN HEB MIXERS

The main noise of concern for the superconducting HEB is that which is directly emitted by the device at the IF frequency. The noise theory does not need to take into account correlation of noise contributions at a number of frequencies, as is the case for the Schottky-barrier mixer, for example. The IF output noise is primarily of two types [10,23]: 1) Nyquist noise at the electron temperature, and 2) temperature fluctuation noise. The latter contribution is not usually important for microwave or millimetre wave mixers, but is likely to be the dominant one in superconducting HEB mixers. Temperature fluctuation noise is well-known from the theory of conventional (lattice) bolometers [24]. The electrons are assumed to form a subsystem in thermal equilibrium at the electron temperature, θ, while conducting heat to the lattice heat reservoir at temperature T. The corresponding thermal conductance is G_e [W/K]. It is known from thermodynamics that the average temperature of the electron subsystem will show spectral RMS fluctuations in a 1 Hz bandwidth given by [24]:

$$\left\langle \Delta\theta^2 \right\rangle = \frac{4k_B\theta^2}{G_e} \quad where \quad G_e = \frac{C_e V}{\tau_\theta} \tag{13}$$

where V is the volume of the superconducting strip. Since the resistance of the HEB depends on θ, and the device is biased with constant DC current, I_0, there will be a noise voltage developed across the device. We can derive the equivalent device output noise temperature, $T_{d,t}$, by equating the noise power due to temperature fluctuations to that from a Nyquist source at $T_{d,t}$:

$$T_{d,t} = \frac{I_o^2 \, (dR/d\theta)^2 \, \theta^2}{R_o G_e} \tag{14}$$

Here $(dR/d\theta)$ must be evaluated for actual mixer operating conditions. The total device *output* noise temperature will be

$$T_d \approx \theta + T_{d,t} \tag{15}$$

This noise temperature can now be used in standard expressions for the mixer noise temperature, given the conversion loss, L. If typical values are inserted in Eq. (14), one finds that this term is likely to dominate. Note that $T_{d,t}$ is proportional to the electron temperature squared. Furthermore, θ is expected to be close to T_c. HEBs based on superconductors with higher T_c then should have higher noise. The θ-dependence may be compensated by the appearance of G_e in the denominator, however. As a rule, higher T_c materials have shorter relaxation times , resulting in larger G_e (See Eq. 15). An optimum material in terms of the output noise temperature of the device can therefore not be identified. One must also consider the conversion loss which can be achieved, in order to minimise the receiver noise temperature. These matters are under investigation, but have not yet been resolved in detail. Another noteworthy consequence of the dominance of $T_{d,t}$ is that the output noise temperature is predicted to fall with IF frequency with a time constant similar to τ_{mix} (see Eq. 12). This has been verified, and we can conclude that the bandwidth over which a given mixer noise temperature can be maintained is actually wider than $1/(2\pi\tau_{mix})$, since the device output noise decreases at the same time that the mixer conversion gain decreases.

In conclusion, one can roughly estimate a device output noise temperature for Nb in the range 25-100 K. This is consistent with measurements, as described in a later section.

3.5. EXPERIMENTS

3.5.1. *Nb Phonon Cooled HEB Mixers*
The most extensive experiments published so far have been performed on phonon cooled HEB mixer at a comparatively low signal frequency (20 GHz) [9]. The HEB devices were made from 90 to 150 Å thick niobium films, DC-magnetron sputtered on silicon substrates held at room temperature and patterned by conventional photolithography. The HEBs consisting of 2 parallel niobium strips, 1.5 μm wide and 7 μm long, had a normal resistance between 40 and 150 Ω and were measured at 20 GHz signal and 1-1000 MHz intermediate frequency. T_c of the 90 Å thick devices was about 4.3 K. This value is expected for thinner films [25]. However, these films were not passivated, and therefore include a 20-30 Å thick oxide layer. The films also have a diffusion constant of about half the value of that given by [26].

The lowest **intrinsic conversion loss** of several investigated devices was around 1 dB at 20 MHz IF and operating temperatures between 4.2 K and 2.1 K (Figure 9). The IF of 20 MHz was chosen for the experiments since it is well below the cut-off frequency for hot electron effects (compare Figure 5). In Figure 14 one can see how the conversion loss depends on bias voltage and temperature. The best conversion gain is found for samples with the narrowest superconducting-normal transition width $\mathit{ÆT}_c$, i.e. dR/dT is large.

In Figure 14 the measured intrinsic conversion loss is compared with the theoretical conversion loss calculated from Eq. (9) for two temperatures, 4.2 K (≈ 0.05 K below T_c of the thin strips) and 2.1 K. P_{LO}, R_o, P_{DC} I_o, I_{oo} and $(dV/dI)_{DC}$ were obtained from the DC IV-plots at different LO powers. Pump and signal powers, of about -44 dBm and -72 dBm respectively, are typical values in the experiments.

The conversion gain has dropped 3 dB at around 80 MHz IF, see Figure 15. For larger bias voltages than for maximum conversion, there is a small increase in bandwidth, i.e. the mixer time constant will be slightly shorter with a larger bias voltage.

Assuming the electron relaxation time constant $\tau_\theta = 1.78$ ns, and calculating $C_o I_o^2$ and R_o from the DC IV-characteristic (were obtained theoretically from (Eq. 12)) $\tau_{mix} = 2.6$, 2.14, and 1.83 ns, respectively, for the three bias voltages 1.1, 1.4 and 3.2 mV, in excellent agreement with the experimental time constants.

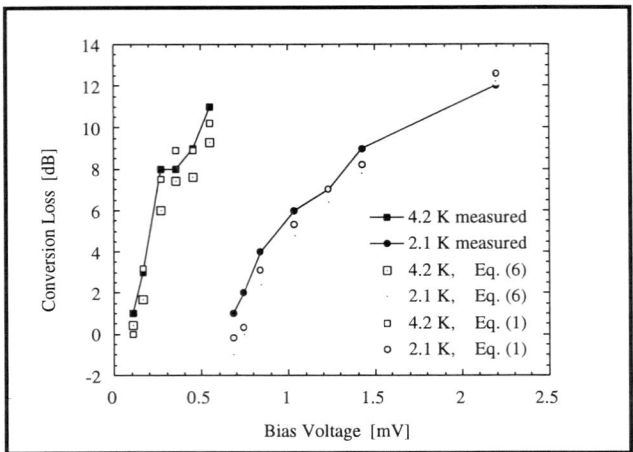

Figure 14. Experimental and calculated intrinsic conversion loss vs. bias voltage and temperature. The LO power is optimised at the lowest bias voltage and kept constant when bias is changed. In the modified Arams expression (Eq. 6), the slope of the unpumped curve is calculated from the slope of the pumped curve.

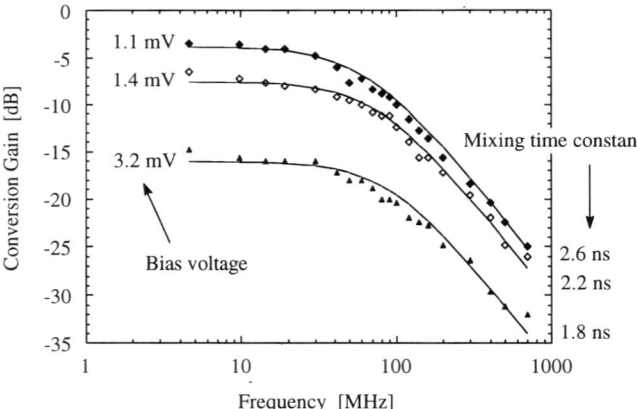

Figure 15. Conversion Gain vs. frequency and bias, showing an increased mixing time constant for low bias at 2.1 K .

The *IF-impedance* of the device has been measured between 1 and 500 MHz (see Figure 16). There is a transition of the impedance in the range 10 to 100 MHz from 90 Ω which is close to the DC differential resistance (108 Ω) at the bias point, to 26 Ω, which is the DC resistance at the bias point. This frequency dependence is in agreement with the assumption that $C_0(\omega)=C_0/(1+j\omega t\theta)$ (see ref. [9]).

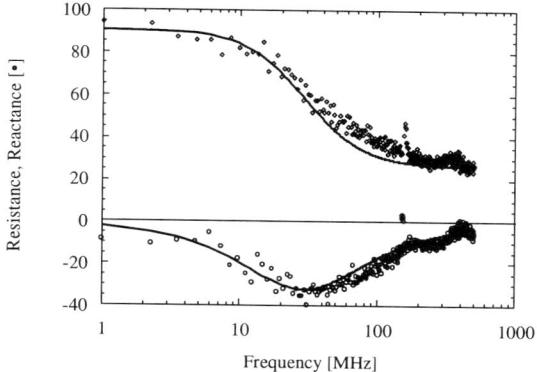

Figure 16. The measured bolometer impedance (dots). The solid lines show the real and imaginary part of the circuit in Figure 14. R_0 is 26 ½ and $(dV/dI)_{DC}$ =108 ½. Optimum LO power is applied. T=2 K.

The *receiver noise* temperature was measured using hot-cold body techniques. In a separate experiment the output noise temperature from the mixer was also measured. Together with conversion loss measurements the conclusion was that the mixer noise temperature at the input was about 200 K. The intrinsic conversion loss (not accounting for attenuation in the input circuits etc.) was 7 dB. Further work is necessary in order to understand the noise mechanisms in the hot electron bolometer mixer better.

3.5.2. Phonon cooled NbN HEB Mixers

Experiments on phonon cooled NbN HEB mixers have been performed at about 400 K (DSB) at, 300 GHz, and 1700 K (DSB) at 700 GHz (recent result at Chalmers). At 2.5 THz a noise temperature of about 15 000 K has been obtained. The conversion loss is high, partly because the devices are not optimised for this frequency. The highest IF bandwidth has been measured to be about 3.5 GHz (recent result at Chalmers).

3.5.3. *Diffusion Cooled Nb mixers*

A noise temperature of 650 K DSB at 533 GHz was measured by the JPL group [11]. These experiments indicate that an IF bandwidth of at least 3 GHz is achievable. Later experiments show that at least 50% higher IF bandwidths should be possible to in practical mixers (private communication)

4. Conclusions

Figure 17 shows recent results on mixers of different types (private communication). It is interesting to see that the HEB mixers in a very short time have reached noise temperatures comparable to the best Schottky mixers. It can be expected that above 1 THz they will outperform both the Schottkys and the SIS mixers in the near future.

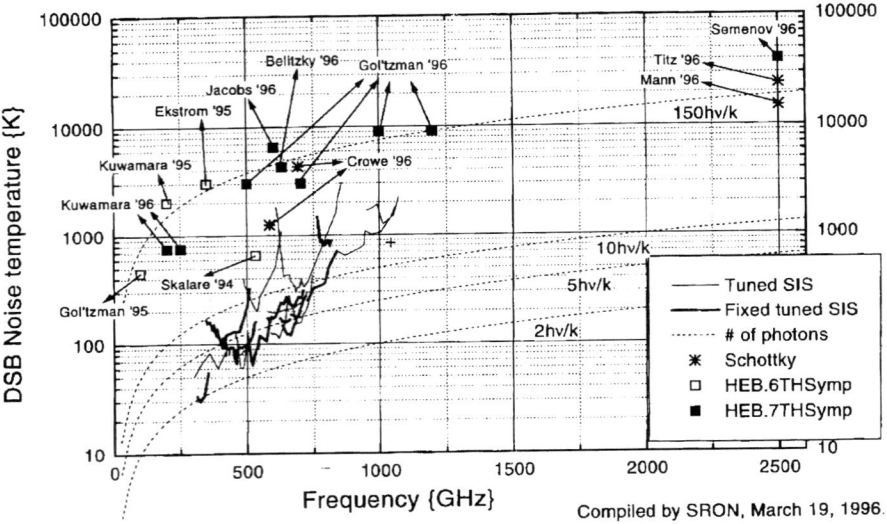

Figure 17. Recently measured noise temperatures. The names indicated are referenced to private communication. Some of the results have been published, and can be found in the reference list.(by courtesy from H. van der Stadt, SRON, The Netherlands).

References

1. Blundell, C.E. Tong (1992), Submillimeter Receivers for Radio Astronomy, *Proceedings of the IEEE*, **80,** 1702-1720.
2. Zimmermann R&R&P (1992) All solid state radiometers for environmental studies to 700 GHz, *Proceedings of the Third International Symposium on Space Terahertz Technology*, 706-723.
3. Mees, S. Crewell, H. Nett, G. de Lange, H.van de Stadt, J. J. Kuipers, R. A. Panhuyzen (1994) An Airborne SIS-Receiver for Atmospheric Measurements at 630 and 720 GHz, *Proceedings of the Fifth International Symposium on Space Terahertz Technology*, 142-155.
4. Zmuidzinas, H. D. LeDuc, J. A. Stern, and S. R. Cypher, (1994) "Two-Junction Tuning Circuit for Submillimeter SIS Mixers," *IEEE Trans. on Microwave Theory and Techniques*, **MTT-42,** 698-706
5. Yu. Belitsky, S. W. Jacobsson, L. V. Filippenko, E. L. Kollberg (1995) Broadband Twin-Junction Tuning Circuit for Submillimeter SIS Mixers, *Microwave and Optical Technology Letters*, **10,** 74-78.
6. de Lange, C. E. Honing, J. J. Kupiers, H. H. A. Schaeffer, R. A. Panhuyzen, T. M. Klapwijk, H. van de Stadt, M. de Graauw (1994) Heterodyne mixing with Nb tunnel junctions above the gap frequency, *Appl. Phys. Lett.* **64,** 3039-3041.

7. Gershenzon, G.N. Gol'tsman, I.G. Gogidze, A.I. Elant'ev, B.S. Karasik, A.D. Semenov (1990) Millimeter and Submillimeter Range Mixer Based on Electronic Heating of Superconducting Films in the Resistive State, *Sov.Phys.Superconductivity*, **3**, 1582-1597.

8. Gol'tsman, A.D.Semenov, Y.P.Gousev, M.A.Zorin, I.G.Gogidze, E.M.Gershenzon, P.T.Lang, W.J.Knott, K.F.Renk (1991) Sensitive Picosecond NbN Detector for Radiation from Millimeter Wavelengths to Visible Light, *Supercond.Science and Technology*, **4**, 453.

9. Ekström, B. Karasik, E. Kollberg, and K.S. Yngvesson (1995) Conversion Gain and Noise of Niobium Superconducting Hot-Eelectron-Mixers, *IEEE Transactions on Microwave Theory and Techniques*, **43**, 938-947.

10. Prober (1993) Superconducting Terahertz mixer using a transition-edge microbolometer, *Appl. Phys. Lett.* **62**, 2119-2121.

11. Skalare, W.R. McGrath, B. Bumble, H.G. LeDuc, P.J. Burke, A.A. Verheijen, and D.E. Prober (1995) A heterodyne receiver at 533 GHz using a diffusion cooled superconducting hot electron mixer,"*IEEE Trans. on Applied Superconductivity*, **5**, March 1995.

12. Maas (1988) *Nonlinear Microwve Circuits*, Artech House Inc., Norwood, Massachusetts.

13. Tucker, M. J. Feldman (1985) Quantum Detection at Millimeter Wavelengths, Review of Modern Physics, **57**, 1055-1113.

14. Yu. Belitsky, E. L. Kollberg (1996) Superconductor-Insulator-Superconductor Tunnel Strip Line: Features and Applications, to be published in *Applied Physics journal*

15. Skocpol, M. R. Beasly, and M. Tinkham (1974) Phase-Slip Centers and Nonequilibrium Processes in Superconducting Tin Microbridges, *J. Low Temp. Phys.*, **16**, 145-167

16. Gurevich and R. G. Mints (1987) Self-heating in normal metals and superconductors, *Reviews of Modern Physics*, **59**, 841-999, .

17. Huebener (1979) *Magnetic flux Structures in Superconductors*. Berlin: Springer.

18. Gershenzon, M.E. Gershenzon, G.N. Gol'tsman, A.M. Lyul'kin, A.D. Semenov, A. V. Sergeev (1990) Electron-phonon interaction in ultrathin Nb films", *Sov. Phys. JETP*, **70**, 505-508.

19. Ekström, B. Karasik, E. Kollberg, G. Gol'tsman, and E. Gershenzon (1995) *350 GHz NbN hot electron bolometer mixer*, 6th Int. Symp. on Space Terahertz Technology, Pasadena, 269-283.

20. Yagoubov, G. Gol'tsman, B. Voronov, and E. Gershenzon (1996) The bandwidth of HEB Mixers Employing Ultrathin NbN Films on Sapphire Substrate, *presented at 7th Int. Symp. on Space Terahertz Technology, Charlottesville, VA*.

21. Andreev (1964), Sov. Phys JEPT, **19**, 1228

22. Arams, C. Allen, B. Peyton, E. Sard (1966) Millimeter Mixing and Detection in Bulk InSb", *Proc. IEEE*, **54**, 308-318.

23. Karasik, A. I. Elantev, (1995) Analysis of the Noise Performance of a Hot-Electron Superconducting Bolometer Mixer, *Sixth International Symposium on Space Terahertz Technology*, 229-246

24. Mather (1982) Bolometer Noise: nonequilibrium Theory, *Appl. Optics*, **21**, 1125-1129.

25. Park, T. H. Geballe (1986) Superconducting Tunnelling in Ultrathin Nb Films, *Phys. Rev. Lett.* **57**, 901-904.

26. Hsu, A. Kapitulnik (1992) Superconducting transition, fluctuation, and vortex motion in a two-dimensional single-crystal Nb film, *Phys. Rev. B*, **45**, 4819-4835.

27. Huggins and M. Gurvitch (1985) Preparation and characteristics of Nb/Al-oxide/Nb tunnel junctions, *J. Appl. Phys.*, **57**, 2103-2109

28. Gerecht, C. F. Musante, Z. Wang, K. S. Yngvesson, E. R. Mueller, J. Waldman, G. N. Gol'tsman, B. M. Voronov, S. I. Cherednichenko, S. I. Svechnikov, P. A. Yagoubov, and E. M. Gershenzon, (1996) Optimization of Hot-Electron Bolometer Mixing Efficiency in NbN at 119 Micrometer Wavelength, *presented at 7th Int. Symp. on Space Terahertz Technology, Charlottesville, VA*.

29. Kawamura, R. Blundel, C.-Y. E. Tong, G. Gol'tsman, E. Gershenzon, and B. Voronov (1996) Superconductive NbN Hot-Electron Bolometric Mixer Performance at 200-250 GHz, *presented at 7th Int. Symp. on Space Terahertz Technology, Charlottesville, VA*.

116

30. Karasik, G. N. Gol'tsman, B. M. Voronov, S. I. Svechnikov, E. M. Gershenzon, H. Ekström, S. Jacobsson, E. Kollberg, and S. K. Yngvesson (1995) Hot Electron Quasioptical NbN Superconducting Mixer, *IEEE Trans. Appl. Superconductivity*, **5**, 2232-2235.

31. Yagoubov, G. Gol'tsman, B. Voronov, S. Svechnikov, S. Cherednichenko, E. Gershenzon, V. Belitsky, H. Ekström, E. Kollberg, A. D. Semenov, Y. P. Gousev, and K. F. Renk (996) Quasioptical Phonon-cooled NbN Hot-Electron Bolometer Mixer at THz Frequencies, *presented at 7th Int. Symp. on Space Terahertz Technology, Charlottesville, VA.*

32. Ekström, B. Karasik (1995) Electron Temperature Fluctuation Noise in Hot Electron Superconducting Mixers, *Applied Phys. lett.*, **66**, 3212-3214

APPENDIX: Relating Conversion Properties to the IV Characteristics.

Let us assume that the resistance is only dependent on the electron temperature and that the temperature increases linearly with dissipated power. Then consider the IV-characteristic shown in Figure 3. For a given device resistance $R_0=V_0/I_0$, the device temperature must be constant, i.e. the total power dissipated must be the same. When the device is pumped with a certain LO-power, the IV-characteristic will change, as shown in Figure 3. The above argument indicates that in points A and a we have the same total dissipated power; in A only DC power, and in a LO power plus DC power. If we increase either the LO power or the DC power, or both, the resistance will increase, and become $R_0+\Delta R$. where $\Delta R=C_0 \cdot \Delta P$

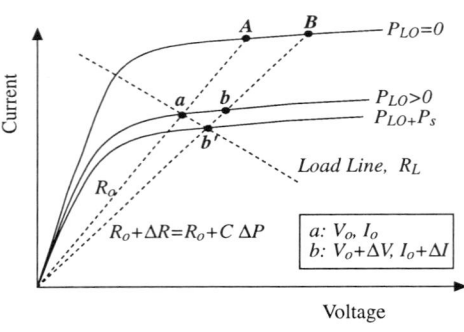

Fig. A1 Schematic illustration of the behaviour of the IV-characteristic and bias-point when LO- and RF-power is applied to the bolometric device.

Figure A1 Schematic illustration of the behaviour of the IV-characteristic and bias-point when LO- and RF-power is applied to the bolometric device.

We may now calculate ΔR and ΔP.

$$\Delta R = \frac{V_0 + \Delta V}{I_0 + \Delta I} - \frac{V_0}{I_0} \approx \frac{V_0}{I_0} \frac{\Delta I}{I_0}\left(\frac{\Delta V}{\Delta I} \frac{I_0}{V_0} - 1 \right)$$

(A1)

$$\Delta P = \left(V_0 + \Delta V \right)\left(I_0 + \Delta I \right) - V_0 I_0 \approx V_0 \Delta I \left(\frac{\Delta V}{\Delta I} \frac{I_0}{V_0} + 1 \right)$$

(A2)

From Eqs. (2) and (3) we get

$$C_o = \frac{dR}{dP} = \frac{1}{I_o^2} \frac{\dfrac{\Delta V}{\Delta I} \dfrac{I_o}{V_o} - 1}{\dfrac{\Delta V}{\Delta I} \dfrac{I_o}{V_o} + 1} = \frac{R_o}{P_{DC}} \frac{\left(\dfrac{dV}{dI}\right)_{DC} - R_o}{\left(\dfrac{dV}{dI}\right)_{DC} + R_o}$$

(A3)

where $(dV/dI)_{DC}$ is the differential resistance of the pumped IV at the bias point determined by DC bias and LO power. Using this value for C_o in Eq. (9), we may derive Eq. (10). It is in fact also possible to derive P_{DC} and P_{LO} from Figure A1 and get an expression for conversion gain that only uses parameters derived from the IV curve [9].

SCHOTTKY BARRIER DEVICES FOR THZ APPLICATIONS

H.P. RÖSER and H.-W. HÜBERS
German Aerospace Research Establishment
Institute of Space Sensor Technology
Rudower Chaussee 5, D-12489 Berlin, Germany

1. Introduction

Recent technological advances in detectors, coupling structures, local oscillators (LO) and spectrometer back ends have made possible the development of heterodyne receivers with high sensitivity and high spectral resolution ($\Delta v/v \cong 10^{-7}$) for frequencies up to 3,000 GHz (100 µm) and higher. Although some of the technical solutions are still not ideal and need further improvement to become a real facility instrument, heterodyne receivers have already been used very successfully in many research areas, for example, astronomy, plasma diagnostic experiments, atmospheric physics and radar techniques [1].

This article describes the state of the art of Schottky barrier diodes which have been used as mixers in heterodyne receivers in the research areas mentioned above. Because the description of the spectral range above 300 GHz is yet not standardized and several notations exist without any exact definition, for this paper the Terahertz region or far infrared (FIR) denotes the range from 1 - 10 THz or 300 - 30 µm, respectively.

2. GaAs Schottky Barrier Diodes

A great deal of research has been required by many individuals to bring GaAs Schottky technology to the point where it is useful at THz frequencies. Although there are only a few research laboratories worldwide fabricating GaAs Schottky diodes the performance of the devices has been significantly improved over the last 10 years [2-5].

The understanding of how Schottky diodes operate at high frequency has progressed roughly in step with the ability to fabricate high quality diodes with reproducible geometrical and electrical parameters. The application of advanced nanotechnology to achieve devices having very small dimensions and capacitances has been of particular importance.

Figure 1 shows the cross-section of a typical Schottky barrier diode with the honeycomb structure contacted by a whisker and an enlarged view of the Schottky contact including the equivalent circuit of the junction. Table 1 summarizes the geometrical and electrical parameters of Schottky diodes which have been investigated so far. The actual design has been determined by the smallest physical size of the Schottky contact which can be realized in the laboratory (≈ 0.25 µm diameter) and therefrom the optimum epilayer doping and epilayer

119

J.M. Chamberlain and R.E. Miles (eds.), New Directions in Terahertz Technology, 119–125.

epilayer thickness. Very recently diodes with sub-quarter-micron anodes have been fabricated using a new fabrication process as shown in Figure 1b [6]. It is expected, and first measurements already indicate, that this type of diode will show better performance than the others.

Table 1. Geometrical and electrical parameters of GaAs Schottky barrier diodes fabricated by the University of Virginia in Charlottesville. The capacitances are measured values.

Diode	1I7	1I12	1T15	1T23
Anode diameter [μm]	0.8	0.45	0.25	~ 0.2
Depletion thickness at zero bias D_0 [Å]	1000	600	~ 300	~ 300
Epitaxial layer doping $N_D \times 10^{17}$ [cm^{-3}]	3	4.5	10	10
Capacitance at zero bias C_{j0} [fF]	0.9	0.45	0.25	0.45-0.65
Series resitance R_s	13	33	~ 20	20-27
Ideality factor η	1.3	1.4	1.5	1.53-1.56

The primary design parameters are the epitaxial layer doping density and thickness, the anode diameter, and the doping density and geometry of the substrate. Together these physical characteristics determine the electrical diode parameters, such as junction capacitance, C_j, series resistance, R_S, figure-of-merit cut-off frequency, $v_{co} = (2\pi R_S C_j)^{-1}$, the plasma resonance frequency, the quality of the IV curve (ideality factor) and the electron transport mechanism through the barrier. Tradeoffs between the parameters are very important for obtaining the optimum diode performance.

At lower frequencies (≤ 300 GHz) the operation of Schottky barrier diodes as detectors and mixers is now very well understood and well described by classical mixer theory. However, in the THz frequency range, the simple circuit models used at lower frequencies are only of limited usefulness. This is because of the great increase in the complexity of the diode model at THz frequencies due to such effects as charge carrier inertia, dielectric relaxation and skin effect, as well as possible transient effects such as ballistic transport. For this reason, device design for THz frequencies has relied on general guidelines and, more importantly, active experimental collaboration between device and receiver researchers.

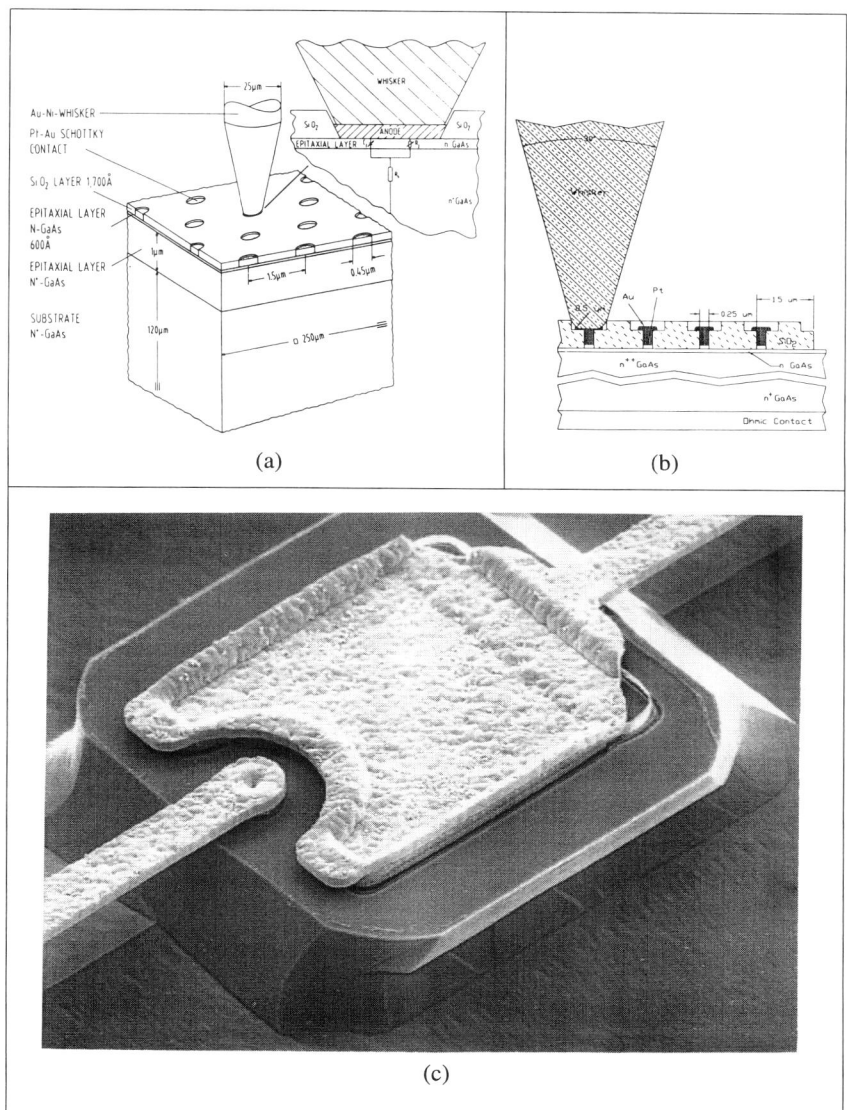

Figure 1. a) Scaled cross section of a typical Schottky barrier diode contacted by a whisker with an enlarged view of the contact including the equivalent circuit of the junction, b) whisker contacted diode with stepped - oxide profile, c) planar diode.

In the THz frequency range whisker-contacted diodes are the most commonly used mixer element. This is primarily due to their ability to yield good noise performance and the relative simple corner cube structure. But when it comes to high reliability applications and in particular space-borne systems a planar (whiskerless) diode would be desirable. Other reasons for planar structures are a) the poor coupling efficiency of corner cubes especially for wavelengths below 100 µm and b) the possibilty to develop diode arrays. But the problem of fabricating submicron anodes in a planar diode package is a matter of process development which is just at the beginning. So far planar diodes have been tested with reasonable results only below 1 THz [7, 8]. The results are still inferior compared to whiskered diodes but are promising enough to expect that planar structures probably will become the structure for the THz frequency range in the future.

3. Heterodyne Sensitivity

Schottky barrier diodes are mostly used as mixer in heterodyne receivers for high resolution spectroscopy. In the THz region these spectrometers consist of an optically pumped submillimeter laser as local oscillator (LO), a wire grid Martin-Puplett diplexer and a GaAs Schottky barrier diode in an open structure corner cube mixer. The signal is then passed through low noise intermediate frequency (IF) preamplifiers (FET, HEMT), microwave components and a second mixer stage followed by a filter or acousto optical spectrometer [1]. Some systems use a ring laser instead of standing wave resonator for the optically pumped laser providing much higher amplitude and frequency stability [9]. The corner cubes are usually used with 4λ long wire antennas but below 100µm up to 25λ antennas are more useful [10]. This type of system meets airborne specifications and has been used in different aircraft for astronomy and atmospheric physics detecting for example different CO transitions, neutral carbon CI at 157 µm, OH at 118 µm and neutral oxygen OI at 63 µm [1, 10, 11].

The sensitivity of the heterodyne system is determined at different wavelengths by measuring the system noise temperature T_{sys} with a hot/cold load. The general experimental set-up for calibration is described in detail in reference [9]. Figure 2 shows the double sideband system noise temperature T_{sys} (DSB) versus frequency for the different diodes listed in table 1. It also shows how the performance has progressed with the ability to fabricate smaller devices in geometry as well as capacitance.

As a consequence of the high cut-off frequency and fast response time of the electrons (see section 4) these devices can also be used as extremely fast video detectors ($t \approx 10^{-13}$ s) operating from room to liquid helium temperatures, with an estimated sensitivity of about 10^{-16} W/Hz$^{-1/2}$. Measurements have already been done up to 30 THz [12].

4. Current-Frequency Behaviour

Recently, it has been shown that when the dc bias and coherent LO radiation are adjusted for optimum mixing performance, the number of electrons, N_e, passing through the Schottky contact each LO cycle is constant for all frequencies. Furthermore, this constant has been shown to be a characteristic of the particular type of diode used and is in the range of 1000 -

5000 electrons [13, 9]. This behaviour is expressed by the experimentally derived equation for the optimum mixing current $I_{opt} = N_e\, e\, \nu$, where ν is the frequency of the radiation and e the electronic charge. This simple equation indicates that the mixing process is governed by a quite modest number of electrons N_e each LO cycle which is independent of frequency. For example, at 1,500 GHz , with the diode 1I12, a typical optimum current of 0.5 mA is made up of bunches of about 2,200 electrons which pass through the Schottky contact at a rate of 1,5 x 10^{12} per second. Knowing N_e and the geometry of the Schottky contact, it is possible to calculate the D_{depl} of an active depletion region. It is important to note that this active depletion thickness, when the diode is under forward current, is much less than the value for an unbiased junction which is approximately equal to the epitaxial layer thickness D_e shown in table 1.

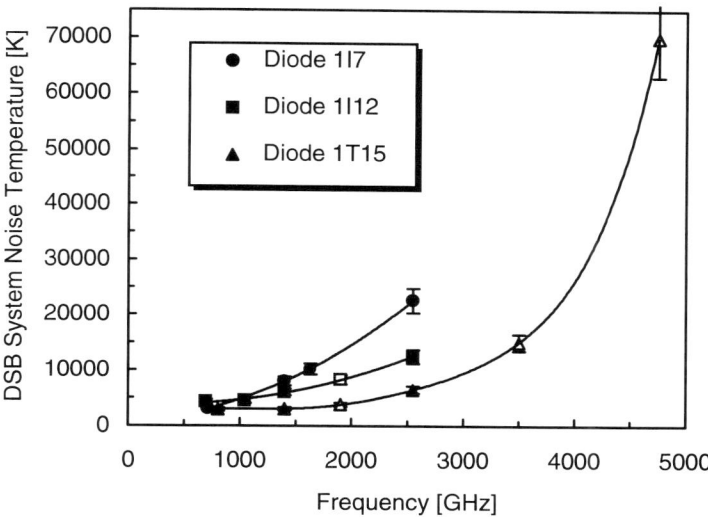

Figure 2. Measured system noise temperature as a function of frequency for different GaAs Schottky barrier diodes. The data points are from references [1, 9, 10].

When the Schottky contact is irradiated by a coherent source the depletion region D_{depl} is filled and emptied with N_e electrons during each LO cycle. Calculation of the mean free path of the electrons shows that this is greater than the active depletion thickness for Schottky diodes used, so one would expect the process to be essentially free of collisions. This is supported by the experimental observation that there is only marginal improvement (<30%) in the noise figure for the diodes when they are cooled to 20 K and part of this improvement is not due to the Schottky contact itself but to the embedding network [9]. The ballistic transport would also explain the low noise figures at room temperature and very high current densities of up to $2x10^6$ Acm^{-2} at the same time.

In addition, it is relevant to note that these diodes have mesoscopic dimensions which make them good candidates for observing coherence effects at room temperature as the anode diameter is only about 20 electron Fermi wavelengths and the DeBroglie wavelength for typical electron velocities is close to the diameter of the contact and larger than the active depletion thickness. Further experiments have already shown first evidence for this [14, 20]. It has been speculated that the coherent LO photons absorbed by the electrons might force N_e electrons to move in phase in a characteristic device length D_{depl} where the minimal scattering allows the electrons to maintain phase coherence [13, 14, 20].

5. Outlook

Over the last few years there has been a great effort in improving the coupling efficiency for signal and LO radiation. To improve the corner cube a quasi-planar Schottky diode design has been investigated leading to more reproducible and defined conditions [15]. Another approach is replacing the corner reflector by a waveguide structure. But generally they are perceived to be difficult to build especially single moded waveguides followed by corrugated horns, because circuit structures dimensions need to be much less than 100 μm. But modern lithographical techniques and micromachining have succeeded in building surprisingly good waveguide structures with excellent radiation patterns. Very first test measurements demonstrate DSB system noise temperatures down to 20,000 K [16, 17].

For the next few years GaAs Schottky barrier diodes will be the detector choice in the THz region for high resolution heterodyne receivers with receiver bandwidths of >1 GHz and they still show a lot of potential for significant improvements in terms of sensitivity and coupling structure. In addition, they operate at ambient temperature, so that cooling requirements are not essential. This is a very important aspect when using heterodyne receivers on small satellites where space and power requirements are strong limitations and a lifetime of one or more years is desirable.

Competing technologies are superconductor-insulator-superconductor (SIS) mixers and so-called hot electron bolometers (HEB). SIS receivers have sensivities close to the quantum noise limit and therefore are much better than Schottky diodes. They are expected to be applicable up to 1.2 THz or eventually 1.5 THz [18]. The most recent detector development is the HEB with promising results at about 0.5 THz. Several groups are now evaluating the performance in the range 1-3 THz (see for example reference [19] and references therein). If they are as sensitive as expected they will replace Schottky diodes. In this respect the future of Schottky diodes will be for all applications where ambient temperature and large IF's are required or is an advantage.

ACKNOWLEDGEMENTS. We would like to gratefully acknowledge the contribution of E. Bründermann, M.F. Kimmitt, A.L. Betz and T.W. Crowe

6. References

1.   Röser, H.P. (1991) Heterodyne spectroscopy for submillimeter and FIR wavelengths from 100µm to 500µm, *IR Phys.* **32**, 385-407.

2.   Crowe, T.W., Mattauch, R.J., Röser, H.P., Bishop, W.L., Peatman, W.C.B., Liu, X. (1992) GaAs Schottky diodes for THz mixing applications, *IEEE Proc., Special Issue on THz Technology* **80**, No.11,1827-1841.

3.   Kelly, W.M., Mackenzie, S., Maaskant, P. (1994) Novel chip geometries for THz Schottky diodes, *Conf. Proc., 5th Int. Symp. on Space THz Technology*, 404-408.

4.   Krozer, V., Grueb, A. (1994) A novel fabrication process and analytical model for Pt/GaAs Schottky barrier mixer diodes, *Solid State Electronics* **37** (1), 169-180.

5.   Nokozido, T., Chang, J.J., Mann, C.M., Suzuki, T., Mizuno, K. (1994) Optimization of a Schottky barrier mixer diode in the submillimeter wave region, *Intl. J. IR and MM Waves*, **15**, No.11, 1851-1865.

6.   Bishop, W.L., Marazita, S.M., Wood, P.A.D., Crowe, T.W. (1996) A novel structure and fabrication process for sub-quarter-micron THz diodes, *Conf. Proc., 7th Intl. Symp. Space THz Technology*, March 12- 14, Charlottesville, 511-524.

7.   Hesler, J.L., Hall, W.R., Crowe, T.W., Weikle, R.M., Deaver, S.D., Bradley R.F., Pan, S.K. (1996) Submillimeter wavelength waveguide mixers using planar Schottky barrier diodes, *Conf. Proc., 7th Intl. Symp. Space THz Technology*, March 12-14, Charlottesville, 462-473.

8.   Molodnyakov, S.P., Shashkin, V., Sukhodoev, L.V., Daniltzev, V.M., Molodnyakov, A.S. (1993) Submicron planar Schottky diodes for submillimeter wavelengths, *Conf. Proc., Intl. Semiconductor Device Research Symp.*, Dec. 1-3, Charlottesville, Virginia, USA, 377-380.

9.   Röser, H.P., Hübers, H.-W., Crowe, T.W., Peatman, W.C.B. (1994) Nanostructured GaAs Schottky diodes for FIR heterodyne receivers, *IR Phys.* **35**, 451-462.

10. Betz, A.L., Boreiko, R.T. (1996) A practical Schottky mixer for 5 THz, *Conf. Proc., 7th Intl. Symp. Space THz Technology*, March 12-14, Charlottesville, 503-510.

11. Titz, R., Birk, M., Hausamann, D., Nitsche, R., Schreier, F. Urban, J., Küllmann, H., Röser, H.P. (1995) Observation of stratospheric OH at 2.5 THz with an airborne heterodyne system, *IR Phys. Technol.* **36**, 883-891.

12. Hübers, H.-W., Schwaab, G.W., Röser, H.P. (1994) Video detection and mixing performance of GaAs Schottky-barrier diodes at 30 THz and comparison with metal-insulator-metal diodes, *J. Appl. Phys.* **75** (8), 4243-4248.

13. Roeser, H.P., Titz, R.U., Schwaab, G.W., Kimmitt, M.F. (1992) Current-frequency characteristics of submicron GaAs Schottky barrier diodes with femtofarad capacitance, *J. Appl. Phys.* **72**, 3194-3197.

14. Röser, H.P., Bründermann, E., Hübers, H.-W. (1995) Interaction of FIR-laser radiation with mesoscopic GaAs structures and the experimental determination of h/e^2 and $h/2e$ at room temperature, *Conf. Proceedings Annual Meeting of the Japanese Physical Society of IR Technology*, **5**, 1-5.

15. Nitsche, R., Titz, R., Biebl, E.M. (1996) Quasi-planar Schottky diode design, *Conf. Proc., 7th Intl. Symp. Space THz Technology*, March 12-14, Charlottesville, 488-493.

16. Brown, D.A., Treen, A.S., Cronin, N.J. (1994) Micromachining of Terahertz waveguide components with integrated active devices, *Proc. 19th Intl. Conf. IR and MM Waves*, Sendai, Japan, 359-360.

17. Ellison, B.N., Maddison, B.J., Mann, C.M., Matheson, D.N., Oldfield, M.L., Marazita, S., Crowe, T.W., Maaskant, P., Kelly, W.M. (1996) First results for a 2.5 THz Schottky diode wavguide mixer, *Conf. Proc., 7th Intl. Symp. Space THz Technology*, March 12-14, Charlottesville, 494-502.

18. *Conf. Proc., 7th Intl. Symp. Space THz Technology* (1996) Session 1 and 9 on SIS Mixers, March 12-14, Charlottesville.

19. Skalare, A., McGrath, W.R., Bumbl, B., LeDuc, H.G., Burke, P.J., Schoelkopf, R.J., Prober, D.E. (1996) Niobium superconducting diffusion-cooled hot-electron bolometer mixers above 1 THz, *Conf. Proc., 7th Intl. Symp. Space THz Technology*, March 12-14, Charlottesville, 561-564.

20. Röser, H.P., Hübers, H.-W., Bründermann, E., Kimmitt, M.F. (1996) Observation of mesoscopic effects in Schottky diodes at 300 K when used as mixer at THz frequencies, accepted for publication in Semiconductor Science and Technology.

DEVICE PHYSICS OF INTERSUBBAND LASERS

P. HARRISON
Department of Electronic and Electrical Engineering,
University of Leeds, LS2 9JT, U.K.

1. Introduction

Solid state microwave oscillators generate electromagnetic radiation by accelerating electric charge and can operate up to frequencies of around 150 GHz (2 mm). Solid state optical sources of electromagnetic radiation work on a fundamentally different principle, with photons generated as a result of electron transitions between discrete states. Devices based on cross-bandgap electron-hole recombination are available typically down to frequencies of around 150 THz (2-3 μm). Recent work on unipolar solid state optical sources has extended this down to around 70 THz (13 μm). The purpose of this work is to give an outline of the physics of these intersubband devices which generate photons via electron transitions between excited conduction band states localised in quantum wells.

Following a brief recap of the basic physical principles of quantum wells, a short introduction will be given to the computational techniques commonly employed to solve the corresponding Schrödinger equation. After a short summary of the recent successes of intersubband lasers operating in the mid-infrared, the discussion moves on to how the device designs may be adapted to the longer wavelengths of the terahertz (1→10 THz) region of the spectrum.

2. Introduction to quantum well physics

As ever, certain approximations and assumptions need to be employed in order to make progress amenable. Stating these explicitly gives the true context and applicability of the theories discussed.

In the effective mass approximation the complex dispersion relations of electrons (or holes) are approximated at small wave vectors by a parabola, $E = \hbar^2 k^2/(2m^*)$[1]. In the description of localized states, such as those that occur in quantum wells, the wave functions are written as a product of two terms, a slowly varying envelope ψ and a rapidly varying function u with a period of the order of the microscopic crystal potential[1], i.e.,

$$\Psi = \psi(\mathbf{r})u(\mathbf{r}) \tag{1}$$

The sets of envelope functions ψ and Bloch functions u are both orthonormal. Generally the properties of quantum wells can be expressed in terms of just the more readily calculated envelope function ψ, thus simplifying the problem considerably. The further assumption will be made that the effective mass is a constant across the quantum well structure,

127

J.M. Chamberlain and R.E. Miles (eds.), New Directions in Terahertz Technology, 127–133.

which is a good approximation in GaAs/Ga$_{1-x}$Al$_x$As structures with low x and avoids the controversy over the form of the kinetic energy operator in the Schrödinger equation[1, 2]. As the *majority* of work to date has been on the physically simpler n-type systems, then for this introduction this is where attention will be focussed.

The Schrödinger equation follows as a linear second order differential equation of the form

$$\left(-\frac{\hbar^2}{2m^*} \frac{\partial^2}{\partial z^2} + V(z) \right) \psi(z) = E\psi(z) \tag{2}$$

This is commonly solved in the one-dimensional potential $V(z)$ of a quantum well structure, by dividing the potential into regions with constant $V(z)$ and choosing general solutions of the form

$$\psi(z) = A\cos(kz) + B\sin(kz), \qquad\qquad E > V(z) \tag{3}$$

$$\psi(z) = A\exp(\kappa z) + B\exp(-\kappa z), \qquad\qquad E < V(z) \tag{4}$$

where

$$k = \sqrt{\frac{2m^*E}{\hbar^2}} \qquad\qquad \kappa = \sqrt{\frac{2m^*(V(z) - E)}{\hbar^2}} \tag{5}$$

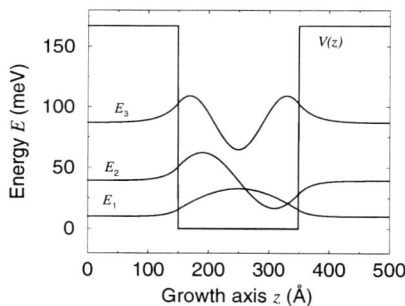

Figure 1. Three lowest quantum well electron wavefunctions

As mentioned above, assuming constant mass across the structure, then the solution (ψ) must be continuous and differentiable. Recalling the standard boundary conditions $\psi(z) \to 0$ as $z \to \pm\infty$ and matching wavefunction and derivative at the boundaries between regions of differing V gives, for the even parity states of a single quantum well

$$k \tan\left(\frac{ka}{2} \right) = \kappa, \qquad\qquad \text{where } a \text{ is the well width} \tag{6}$$

and can be solved for the energy E by Newton-Raphson techniques etc.. A similar expression follows for the odd parity states. The solutions for a single 100Å GaAs/Ga$_{0.9}$Al$_{0.1}$As quantum well are illustrated in Figure 1.

In principle, any structure can be solved with this 'Transfer Matrix Technique' by discretizing the potential into single monolayer steps. However a more general and powerful technique is available. It involves expanding the derivatives within the Schrödinger equation in terms of finite differences, in order to produce an iterative shooting equation[3]. This has been used successfully in the solution of complex potential profiles[4, 5] and lends itself better to the inclusion of an electric field[6], as usually encountered in real devices, than the transfer matrix technique whose general solutions would now be linear combinations of Airy functions[7].

Within the plane $(x$-$y)$ of the quantum well the potential is zero and the solutions are plane waves. Again parabolic dispersion curves are assumed (see Figure 2) hence the total energy is given by

$$E = E_n + \frac{\hbar^2 k_{\parallel}}{2m^*}, \qquad n = 1, 2, 3, \ldots \qquad (7)$$

The ground state E_1 and higher energy solutions E_2, E_3, etc. are the minima of continuous energy bands and are referrred to as 'subbands', as they originate from the conduction band of the bulk crystal.

Optical transitions between the conduction subbands of a GaAs quantum well were first observed by West[8], and have since become known simply as 'intersubband transitions'.

3. Physics of intersubband transitions

The word 'opto-electronic' refers to devices in which the technologies of optics and electronics are merged. On a fundamental physics level these devices are based on the interactions of electrons with photons. Generally a classical description of an electromagnetic field is used to describe the photons and hence the Schrödinger equation must contain an additional term which represents the energy of the electron in the electromagnetic field.

Considering the electromagnetic field as a time dependent perturbation H'_{if} causing transitions between electronic energy levels[9] then the transition rate from the initial electronic state Ψ_i to the final state Ψ_f is given by Fermi's golden rule[10]

$$W_{i \to f} = \frac{2\pi}{\hbar} \left| H'_{if} \right|^2 \delta(E_i - E_f - \hbar\omega) \qquad (8)$$

Using the envelope function formalism, i.e. $\Psi(\mathbf{r}) = u\psi$, gives[11]

$$H'_{if} = \frac{eA_0}{2m_0} \left\{ \langle u_f | \hat{\mathbf{e}}.\mathbf{p} | u_i \rangle_{\text{cell}} \langle \psi_f | \psi_i \rangle + \langle u_f | u_i \rangle_{\text{cell}} \langle \psi_f | \hat{\mathbf{e}}.\mathbf{p} | \psi_i \rangle \right\} \qquad (9)$$

where $\hat{\mathbf{e}}$ is the polarization and $\mathbf{p} = -i\hbar\nabla$ is the momentum operator.

In the case of interband transitions between the conduction and valence band, the second term in equation 9 gives zero since the Bloch functions at the same point in the Brillouin zone, in two different bands are orthogonal, i.e. $\langle u_f | u_i \rangle_{\text{cell}} = \int_{\text{cell}} u_f(\mathbf{r}) u_i(\mathbf{r}) \; d\mathbf{r} = 0$. Hence the envelope function overlap integral $\langle \psi_f | \psi_i \rangle$ in the first term, determines which transitions are allowed and which are forbidden.

Lasers based on interband transitions have been very successful[10] and give access to a frequency range from the near-infrared[12] to the blue[13]. For a review of II-VI devices see Nurmikko and Gunshor[14].

For *intersubband* transitions the first term on the right hand side of equation 9 is zero, since the conduction subband envelope functions ψ_f and ψ_i are both eigenfunctions of the same Hermitian operator (conduction band Hamiltonian), and are therefore orthogonal, i.e. $\langle \psi_f | \psi_i \rangle = \int_{\text{all space}} \psi_f(\mathbf{r}) \psi_i(\mathbf{r}) \ d\mathbf{r} = 0$. Therefore

$$H'_{if} = \frac{eA_0}{2m_0} \langle u_f | u_i \rangle_{\text{cell}} \langle \psi_f | \hat{\mathbf{e}}.\mathbf{p} | \psi_i \rangle \tag{10}$$

For intersubband absorption and emission from electrons with low in-plane momenta the envelope functions ψ are functions of z only, hence the linear momentum operator $\mathbf{p} = -i\hbar \partial/\partial z$. Therefore

$$\langle \psi_f | \hat{\mathbf{e}}.\mathbf{p} | \psi_i \rangle = -i\hbar \hat{\mathbf{e}}_z \left\langle \psi_f \left| \frac{\partial \psi_i}{\partial z} \right. \right\rangle \tag{11}$$

This implies that transitions are only allowed when there is a component of the polarization vector $\hat{\mathbf{e}}$ along the growth z-axis[8, 15]. Which means that no intersubband absorption occurs for normal (along the growth z-axis) incident light. This is a major difference between intersubband and interband transitions.

The natural device geometry for optical detectors, and optically stimulated lasers would be based upon normal incidence excitation. However the restriction that in-plane (x-y) polarized optical intersubband transitions are forbidden has led to the adoption of the Brewster angle or other more complex configurations[16]. Another alternative has been sought, normal incidence infrared photodetectors using *'intervalence'* subband transitions in p-type quantum wells have been demonstrated[17].

Emitting devices based on electrical injection with a perpendicular electric field[18], naturally produce polarized radiation which leaves the sample at the edges, this allows for fabrication into simple edge emitters[19].

In a symmetric system $\partial \psi/\partial z$ is of opposite parity to ψ, hence the right hand side of equation 11 becomes zero for ψ_i and ψ_f of the same parity and these transitions are forbidden[20]. Representing $|i - f|$ as Δn then this selection rule can be summarized as $\Delta n = 1, 3, \ldots$ This can be overcome by introducing an asymmetry into the system, either with the application of an electric field, or structurally[19].

In an intersubband quantum well device, as well as radiative (photon emitting) transitions between electron energy levels, non-radiative (phonon emitting and absorbing) transitions also occur, see Figure 2. These phonon scattering mechanisms can be either beneficial or detrimental to the production of a population inversion and hence a laser.

There are two types of phonon that are of interest here. Scattering events involving the Longitudinal Optical (LO) phonon typically occur on the time scale <1 ps. As the energy of the LO phonon is large compared to the acoustic phonons (36 meV for GaAs), LO phonon scattering is particularly important for intersubband events.

In contrast to this, the characteristic scattering time of the acoustic branch is typically >100 ps. Generally the momenta and energy of the acoustic phonons are small (~ 1 meV), and hence they are important for intrasubband relaxation of 'warm' carriers.

In the example in Figure 2, an electron initially in the 2nd subband can relax (lose in-plane momentum) by emitting both LO and acoustic (AC) phonons. At any point the

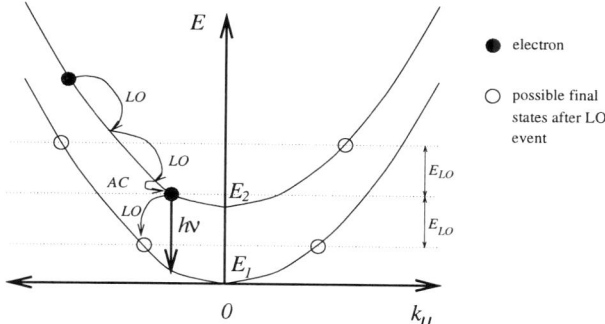

Figure 2. Acoustic (AC), LO phonon and optical ($h\nu$) scattering processes for a two conduction subband system.

electron is able to emit and absorb an LO phonon which can lead to four possible final states.

For optical emission the desired photon $h\nu$ emission must compete with the undesired intersubband LO phonon emission and this is an example where the non-radiative LO emission is detrimental to the production of a population inversion. For a full theoretical derivation of the scattering rates see Lundstrom[21] or Piorek[22].

4. Infrared and terahertz intersubband lasers

If a population inversion is not attained then a device can still give incoherent radiation. This is called electroluminescence, and was first demonstrated by Helm *et al.*[18, 23] at low temperatures and from a subband spacing below the optical phonon energy. The latter eliminates the non-radiative LO phonon emission mechanism and hence makes an optical transition more favourable. But perhaps a larger leap in technology came from Faist *at al.*[24] who demonstrated the first electroluminescence in the mid-infrared (5μm), generated by a subband spacing greater than the LO phonon energy and at room temperature.

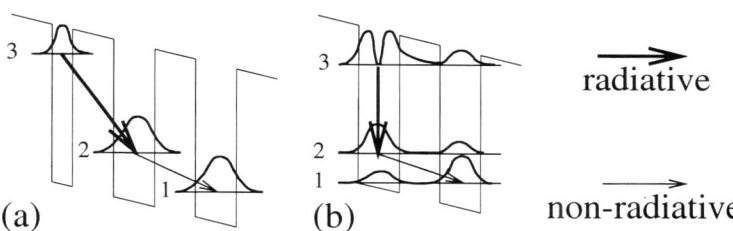

Figure 3. Schematic diagram of the three level diagonal and vertical optical transition device designs.

The GaInAs/AlInAs multiple quantum well structure, as illustrated in Figure 3(a), used a graded gap injector (which becomes almost flat under the applied bias) and generated nanowatt power output. Soon after, came the first report of lasing in such a structure, referred to as the *'Quantum Cascade Laser'*[25]. This gave a peak power of 8 mW at a wavelength of 4.2 μm (71 THz).

The upper level was populated by carrier injection through a barrier into the $n=3$ state which had a reduced spatial overlap with the $n=2$ state in the next well. In turn this had a strong coupling to the $n=1$ level. Hence the lifetime in the E_3 level was long, this was followed by the photon assisted tunnelling transition to the E_2 level which constituted the lasing transition. The lower E_2 level was rapidly depopulated by non-radiative mechanisms to the strongly coupled E_1 level, which ensured a population inversion.

This can be demonstrated by using standard rate (kinetic) equation analysis. The change in population of level 2 is equal to the rate at which carriers populate the level minus the rate at which they leave the level, i.e.

$$\frac{dN_2}{dt} = +\frac{N_3}{\tau_{3\to2}} - \frac{N_2}{\tau_{2\to1}} \tag{12}$$

where the mean relaxation times τ are inversely proportional to the total scattering rates which include both radiative (optical) and non-radiative (phonon) mechanisms. At equilibrium $\frac{dN_2}{dt} = 0$, if $\tau_{3\to2} > \tau_{2\to1}$, then $N_3 > N_2$. This is one of the key requirements for the achievement of a population inversion, the lower laser level must be depopulated faster than it is repopulated.

A refinement of the design has led to devices in which the lasing transition is vertical, i.e. within the same well, rather than the photon-assisted-tunnelling diagonal transition, see Figure 3(b). The reasoning behind such a move was that the device would be expected to be less sensitive to interface roughness and impurity fluctuations, and would therefore exhibit a narrower gain spectrum and thus a lower threshold[26]. This has led to pulsed operation up to 100 K, with a threshold current density of 3 kA/cm^2 and a slope efficiency of 300 mW/A. Although intersubband light-emitting-diodes have been reported in the 8-13 μm wavelength region[27], the longest wavelength intersubband *laser* to date is only 8.4 μm[28]. For an early review of infrared intersubband lasers see Yang[29].

Recent theoretical work[19, 30] has discussed the possibility of extending the wavelength region into the far-infrared ($>30\,\mu$m) or terahertz (1-10 THz) region of the spectrum. There is a dynamical advantage to be gained with such a move, namely that the photon energy of terahertz frequencies is below that of the LO phonon. Hence the required subband spacing will automatically lead to a suppression of the dominant loss mechanism, i.e. the LO phonon scattering from the upper to lower laser levels.

In GaAs the LO phonon energy is 36 meV which is equivalent to a photon frequency of 8.7 THz (34 μm). This is a physical limit on the maximum operating frequency of GaAs based devices. However as the LO phonon energy in AlAs is 50 meV, higher frequencies could be attained in this system. Balanced with this gain is the increased problem of temperature which could induce non-radiative loss from the high energy tail of the thermalized carrier distribution in the upper laser level and scattering from the lower to upper laser levels[30].

References

1. Bastard, G. (1988) *Wave Mechanics Applied to Heterostructures*, Editions de Physique, Paris.

2. Hagston, W. E., Harrison, P., Piorek, T. and Stirner, T. (1994) *Superlatt. Microstruct* **15** 199.
3. Killingbeck, J.P. (1992) *Microcomputer Algorithms* Hilger, Bristol.
4. Harrison, P., Hagston, W. E. and Stirner, T. (1993) *Phys. Rev.B* **47** 16404.
5. Harrison, P., Piorek, T., Stirner, T. and Hagston, W. E. (1996) *Phys. Rev. B* **53** 11060.
6. Roberts, R. G., Harrison, P., Stirner, T. and Hagston, W. E. (1993) *J. de Physique IV* **3** C 5-203.
7. Allen, S. S. (1996) *J. Appl. Phys.* **79** 886.
8. West, L. C. and Eglash, S. J. (1985) *Appl. Phys. Lett.* **46** 1156.
9. Liboff, R. L. (1980) *Introductory Quantum Mechanics*, Holden-Day, San Francisco.
10. Zory, P. S. (1993) *Quantum Well Lasers*, Academic Press, Boston.
11. Adachi, S. (1994) *GaAs and Related Materials*, World Scientific, Singapore.
12. Baranov, A. N., Sherstnev, V. V., Alibert, C. and Krier, A (1996) *J. Appl. Phys.* **79** 3354.
13. Nakamura, S. Senoh. M. Nagahama, S-I., Iwasa, N., Yamada, T., Matsushita, T., Kiyoku, H. and Sugimoto, Y. (1996) *Jpn. J. Appl. Phys.* **35** (2) L74.
14. Nurmikko, A. V. and Gunshor, R. L. (1995) Physics and Device Science in II-VI Semiconductor Visible Light Emitters, *Solid State Physics*, **49**.
15. Kaufman, D., Sa'ar, A. and Kuze, N. (1994) *Appl. Phys. Lett.* **64** 2543.
16. (1992) *Intersubband Transitions in Quantum Wells* H. C. Liu, B. F. Levine, and J. Y. Andersson (eds.), Plenum , New York.
17. Katz, J., Zhang, Y. and Wang, W. I. (1992) *Electron. Lett.* **28** 932.
18. Helm, M., Colas, E., England, P., DeRosa, F. and Allen, S. J. (1988) *Appl. Phys. Lett.* **53** 1714.
19. Berger, V. (1993) *Semicond. Sci. Technol.* **9** 1493.
20. Saar, A. (1993) *J. Appl. Phys.* **74** 5263.
21. Lundstrom, M. (1990) *Fundamentals of Carrier Transport*, Addison-Wesley.
22. Piorek, T. (1996) Quantum Theory of Low Dimensional Semimagnetic Structures, PhD Thesis, University of Hull, U.K.
23. Helm, M., England, P., Colas, E., DeRosa, F., and Allen, S. J. (1989) *Phys. Rev. Lett.* **63** 74.
24. Faist, J., Capasso, F., Sirtori, C., Sivco, D. L., Hutchinson, A. L., Chu, S. N. G. and Cho, A. Y. (1993) *Electronics Lett.* **29** 2230.
25. Faist, J., Capasso, F., Sivco, D. L., Sirtori, C., Hutchinson, A. L. and Cho, A. Y. (1994) *Science* **264** 553.
26. Faist, J., Capasso, F., Sirtori, C., Sivco, D. L. and Hutchinson, A. L. (1995) *Appl. Phys. Lett.* **66** 538.
27. Sirtori, C., Capasso, F., Faist, J., Sivco, D. L., Hutchinson, A. L. and Cho, A. Y. (1995) *Appl. Phys. Lett.* **66** 4.
28. Sirtori, C., Faist, J., Capasso, F. and Sivco, D. L. (1995) *Appl Phys. Lett.* **66** 3242.
29. Yang, R. Q. (1995) *Superlatt. Microstruct.* **17** 77.
30. Harrison, P. and Kelsall, R. W. (1996) to be published in Superlatt. Microstruct..

FROM QUANTUM MECHANICS TO S-PARAMETERS?

W.S. TRUSCOTT
Department of Electrical Engineering and Electronics,
University of Manchester Institute of Science and Technology,
PO Box 88, Manchester M60 1QD, U. K.

1. Introduction

This paper answers two questions: first, are there transport devices, such as diodes and transistors, whose operation can only be described by quantum mechanics and second, is it possible to calculate their electrical circuit characteristics, for example s-parameters, from quantum mechanics? Both answers are "yes": in addition there are designs for devices that are able to use uniquely quantum features to improve operation at terahertz frequencies. Another conclusion is, unfortunately, that only time-consuming self-consistent calculations can give accurate values for quantities like s-parameters.

This paper first establishes where a quantum mechanical description of carrier transport differs from drift-diffusion theory or classical ballistic transport. Exact solutions to Schrodinger's equation with a time-dependent potential are presented for two cases of importance in transport devices. The combination of locally exact solutions giving time-dependent charge densities and currents is discussed. At low frequencies these results are identical with the ballistic model; however they show that devices can be designed whose performance at terahertz frequencies is enhanced through the exploitation of quantum resonances. The paper then presents calculated resistive and reactive terminal currents for possible "real" device structures with this quantum mechanical enhancement. Finally, self-consistent solutions are discussed.

2. Why bother with Quantum Mechanics in real devices?

The standard physical description of semiconductor transport devices, for example diodes and transistors, uses Poisson's equation for the electrostatics and the drift-diffusion-recombination equation to give the carrier dynamics. This approach is highly successful and justified whenever the carrier scattering time is much shorter than any other relevant time such as the transit time for carriers across the active region or the period of any external electric waveform. Carrier scattering times are dependent on temperature, doping density, carrier type and material system. The room temperature electron mobility in low-doped GaAs is 0.7 m^2/Vs corresponding to a scattering time of

135

J.M. Chamberlain and R.E. Miles (eds.), New Directions in Terahertz Technology, 135–142.
© 1997 *Kluwer Academic Publishers. Printed in the Netherlands.*

0.3 ps for $m^* = 0.067m_e$; doped materials have significantly shorter times. These times are short enough for the assumptions of the drift-diffusion description to be valid at all frequencies below the terahertz region. Quantum mechanical effects cannot be observed in the drift-diffusion region because carrier scattering randomises the phase of the wave function, stopping the time development described by quantum mechanics.

At terahertz frequencies and in devices with very short active regions an alternative to drift-diffusion is needed and the ballistic carrier model based on Newtonian mechanics is usually used. The quantum mechanical equivalent is a wave packet: wave packets travel at the group velocity, $\partial\omega/\partial\mathbf{k} = \mathbf{p}/m$ in any parabolic band. A wave packet behaves as a Newtonian particle if it does not spread out significantly in space over the time that it crosses the active region of a device. The broadening, Δx, in a time t equals $t\Delta v$, where Δv is the range of velocities in the packet. The distance travelled in this time $d = tv$ and $\Delta x = d\Delta v/v$. Hence the relative broadening of the wave packet, $\Delta x/d = \Delta v/v = 2\Delta E/E$, where ΔE is the range of energies making up the wave packet. This ballistic description can be used if $\Delta E/E$ is small, which implies that the wave packet is large compared to λ, the carrier wavelength. The ballistic model therefore fails, and quantum mechanics is required, for device sizes that approach is the wavelength of carriers ($h/(2mE)^{\frac{1}{2}}$ 30 nm for electrons at room temperature in GaAs). The active regions of many terahertz devices are this size and they require a fully quantum mechanical model to predict their electrical behaviour.

3. Solutions to Schrodinger's equation in time-dependent potentials

One might expect that the complexity of a time-dependent component of the potential would make Schrodinger's equation analytically insoluble. Surprisingly, there are cases appropriate to electronic devices where such a component does not make the solution significantly harder. The simplest case is a sinusoidally time-dependent potential that is independent of position, $V_1\cos(\omega t)$, added to any time-independent potential $V_0(x)$. If $\Psi_0(x,t)$ is a solution to the time-dependent Schrodinger equation:

$$-\frac{\hbar^2}{2m}\frac{\partial^2\Psi_0(x,t)}{\partial x^2} + V_0(x)\Psi_0(x,t) = -i\hbar\frac{\partial\Psi_0(x,t)}{\partial t}$$

then the solution for the potential $V(x,t) = V_0(x) + V_1\cos(\omega t)$ is [1]:

$$\Psi(x,t) = \Psi_0(x,t)\exp\left[\frac{iV_1\sin(\omega t)}{\hbar\omega}\right]$$

The time-dependence in these equations is the complex conjugate of that normally adopted; it conforms with the electrical engineering convention of $\exp(j\omega t)$ to describe a.c. waveforms and will lead to an easier interpretation in terms of effective capacitance or inductance. Exact solutions can be written even when the time dependence of the potential is not sinusoidal, but it is the variation used in the electrical description of devices. This exact solution corresponds to a phase modulation of the wave function whose evolution in time and space is otherwise unaltered by the time dependence of the potential. The potential, $V_0(x) + V_1\cos(\omega t)$, can describe the part of the channel of a FET that is under the gate; the static potential is produced by the distribution of carriers and the bias on the terminals, it is modulated by a sinusoidal voltage applied to the gate which causes a similar time-dependence of the channel potential.

The second case where Schrodinger's equation has exact solutions with a time-dependent part of the potential is when it has the form of a field that is uniform in space but varies sinusoidally with time: $-F_1 x\cos(\omega t)$. If the static part of the potential corresponds to a constant gradient in space, $-F_0$, giving a combined potential $V(x,t) = V_0 - F_0 x - F_1 x\cos(\omega t)$, then the solution is given by [2]:

$$\Psi(x,t) = \Psi_0\left[x + \frac{F_1\cos(\omega t)}{m\,\omega^2}, t\right]\exp\left[\left(-\frac{i\,F_1\,x}{\hbar\omega} + \frac{i F_0 F_1}{m\hbar\,\omega^3}\right)\sin(\omega t)\right]$$

where $\Psi_0(x,t)$ is a wave function for $V_0(x,t) = V_0 - F_0 x$. This expression omits a spatially-uniform phase modulation that is second order in F_1 which is not significant in this paper as we are concerned only with linear device behaviour. The expression shows the two effects of the time-dependent field: $\Psi_0(x,t)$ is phase modulated by a spatially-dependent factor and, in addition, the envelope of the wave function oscillates in time as $-F_1 x\cos(\omega t)/m\omega^2$; this displacement is exactly the same as that of a Newtonian particle, mass m, subject to the time-dependent field. This solution applies to the static potentials that are constant in space, $F_0 = 0$, for which the static solutions are analytical. The solution can be extended to static potentials that are parabolic in space: $V(x,t) = \frac{1}{2}m\omega_0^2 x^2 - F_1 x\cos(\omega t)$; but in this case the field F_1 is replaced by the scaled field, $F'=F_1/(1-\omega_0^2/\omega^2)$, this again describes the behaviour of a Newtonian particle. Linear and parabolic static potentials are found in undoped and uniformly-doped regions of two-terminal electronic devices where an external sinusoidal bias adds a uniform a.c. field. The analytical solutions for $F_0 = 0$ lead to the most rapid calculation of the behaviour of entire structures since numerical integration is not needed; they are therefore useful approximations if the variation in the static potential across the active region of the device is small.

4. Linking exact solutions.

The solutions in the previous section are exact in any region with a potential of the appropriate form. Any physically possible behaviour within such a region can be described by a complete set of such solutions. In a real device the forms of the time-dependent and time-independent parts of the potential vary in space, being different in the active region and the contacts; there are also potential steps in heterostructure devices. Since the aim of this work is to calculate terminal characteristics, it is necessary to link the solutions in the active region with solutions in the contacts that act as the physical boundaries of the part of the device requiring a quantum mechanical description. Linking solutions in different regions is achieved by matching $\Psi(x,t)$ and $\partial\Psi(x,t)/\partial x$ at any boundary between different exact solutions for all times [1]. In this paper such a matching will be outlined for the case of a simplified single heterojunction barrier tunnelling device. A barrier exists between $x = 0$ and $x = d$; a steady flux of particles is incident on the barrier from $x < 0$ with energy E; the barrier is subject to a time-dependent field $F\cos(\omega t)$; the potential in the region $x<0$ is 0; for $0<x<d$, $V(x,t) = V_0 - Fx\cos(\omega t)$; for $x>d$, $V(x,t) = -Fd\cos(\omega t)$. General wave functions without the time-dependent field in the three rgions are: $A\exp(-ikx+iEt/h)+B\exp(ikx+iEt/h)$, $C\exp(-\kappa x+iEt/h)+D\exp(\kappa x+iEt/h)$, and $P\exp(-ikx+iEt/h)$. The time-dependent field does not affect the first region, but the wave function in the second becomes:

$$\left\{ C\exp\left[-\kappa\left(x+\frac{F\cos(\omega t)}{m\,\omega^2}\right)\right] + D\exp\left[\kappa\left(x+\frac{F\cos(\omega t)}{m\,\omega^2}\right)\right]\right\}\exp\left[\frac{iEt}{\hbar} - \frac{iFx\sin(\omega t)}{\hbar\omega}\right] =$$

$$\sum_{n,m=-\infty}^{\infty}\left[i^n\,C\exp(-\kappa x) + (-i)^n\,D\exp(\kappa x)\right]J_n\left(\frac{i\kappa F}{m\,\omega^2}\right)J_m\left(\frac{-Fx}{\hbar\omega}\right)\exp\left[i\left(\frac{E}{\hbar} + n\omega + m\omega\right)t\right]$$

where functions of the form $\exp[\alpha\cos(\omega t)]$ and $\exp[i\beta\sin(\omega t)]$ have been expanded as power series in $\exp(in\omega t)$ for integer n. For $x>d$ the wavefunction becomes:

$$P\exp\left[-ikx + \frac{iEt}{\hbar} - \frac{iFd\sin(\omega t)}{\hbar\omega}\right] = P\sum_{n=-\infty}^{\infty}J_n\left(\frac{-Fd}{\hbar\omega}\right)\exp\left(-ikx + \frac{iEt}{\hbar} + in\omega t\right)$$

Matching the wavefunctions at the boundaries $x = 0$ and $x = d$ for all times is achieved by separately matching all terms in the continuity equations that have a time dependence $\exp[i(E/h+n\omega)t]$ for each significant value of n. Since the set of time dependences is orthogonal over time this gives the unique solution to these equations. At $x = 0$ the equations are: $A_n + B_n = \Sigma\,[i^{n-m}C_m + (-i)^{n-m}D_m]\,J_{n-m}(i\kappa_m F/m\omega^2)$ and $ik_n[B_n-A_n] = \Sigma\kappa_m[-i^{n-m}C_m+(-i)^{n-m}D_m]J_{n-m}(i\kappa_m F/m\omega^2)+F/\hbar\omega[C_{n+1}-C_{n-1}+D_{n+1}-D_{n-1}]$.

The final term in the second equation comes from the only $J_m(x)$ with derivatives not equal to zero at $x = 0$, namely $n = 0$, $m = \pm1$. Since the incoming particles have the energy E, $A_n = 0$ for $n \neq 0$. Similarly at $x = d$ two further sets of equations connecting C_m and D_m with P_n can be written. These four sets of equations allow all B_n, C_n, D_n, and P_n to be found in terms of A_0. This general analysis shows that the expansion parameter determining the amplitudes of the various components of the tunnelling current is $i\kappa F/m\omega^2$; this is also true for multi-barrier structures. This paper is concerned with the linear response of devices at high frequencies; the continuity equations can therefore be simplified by substituting $J_n(i\kappa F/m\omega^2)$ by 1 for $n = 0$, $\pm i\kappa F/2m\omega^2$ for $n = \pm1$ and 0 for $|n| > 1$. There are now only six equations to solve at each boundary; in two of these solely the $n=0$ terms are significant because those for $n = \pm1$ are second order in F. Finding the amplitudes is now a manageable task: B_0, C_0, D_0, and P_0 are unchanged by the time-dependent field and are found from a pair of equations at each boundary; those with $n = \pm1$ are found from two more pairs once B_0, C_0, D_0, and P_0 are known.

It might seem that the amplitudes of the terms in the wave function that are first order in κF vary with the frequency as ω^{-2}: however, the portions of the waves $C_{\pm1}$ and $D_{\pm1}$ generated by C_0 and D_0 at $x = 0$ and those generated at $x = d$ are of opposite sign and a cancellation occurs. The resultant includes a factor $\kappa - \kappa_{\pm1}$ that is proportional to ω. The first order terms therefore vary as ω^{-1}.

The conclusion of the complete analysis is that the form of the solution depends on the value of $i\kappa F/m\omega^2$: if this is small, then only the additional wave functions with time-dependence $\exp(iEt/h\pm\omega t)$ are significant and the behaviour is linear in F, but if it is large many wave functions must be considered in the complete solution, and the behaviour is no longer necessarily linear in F. This analysis can be applied to multi-barrier tunnelling structures, at least in the case where $\kappa F/m\omega^2$ is small, and the full wave functions found throughout the structure including the contacts.

5. Calculation of Electrical Properties

This work aims to find the amplitude and phase of the currents that flow in a device as a result of a high frequency bias. There are two sources of such currents; first, the number of charge carriers traversing the active region varies in time, corresponding directly to a time-dependent current, and second, charges within the active region can redistribute themselves in response to the external bias; this implies a displacement current that also produces currents at the contacts where the time-dependent field lines terminate. Both sources must be calculated if the a.c. device current is to be found.

The time dependence of the flux of carriers at any point is readily calculated. To show the principles of the calculation, initially only one of the two time-dependent terms will be considered. The wave function at the last barrier which is now chosen as the origin of x, may be written as: $[P_0+P_1\exp(-i\Delta kx+i\omega t)]\exp(-ikx+iEt/\omega)$ where $\Delta k = k_1-k$. The particle flux is given by $-(ih/m)\Psi^*\partial\Psi/\partial x$. The terms Ψ^* and $\partial\Psi/\partial x$ can be represented by phasors. If, at $x = 0$, $P_0^*\exp(-iEt/h)$ is the reference phasor directed along the real axis, then the second term in Ψ^* has an amplitude $|P_1|$ and rotates with a relative angular velocity $-\omega$. The representation of $\partial\Psi/\partial x$ comprises ikP_0 as the reference phasor with a second phasor of magnitude $|k_1P_1|$ rotating at ω. The product $-i\Psi^*\partial\Psi/\partial x$ equals: $kP_0^2+(k+k_1)P_0P_1\cos(\phi+\omega t)+k_1P_1^2+ i(k_1-k)P_0P_1\sin(\phi+\omega t)$ where ϕ is the phase angle of P_1 relative to P_0 at time $t = 0$. The second term in this expression describes a fractional modulation of the current by approximately $2P_1/P_0$ at a frequency ω and with a phase angle ϕ. This current has a resistive part $2P_1\cos(\phi)/P_0$ in phase with the applied bias and a capacitive part $2P_1\sin(\phi)/P_0$ leading the applied bias by 90°. The term, $P_{-1}\exp(-ik_{-1}x+iEt/h-i\omega t)$, produces a similar modulation in the current; the total is given by their sum. In tunnelling devices at the low frequency limit the amplitudes P_1 and P_{-1} vary as ω^{-1}; they have opposite signs and the a.c. current depends on the difference $|P_1|-|P_{-1}|$. At energies where such a structure has a positive conductance, $(|P_1|-|P_{-1}|)/(|P_1|+|P_{-1}|)$ is positive and proportional to ω; giving a frequency-independent in-phase part of the a.c. current.

The calculation of the charge density within the active region of the device and its dipole moment is similar. In any region subject to a field the wave functions have an explicit time dependence like those in the barrier considered in section 4. The calculation of the time dependence of the charge density must therefore include both such terms as C_1 and those like $C_0 J_1(i\kappa F/m\omega^2)$; the latter are easily included in the calculation. As an example, in a double barrier tunnelling device with an incident energy below the resonant level the variation of the charge density in the well is found, as $\omega \to 0$ to be in-phase with the applied field with a frequency independent magnitude.

6. Can Quantum Mechanics help with Device Design?

The reader may feel that the full quantum mechanical calculation of, for example, the currents in a tunnelling structure, is an excessively difficult task with little, if any, reward. However the essential result is that both the current and charge modulation are given by terms like P_1+P_{-1} resulting in a frequency independent magnitude for the current at low frequencies. P_1 and P_{-1} are determined by the properties of the structure at energies $E+h\omega$ and $E-h\omega$; as ω increases the value of P_1+P_{-1} will depart from its low frequency behaviour in a way that depends on the structure and, in particular, on any resonances in the region of $E+h\omega$ and $E-h\omega$. In a double barrier structure there exists a resonance at quasi-bound state energy, E_R. If $E < E_R$ then, as ω is increased $E+h\omega$ will pass through E_R; two factors will then affect P_1+P_{-1} and hence the a.c. current. First, the phase of the P_1 changes by $180°$ through the resonance, above E_R $|P_1|$ and $|P_{-1}|$ therefore add to give the a.c. current and, second, P_1 will be enhanced by the resonance over the value extrapolated from low frequencies.

Quantum mechanics predicts very intersting possibilites for any terahertz transport device in which the carriers in the active region have an energy, E, that is well defined on the scale of the applied frequency, $\Delta E << h\Im$. If such a device has a resonance at an energy, E_R, that is about $h\omega$ away from E then the device conductance at terahertz frequencies may be significant; resonances near both $E+h\omega$ and $E-h\omega$ can double this effect. The author has suggested that this effect should be called quantum modulation in scattering and tunnelling (the QMIST effect).

7. Calculated Terahertz responses for QMIST structures

The simplest calculation is that for an ideal double-barrier device with a single incident energy and resonance energy. The admittance, $\partial J/\partial V$, is real for $\omega \to 0$, its magnitude increases as ω increases and it lags in phase. The maximum admittance is reached when $h\omega = |E-E_R|$, when its phase lags by $90°$, corresponding to a pure susceptance. If $E < E_R$, then the low-frequency admittance is positive and the resonant admittance is inductive, but for $E > E_R$ the sign of the admittance is reversed and the resonant admittance is capacitive. The phase evolution continues for ω above resonance, with a conductance opposite in sign to that at low frequencies. A device has power loss or gain for positive and negative conductance respectively. The phase and amplitude variation of the admittance fits a classic resonance, at maximum it is magnified by $E/\Delta E$; the maximum conductance occurs at frequencies corresponding to the half-width ΔE (or inverse lifetime) of the quasi-bound quantum state. If this device were realisable it could be used with a positive low-frequency conductance, giving stable biasing, and a resonantly enhanced power gain at frequencies above the value for resonance.

For a device with "real" contacts, the admittance must be integrated over the distribution of electron energies and velocities in the cathode at that bias. The broad range of incident energies associated with the highly-doped emitter of a double-barrier structure masks the peaks in the response, since resonance occurs over a wide range of frequencies. Characteristics calculated for "real" devices are similar to those given by a simpler picture based on tunnelling time [3]. A minimum of three barriers is required to observe a resonant enhancement. The quasi-bound state between the first two barriers acts as an energy filter for the second well. If the first two barriers have the greater effective thicknesses then the majority of the electrons tunnel via the first well state which has the smaller energy width but higher peak transmission. This current shows a resonance in admittance that is very similar to that described in the paragraph above. The smaller current tunnelling via the second well state has an opposing admittance. The resonant frequency is given by the energy difference, E_1-E_2, between the quasi-bound states in the two wells which reduces as bias is applied to the device. Increasing the bias also reduces E_1 with respect to the emitter giving increases in the electron density in the first state and the d.c. current. As the frequency increases through resonance at low bias, the admittance passes through an inductive peak from a positive to a negative conductance; all are small, since the d.c. current is small. Initial increases in the bias reduce the resonant frequency and increase the high-frequency admittance with the d.c. current. The resonant frequency is a minimum when the quasi-bound energies coincide and the quantum states combine. As this point is approached the d.c. current increases but the high frequency admittance falls to zero. Additional bias still increases the d.c. current but causes the sequence to be reversed; the resonant frequency again increases, and the admittance changes from negative through a capacitive peak to positive as the frequency increases for fixed bias. Four barrier structures can be designed to exploit resonances at both $E+h\omega$ and $E-h\omega$ and these will have significant advantages as terahertz devices over triple-barrier structures; their characterisitcs have been presented elsewhere [4].

8. The problem of internal self-consistency

These calculations are based on a.c. fields that are uniform across the active region of the device. For a triple barrier device they predict that the charges in the first and second wells vary at the frequency of the a.c. field, particularly when this corresponds to E_1-E_2. If the device has negative conductance (power gain), the variation of charge is out-of-phase with the a.c. field in the second well, but in-phase in the first well. This a.c. polarisation generates an extra field reducing those fields on the contacts and increases the field between the two wells. A device with positive conductance has an opposite charge response and the field is enhanced near the contacts and reduced between the two wells. If the consequences of these charge oscillations are to be included in the model, two effects must be considered: first, a displacement current that leads the charge oscillation by $90°$ is added to the current associated with the flow of carriers, and second, the complete a.c. analysis of the device response requires an iterative solution. In this the charge response to a uniform a.c field is first calculated, then the resultant non-uniform a.c. field found and the calculation repeated for the latter using a stepped approximation. This process may have to be iterated several times for stronly non-uniform a.c. fields.

9. Conclusions

Although quantum mechanical effects are negligible in most devices, this paper shows that, at terahertz frequencies, the conventional models may not describe their behaviour well, and that a quantum treatment is essential. This requires lengthy and difficult calculations; however, new classes of transport devices are predicted that use resonances to overcome the limitations of carrier inertia and structural capacitance.

10. References

1   Buttiker, M. and Landauer, R. (1982) Traversal Time for Tunneling, *Phys. Rev. Lett.* **49**, 1739-1742
2   Truscott, W.S. (1993) Wave Functions in Time-Dependent Fields: (i) Exact Solutions *Phys. Rev. Lett.* **70**, 1900-1903
3   Truscott, W.S. (1994) A Physical High Frequency Equivalent Circuit of Resonant Tunnlling Diodes, in A.A. Rezazadeh (ed.) *EDMO'94 2nd International Workshop on High Performance Electron Devices for Microwave and Optoelectronic Applications*,King's College, London, pp 60-65
4   Truscott, W.S. (1994) Negative Conductance at THz Frequencies in Multi-Well Structures, *Solid State Electronics*, **37**, 1235-1238

TRAVELLING WAVE DETECTORS –
A PRINCIPLE FOR TERAHERTZ OPERATION

M.H. KWAKERNAAK, D. ERNI
Electronics Laboratory, Laboratory for Electromagnetic Fields
and Microwave Electronics,
Swiss Federal Institute of Technology, ETH Zürich
Gloriastrasse 35, CH-8092 Zürich

H. SIGG
Paul Scherrer Institute Zürich
Badenerstrasse 569, CH-8048 Zürich

Abstract

We discuss the promising application of the travelling wave concept for ultrahigh frequency detectors. An extensive introduction on the general operation principles and design considerations, including coupling strategies, losses, the velocity and the impedance matching of such detectors is presented. We report on the realisation of a velocity matched detector with the so-called transmission line integrated photon drag (TIP-)detector. We aim to revive the travelling wave concept, especially in the context of optical THz-signal detection.

1. Introduction

The practical operation of devices in ultra-high frequency applications becomes increasingly difficult because every decrease in the corresponding wavelength must be followed by all the geometrical dimensions of the device. Usually shrinkage is applied until the lumped equivalent circuit models are no longer valid and the device parasitics start to dominate. Therefore, as one ventures into the THz-range, novel device concepts are needed. A powerful method to avoid such geometrical restrictions is addressed here. The principle is that instead of localizing the interaction of the optical field with the microwave field in a restricted area, one transforms the whole interaction process into a travelling wave (TW) event. All of these geometrical restrictions are then converted to an easily satisfied velocity-matching condition between both fields. The interaction length is governed only by the interaction process itself and may be orders of magnitude larger than the wavelengths involved. This relaxation of the predominant wavelength dependence of the device dimensions by a travelling wave (TW) principle represents an attractive potential for detector technologies, since the efficiency of a detector is usually determined by its area. It is therefore our aim to revive this promising concept, especially in the context of optical THz-signal detection.

J.M. Chamberlain and R.E. Miles (eds.), New Directions in Terahertz Technology, 143–153.

144

The paper is organized as follows. In section 2, the travelling wave (TW) concept is elucidated using the analogy with electro-optical modulators (EOM), for which this concept has been known for many years. Section 3 gives the basic equations needed to design TW detectors. A general approach is taken which allows the description of several kinds of detectors, such as TW photodiodes [1, 2], optical-to-microwave transformer [3] and TW photon drag detectors [4]. For the latter system, since it is potentially among the fastest detectors and of conceptual simplicity, the detailed design criteria are given in sections 4 and 5. These considerations have led to the development of the recently presented transmission line integrated photon drag (TIP) detector[4].

2. Travelling Wave Modulators

In a so-called parallel plate electro optical modulator (EOM), the electrical field is excited by a microwave generator oscillating at frequency ω and at constant phase, $E(z, t) = E_0 \cdot cos(\omega t)$. The speed of light and the phase retardation per length of the crystal are defined respectively as $v_{opt} = c/n$ and $d\Gamma = aE_0 dz$. Ignoring the phase dependence of the microwave field, i.e. neglecting its propagation, the total retardation $\Delta\phi$ imposed on an optical wave travelling along the crystal of length l becomes:

$$\Delta\phi = \int_{z=-L/2}^{z=+L/2} d\Gamma(z, t) = \Gamma_0 cos(\omega(t + T)) \cdot \frac{sin(x)}{x} \tag{1}$$

where $T = l \cdot v_{opt}^{-1}$ is the traversal time, $x = T\omega/2$ and $\Gamma_0 = aE_0 l$. Equation (1) shows that because of the non-matched velocities, the retardation is limited to $\Gamma_{Max} < \Gamma_0/2$ at the modulator length given by $l = 2\pi c(\omega n)^{-1}$.

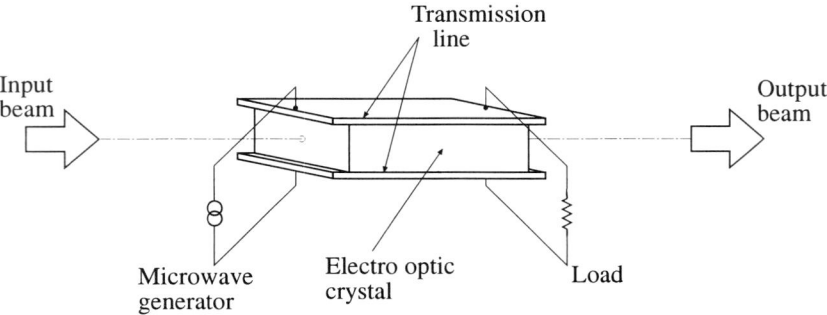

Figure 1.   Travelling wave electro optic modulator

The concept which allows one to overcome this fundamental limit is depicted in Figure 1, showing the scheme of a TW EOM [5]. Here the microwave field is coupled into a transmission line and travels at approximately the same speed as the light: $v_{opt} \approx v_{el}$. One obtains that $\Delta\phi$ is given by an equation similar to equation (1), where x is replaced by $x' = x(1 - v_{opt}/v_{el})$, i.e. the limitation in modulation strength is relaxed by the same factor $(1 - v_{opt}/v_{el})$.

This result can be applied to ultra fast photodetectors. Conceptually, the main difference is that the microwave field becomes the signal carrier while the optical field and the excited currents become the source carriers.

It might be useful to note that the commonly quoted bandwidth limitation for photo-diodes, i.e. the RC-time constant, is identical to the bandwidth limit obtained for entirely non velocity matched systems in a travelling wave picture [1]. By analogy with the un-matched EOM given by equation (1) the RC-time thus merely gives a lower bound for the achievable bandwidth. Proper matching of the propagation, as we will show, improves the detector performance considerably.

3. Principles of Operation of Travelling Wave Detectors

We now discuss the TW principle applied to photodetectors and highlight the major differences to TW modulators. In a TW detector an optical signal is converted into a microwave signal according to the travelling wave principle. As the microwave signal wavelength is comparable to the detector size it is necessary to couple out these signals in a controlled manner. When the geometry of the detector is chosen such that the microwave signal propagates into a controlled direction with a defined velocity the prerequisite for a TW detector is satisfied, where bandwidths can be reached which are no longer limited by the detector size. An obvious geometry is a microwave transmission line in any of its variations. Maxwell's laws can then be described in terms of the propagation properties of the transmission line.

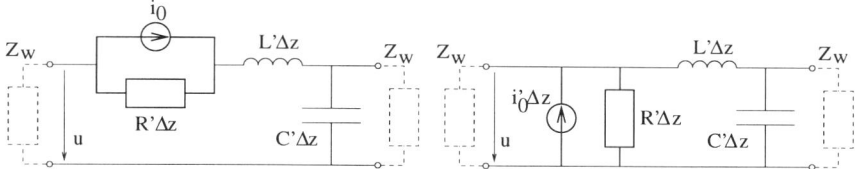

Figure 2. Equivalent circuits of an element of length Δz of TW detectors: serial topology (left) and parallel topology (right)

There are two different ways of introducing an electrical signal into a transmission line. One is to place the source between the two branches of the transmission line, i.e to add parallel current sources between the two branches, Figure 2, right. The other is to disturb the lateral momentum equilibrium of the carriers inside one of the branches, i.e. to add current sources in series into one branch, Figure 2 left. L', R', and C' are the induction, resistance and capacitance per unit length, respectively. The voltage across the the two branches is $u(z, t)$ and the optically driven excitation is represented by the sources i_0 and $i_0' \Delta z$.

Examples of TW photodetectors discussed in the literature are photon drag detectors [4] and photodiodes [1, 2, 3, 6]. In photon drag detectors a lateral current is excited in a 2-D electron gas and the conductance of the 2-D electron gas is used to form a transmission line. In photodiodes carriers are photo-excited in the depletion zone of a p-i-n junction between the two transmission line branches, and the junction capacitance is part of the total transmission line capacitance C'. The major drawback of the latter example is that the detector response time is still limited by the sweep-out time of the carriers out of the depletion zone, whereas in photon drag detectors a current is directly excited in the transmission line branch, and only the intrinsic material response dynamics limit the detector bandwidth (see nextsection).

In the following we restrict our discussion to the serial topology. The parallel topology can be treated in the same manner.

By using Kirchoff's rules the propagation equations are found:

$$\frac{\partial^2}{\partial z^2} u(z,t) = R'C' \frac{\partial}{\partial t} u(z,t) + L'C' \frac{\partial^2}{\partial t^2} u(z,t) + R' \frac{\partial}{\partial t} i_0(z,t) \tag{2}$$

This is simply the well known telegraphist's equation where the inhomogeneous term describes the excitation. The unexcited transmission line is described by the line impedance $Z_w = (j\omega L' + R')^{1/2}(j\omega C')^{-1/2}$ and the propagation constants are $\pm\gamma = \sqrt{(R' + j\omega L') \cdot j\omega C'}$.

Equation (2) is solved in a similar manner to the Green's function method. We calculate the signal occuring in the transmission line caused by the source in one element of length $\Delta z \to 0$ and integrate over all excitations. The voltage per unit detector length at position z_0 is with Kirchhoff's rules:

$$
\begin{aligned}
u_{z_0}(z_0)' &= \lim_{\Delta z \to 0} \frac{u(z_0)}{\Delta z} \\
&= \lim_{\Delta z \to 0} \frac{Z_w R' i_0(z_0,\omega)}{Z_w + j\omega L'\Delta z + \frac{1}{j\omega\Delta z + 1/Z_w} + R'\Delta z} \\
&= \frac{R'}{2} i_0(z_0,\omega)
\end{aligned}
\tag{3}
$$

The overall solution is given by

$$u(z,\omega) = \frac{R'}{2} \int_{-l/2}^{z} i_0(z_0,\omega)e^{\gamma(z-z_0)}dz_0 - \frac{R'}{2} \int_{z}^{l/1} i_0(z_0,\omega)e^{-\gamma(z-z_0)}dz_0 \tag{4}$$

More insight can be gained by neglecting the frequency dependence of the propagation constant γ and of Z_w, which is a good approximation in the example to be discussed. We obtain:

$$
\begin{aligned}
u(z,t) &= \frac{R'}{2} \int_{-l/2}^{z} i_0\left(z_0, t - \frac{z - z_0}{v_{el}}\right)e^{-\alpha_{el}(z-z_0)}dz_0 \\
&\quad - \frac{R'}{2} \int_{z}^{l/2} i_0\left(z_0, t - \frac{z_0 - z}{v_{el}}\right)e^{\alpha_{el}(z-z_0)}dz_0
\end{aligned}
\tag{5}
$$

where the propagation velocity is $v_{el} = (L'C')^{-1/2}$ and the attenuation is $\alpha_{el} \approx R'/(2Z_w)$.

Any source at location z_0 excites one wave travelling to the left and another travelling to the right. In order to observe the total signal one has to integrate over all excitations.

The first integral of equation (5) contains all sources on the left of the observation point z , and the second integral describes all sources to the right of this point. The temporal delay $(z - z_0)v_{el}^{-1}$ represents the propagation behavior, $e^{-\alpha_{el}(z-z_0)}$ the attenuation. An important point to note is that only one of the two waves can be velocity-matched to the optical signal.

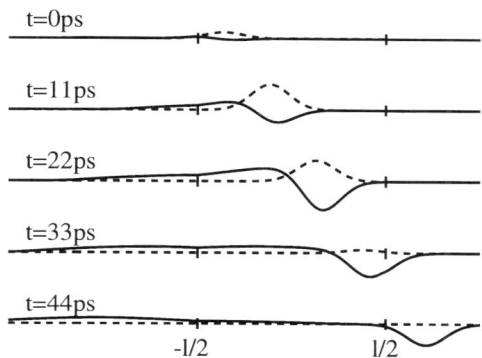

Figure 3. The dashed line shows the intensity $S(z, t)$ of an optical pulse coming from the left. It generates two electrical pulses $u(z, t)$ (solid lines): a short one travelling forward and a wide one travelling backward.

Figure (3 illustrates a scenario in a TW detector. An optical intensity profile of a short pulse travels along the detector. The velocity-matched electrical pulse profile is determined by the optical pulse shape, whereas the backwards electrical waveform is given by the detector length. In this simulation, an instantaneous material response is assumed, i.e. $i_0(z, t)$ is proportional to the radiation intensity $S(z, t)$.

It is important that the backward travelling wave is not reflected somewhere and become coupled into the forward direction, which would decrease the bandwidth dramatically. One should couple the unused wave out of the detector and drain it into an appropriate load.

4. Travelling Wave Photon Drag Detectors

In the following, the realisation of a TW detector with serial in-coupling is described. The two-dimensional electron gas (2DEG) formed by a multi-quantum well system serves as the active conductor. The photo-excitation of the current is due to the so called photon drag effect. Since the TW principle, not the actual detection mechanism, is discussed here only a brief description of the photon drag effect is presented.

The carriers trapped in a multi-quantum well system grown on top of a semi-insulating GaAs substrate are free to move in the x-y-plane. Their motion along the y-direction is confined by 30 GaAs quantum wells each approximately 80 nm thick and separated from one another by centrally-doped 26 nm thick $Al_xGa_{1-x}As$ $(x = 0.35)$ barriers. Room temperature electron density and mobility are respectively, $N_{2D} \approx 0.92 \; 10^{12} \, cm^{-2}$ and

$\mu_e \approx 7'200\,\mathrm{cm^2 V^{-1} s^{-1}}$. Due to the confinement, the 2DEG accommodates a strong dipole transition at the intersubband transition energy, $E_{12} \approx 120\mathrm{meV}$, for excitations polarized along the y-direction. Unexcited, the 2DEG is at equilibrium, and no macroscopic current is flowing, i.e. the sum of the electron velocities $v_{\vec{k},i} = \hbar\vec{k}/m^*$ over all occupied states $\rho_{\vec{k},i}$ vanishes ($\vec{k}$: is the electron wave vector; $i = 1, 2$: for subbands 1 and 2).

$$J_{2D} = -e \sum_{\vec{k},i} \vec{v}_{\vec{k},i} \rho_{\vec{k},i} = 0 \quad (i = 1, 2) \tag{6}$$

Excitations at energies close to E_{12}, however, drive the electron distribution by $\delta\rho$ away from its equilibrium, which results in a macroscopic current composed of the contributions from both subbands.

$$J_{2D} = -e \sum_{\vec{k},i} \vec{v}_{\vec{k},i} (\rho_{\vec{k},i} + \delta\rho_{\vec{k},i}) \neq 0; \quad (i = 1, 2) \tag{7}$$

For the evaluation of $\delta\rho$, we refer to the literature [7, 8, 9, 10]. We find that our experiments are in good agreement with calculations based on the Boltzmann rate equation, in an approximation where the relaxation in the two subbands is parameterized by two individual current relaxation times, τ_1 and τ_2:

$$i_0 = \int_{-b/2}^{b/2} J_{2D} dx = -bd \frac{Se}{v_{opt} m^*} (\alpha_{opt}(2\pi\nu) \cdot \tau_2 + \frac{\partial \alpha_{opt}}{\partial \nu} \frac{E_F}{2h} \cdot (\tau_1 - \tau_2)) \tag{8}$$

E_F is the Fermi level, and $\alpha_{opt}(2\pi\nu)$ is the wavelength dependent absorption coefficient.

As is shown in Figure 4, good agreement with our experiment is obtained for $\tau_1/\tau_2 \approx 2$ [4]. The second term in equation (8) is proportional to $\partial\alpha_{opt}/\partial\nu$, which changes sign at the center frequency of the intersubband resonance E_{12}. This so called resonant term can be understood from the fact that only the 2D electrons at rest are in perfect resonance with an excitation at $h\nu = E_{12}$. The resonance energy shifts slightly towards lower (higher) energies for electrons moving parallel (anti-parallel) to the incident radiation, making this term nonzero for slight detuning from E_{12}.

The first, so-called direct term in equation (8), is a consequence of the momentum transfer of photons to the electron system and is proportional to the absorbed radiation intensity $dS\alpha_{opt}$. The proportionality factor, i.e. the current per width and radiation intensity, is independent of the number of 2D layers, $R \approx e\tau/(m^* v_{opt}) \approx \mu_e n/c \approx 1$.

Figure 5 depict our realisation of a TW photon drag detector [4]. The microwave field is guided by the microstrip line formed by the MQW layers and the metalized backside of the GaAs chip (ground plane), whereas the optical wave is incident via a Ge-prism at the critical angle of total reflection [11].

The optical wave being excited below the surface propagates parallel to the surface, and its velocity is given by the refractive index of GaAs: $v_{opt} = c/n_{GaAs}$.

It can be seen that at resonance, where absorption is high, only a very small fraction of the light is reflected, Figure 6. This optical coupling scheme not only guarantees well-defined irradiance of the surface, but it is also very efficient.

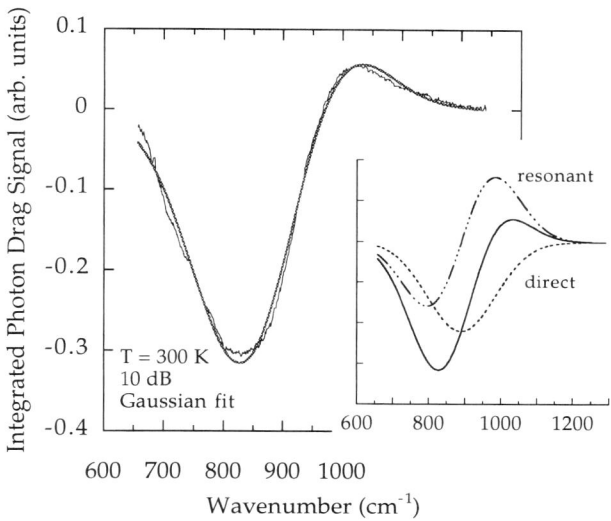

Figure 4. Measured photon drag current as a function of wavelength. Inset: Decomposition of the photon drag current into a direct and a resonant term.

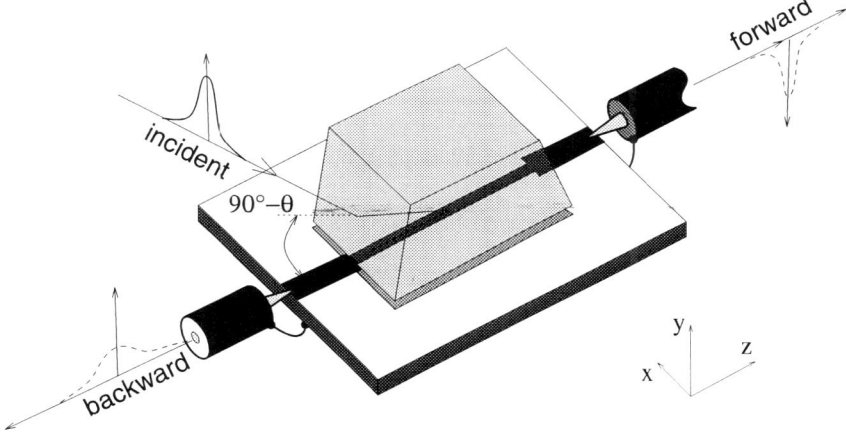

Figure 5. Optical and electrical coupling of the detector

The electrical signal is guided by the detector-microstrip-line. The propagation velocity of the microwave signal is determined by the effective index of the microstrip line $v_{el} = c/n_{eff}$. This effective index is given by the overlap of the electric field with the index distribution and is thus dominated by the (low frequency) refractive index of the material between the microstrip line and the ground plane. We obtain $n_{eff} = 3.7$. The Ge-prism on

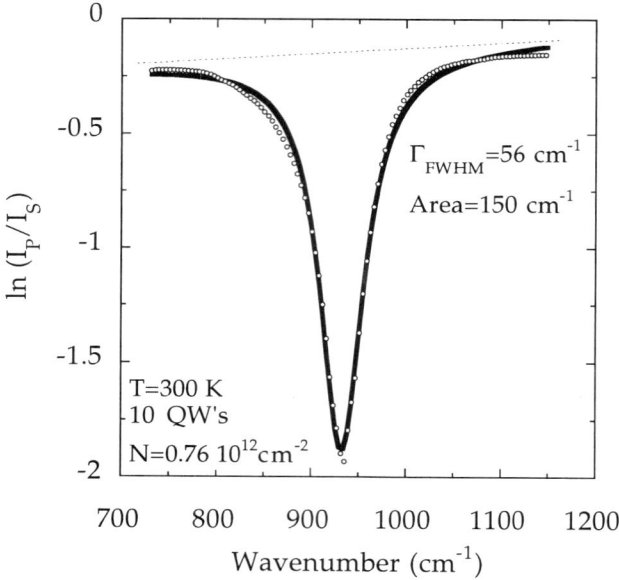

Figure 6. Reflection at the Ge-GaAs surface as a function of wavelength, measured (dots) and calculated (solid)

top of the microstrip line produces, due to its high refractive index $n_{Ge} = 4$, a significant distortion of the line impedance. Care must be taken to properly chose the widths of the microstrip line, both under and beside the prism, c.f. Figure 5. For a 50Ω transmission line on a 25 mil substrate we chose a width of $220\mu m$ under the prism and $450\mu m$ beside the prism.

The velocity matching in our design is relatively good: $v_{el}/v_{opt} = 0.89$. Since the two propagation velocities of the radiation and the microwave signal are determined by the material properties of different regions, velocity-matching can be perfectly achieved with a proper choice of materials.

In Figure 7 the observed electrical response of the detector to a 1 ps radiation pulse generated by a free-electron laser is shown.

The forward response is significantly faster compared to the backwards signal. Note that in this measurement much of the pulsewidth difference is masked by the sampling head bandwidth of 34 GHz.

5. Limitations and Advanced Concepts

The photon drag detector presented suffers from a relatively large electrical signal attenuation due to the high surface resistance of the detector material; we obtained a resistance per length of $R' = 1.43\text{k}\Omega\text{cm}^{-1}$ on a 50Ω transmission line detector on a 25 mil substrate. Therefore, to improve the detector performance, one must overcome the electrical attenuation, for example by enlarging the detector width. The resulting reduced line impedance Z_w should be matched with carefully designed taper structures, to guarantee a smooth

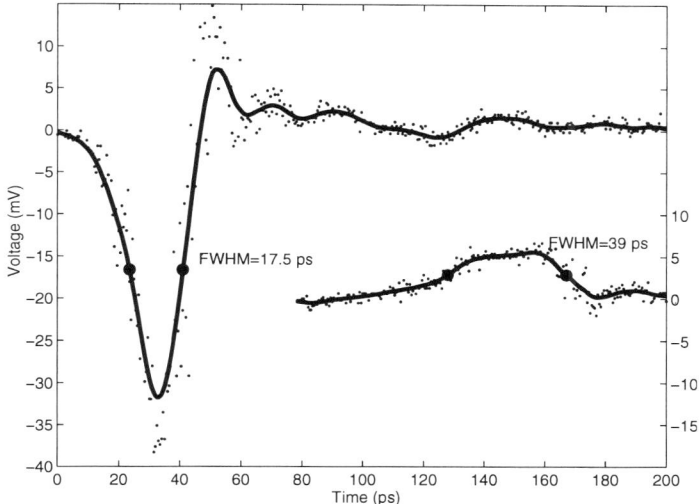

Figure 7. Forward (left) and backward (right) response of the detector to a $1ps$ laser pulse at $11.6\mu m$, measured with a 34 GHz sampling head.

signal transition to the measurement environment of 50Ω while maintaining the high bandwidth characteristics.

The taper structures produce minor pulse distortion, and it is interesting to note that the reduced resistance decreases the frequency dispersion induced pulse distortion, Figure 8. Nevertheless, in this TW detector concept, loss remains the major concern, since it lowers the electrical signal output.

In order to overcome the strong attenuation of the electrical signal within the TIP-detector, one may be tempted to make use of a parallel-line directional coupler as an alternative configuration. Here, one of the coupler arms is represented by the lossy transmission line of the TIP-detector itself. The other arm acts as a parallel-coupled lossless transmission line. Such a configuration would have the advantage that an emerging electrical signal in the lossy detector part is immediately coupled to the lossless coupler arm, before attenuation plays its unfavorable role. Despite the appeal to the ideas described here, there are two fundamental limiting mechanisms to be considered. First, recalling the Bode-Fano theorem in a broad-band matching problem [12], a similar relation can be intuitively postulated for the overall coupling coefficient of a parallel-line coupler: generally speaking, there exists a trade-off between the coupler bandwidth and the coupling coefficient. Therefore, if one intends to extract a short pulse from the TIP-detector in the manner described here, a weak coupling coefficient will be the main limiting factor. The second deficiency in our coupler concept is associated with the coupling process itself. For this reason one has to look into the coupled mode formalism, where coupling arises between the two fundamental eigenmodes of the corresponding coupler arms. Maximum energy exchange between these two arms implies equal propagation constants for both transmission lines. It can easily be shown that the loss introduced by the TIP-detector severely violates this equality.

152

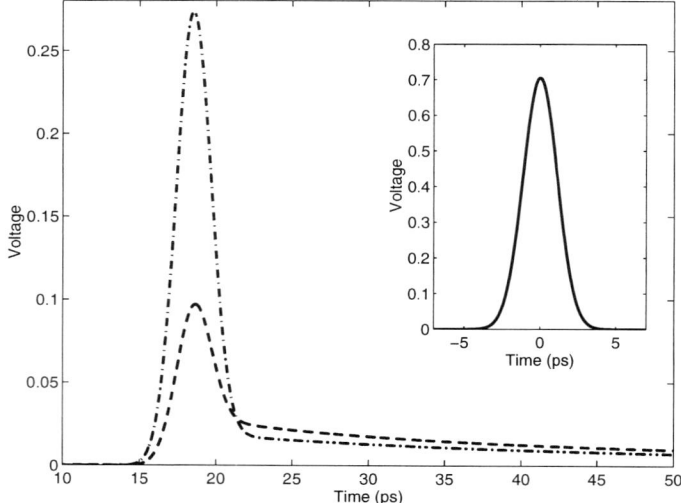

Figure 8. Pulse distortion due to frequency dispersion caused by resistance per length of the detector of width $220\mu m$ (dashed) and $800\mu m$ (dash-dotted) on a $2ps$ Gaussian pulse (inset).

6. Conclusion

The travelling wave principle is a powerful scheme which is applicable to ultrafast detectors. With this principle, the bandwidth versus sensitivity tradeoffs of classical detectors can be circumvented. Travelling wave detectors have been shown to operate beyond the 100 GHz range.They have been successfully realized as photodiodes as well as photon drag detectors.In particular, with photon drag detectors, where large detector sizes are necessary, a dramatic bandwidth improvement could be achieved. We believe that with the travelling wave principle, detector performance can reach the intrinsic material response bandwidth.

Acknowledgments

It is a pleasure to acknowledge the continued collaboration with P. van Son, Delft University, for his substantial contribution to the formulation of the problem of large photon drag detectors and ultra fast response. We greatly appreciate B. Margotte's contribution to the travelling wave solution, which has been part of his Diploma work at the electrical engineering department, ETHZ. We thank H. Siegwart and S. Graf, PSIZ and H. Benedickter, ETHZ for their help in the realisation of the photon drag detectors. Finally, we gratefully acknowledge the support by the stichting voor Fundamenteel Onderzoek der Materie (FOM) in providing the required beam time on FELIX, and highly appreciate the skillful assistance by the FELIX staff.

References

1.   Giboney, K.S., Rodwell, M.J.W. and Bowers, J.E. (1992) Traveling-Wave Photodetectors *Pho-*

tonics Technol. Lett. **4**, 1363-1365.

2. Hietala, V.M., Vawter, G. A., Brennan, T.M. and Hammons, B.E. (1995) Travelling-wave photodetectors for high-power large bandwidth applications,*IEEE Trans. Microwave Theory Tech.* **43**, 2291-2298.

3. Wu, M.C., Lin, L. and Itoh, T. (1994) A new ultrafast photodetector: optical-to-microwave transformer *Proceedings of the SPIE* **2116**, 228.

4. Sigg, H., Kwakernaak, M.H., Margotte, B., Erni,D. van Son, P.. and Köhler, K. (1995) Ultrafast Far-Infrared GaAs/AlGaAs Photon drag detector in microwave transmission line topology, *Appl. Phys. Lett* . **67** (19), 2827-2829.

5. Peters, L.C. (1963) Giga cycle bandwidth coherent light wave, *Proc. IEEE* **51**, 147.

6. Giboney, K.S., Nagarajan, R. L, Reynolds, T. E., Allen, S. T., Mirin, R. P., Rodwell, M.J.W. and Bowers, J.E. (1995) Travelling-wave photodetectors with 172 GHz bandwidth and 76GHz bandwidth efficiency product,*IEEE Photonics Technol. Lett.* **7**, 412-414.

7. Luryi, S. (1987) Photon-drag effect in intersubband absorption by a two dimensional electron gas, *Phys. Rev. Lett.* **58**, 2263-2266.

8. Wieck, A.D., Sigg, H. and Ploog, K. (1990) Observation of resonant Photon Drag in a two-dimensional electron gas, *Phys. Rev. Lett.* **64**, 463.

9. Stockman, M.I., Pandey, L.N., George, T.F. (1990) Light induced drift of quantum confined electrons in semiconductor heterostructures, *Phys. Rev. Lett.* **65** 3433.

10. Keller, O. (1993) Photon drag effect in a single level metallic quantum well, *Phys. Rev.* **B48**, 48.

11. Keilmann, F. (1994) Critical incidence coupling to intersubband excitations, *Solid State Comm.* **92**, 223.

12. Fano, R.M. (Jan 1950) Theoretical limitations on the broadband matching of arbitrary impedances, *Journal of the Franklin Institute*.

INTEGRATED ACTIVE DEVICES

D. LIPPENS
Institut d'Electronique et de Microélectronique du Nord
U.M.R.S. C.N.R.S. 9929, Université des Sciences de Lille 1
Avenue Poincaré, BP 69, 59652 Villeneuve d'Ascq Cedex, France

Abstract

This paper is a review of the processing techniques used to integrate devices for terahertz operation. The main source of difficulty comes from the very small cross sectional area of the devices which leads to stringent conditions in their fabrication. After dealing with epitaxy techniques paying special attention to the growth of low-dimensional semiconductor heterostructures, recent progress in lithography techniques notably by electron beam will be considered. Finally the various techniques for fabricating low parasitic contacts to the devices including free-standing beam leads which interconnect the active devices to the circuit will be addressed.

1. Introduction

For a device to operate at very high frequencies the physical mechanisms involved in the operation characterised by ultra-short time responses. These relevant times are inherent to the material systems but also depends on the geometry of devices. For a confined geometry, such as a quantum well heterostructure, the situation is a little more complicated because we have to substitute the notion of life time, namely the average time spent by electrons on the quantum state, for that of a transit process. The geometry of the devices is that of extremely thin layers in order to take from the quantum effects which can be seen only on a short scale compared to the electron wavelength. While the physics underlying the operation of the device is one thing, the way to integrate the device to the external circuit is also extremely important particularly at terahertz frequencies where practically all orders of magnitude have to be revisited. To illustrate this point, let us consider the very simple case of an ohmic contact. If we assume a contact resistance of $1 \times 10^{-6} \Omega.cm^2$ which is the standard value for a contact to n-type GaAs, the resistance of this contact is 1Ω for a device on the $10 \times 10 \mu m^2$ scale, but $100 \ \Omega$ for a device of $1 \mu m^2$ area, the typical dimension for a terahertz operation. With this order of magnitude in mind, it can be easily understood that the fabrication of interconnects is a challenging issue where parasitic elements have to be as low as possible. Fortunately, owing to the remarkable progress in the growth of high quality

155

J.M. Chamberlain and R.E. Miles (eds.), New Directions in Terahertz Technology, 155–162.

epilayers, in the submicron patterning, using reactive ion etching and in the fabrication of free standing connecting elements, most of the difficulties aforementioned can today be overcome. In this paper, I will try to illustrate each of those issues. To this aim, in the first section, I will report on epitaxy techniques. In the second section, the process techniques involved in the fabrication of small area devices will be illustrated paying attention to some recent ideas.

2. Epitaxial growth

There are two techniques which are currently used for growing III-V epitaxial heterostructures with layer thickness of a few monolayers (where a monolayer is a/2 and hence ~3Å for III-V materials). These are Molecular Beam Epitaxy (MBE) and the Metal Organic Chemical Vapor Deposition (MOCVD). Schematically GaAs-based compounds can be grown in a MBE system whereas MOCVD is employed essentially to grow epilayers which incorporate Indium Phosphide layers. This distinction tending to disappear with recent developments of Gas Source MBE systems which permit the growth of layers with Aluminium and Phosphorus content. MBE is now well understood. In short, it can be compared to an evaporation technique performed in ultra-high vacuum. The substrate is placed in a high vacuum chamber and elemental species (As, Ga, In, Al...for the growth material and also Si, Be for n-type and p-type doping respectively) are evaporated from ovens and impinge the heated substrate. The typical growth temperature is between 500 and 600 ° C. It is also possible to grow material at much lower temperature while preserving the crystal quality typically between 200° C and 250 ° C. Under these conditions the material contains a high concentration of deep defects which promotes carrier recombination. Optically generated carriers thus recombine at rates an order of magnitude faster than for higher temperature grown materials. Such Low Temperature Grown (LTG) materials in particular LTG GaAs are used in electro-optic sampling for detecting terahertz radiation.

Figure 1. Transmission Electron Microscopy of a $Al_{0.4}Ga_{0.6}As/GaAs$ Double Barrier Heterostructure.

For heterojunctions and related superlattices, the accuracy required in the growth of the epilayers strongly depends on the application and particularly on whether transport is along or parallel to the direction of growth. Therefore, for devices whose operation relies on tunnelling through potential barriers and quantum confinement any difference in the nominal thickness lead to dramatic changes in the current-voltage characteristics because quantum transmission is an exponential function of the barrier thickness. This is particularly true for double barrier resonant tunnelling structure whose transmission electron microscopy photograph is shown in Figure 1. For devices intended to operate at terahertz frequencies it is imperative to operate at very high current density and hence with barrier dimensions ranging between 10 and 20 Å.

On the other hand, the quality of the epitaxy is also of prime importance notably of interfaces. At such short dimensions it is clear that in situ analysis techniques have to be employed for the control of the epitaxy parameters. Up to now the technique routinely employed is Reflection High Energy Electron Diffraction (RHEED) which can be used to monitor the growth of the material as a function of time. The diffracted signal shows an oscillating nature characteristic of the two-dimensional growth of the material layer by layer. Also one can make use of growth interruption in order to minimise the interface roughness which is one of the main scattering mechanisms in ultra thin heterojunctions. Very recently, other 'in situ' advanced techniques have been used to control the quality of interfaces notably Scanning Tunnelling Microscopy which can be mounted close to the MBE system. MOCVD involves transport of species using organic molecules. The precursor for group V elements are Arsine (AsH_3) and Phosphine (PH_3) which are highly toxic gases and therefore a major difficulty of this technique is in the necessary safety requirements and precautions which must be implemented. Compositional and layer uniformity are not ideal and it appears relatively difficult to grow quantum well-barrier structures with good reproducibility. Gas Source MBE [1] alleviates most of these drawbacks but is generally more expensive.

3. General process techniques

3.1. LITHOGRAPHY TECHNIQUES.

Schematically there are two techniques which are currently used for patterning the devices namely *photolithography* and direct writing by *electron beam*. Other non optical procedures also exist notably X ray lithography which is used for micromachining microstructures with high aspect ratio. Photolithography proceeds by exposing resists to uv light through a mask. In general, the mask is placed in close proximity to the wafer for alignment. This technique is capable of defining geometries near one micron. Below this critical dimension, researchers have to take advantage of sophisticated strategies. Therefore, workers at the University of Virginia have fabricated submicron anodes by means of the so called Electroplate Window Shrink (EWS)[2]. In this process an initial hole, written on the 1μm scale is subsequently shrunk as metal plates the periphery, narrowing its diameter. Otherwise, electron beam lithography is employed with practically no restriction on pattern size down to 50 nm. This is the preferred technique for writing the submicron gate lengths necessary for high mobility transistors which exhibit a maximum frequency of oscillation at submillimetre wave lengths. The main advantages of e-beam lithography include high resolution, easy pattern modification and absence of mask defects due to the direct writing in e-beam resists such as polymethylmethacrylate (PMMA). Accelerating voltages range from 20kV to 100kV. The development is usually performed with Methylisobutylketone (MIBK). One of the main issues in writing such closely spaced small size patterns is the proximity effect. In practice, the resists are exposed over an area wider than that of the incoming beam due to electron scattering within the resist, secondary electron emission and backscattered electrons. Proximity correction code can however be used to solve this problem. To illustrate the e-beam lithography technique, figure 2 shows a completed submicron Y-shaped gate in order to decrease the gate resistance. In practice such shape is achieved by patterning multilevel resists notably PMMA-P[MMA-MAA] [3] The other advantages which stem from the use of bi- or tri-layer resists concern the lift-off process, which needs an overhanging resist profile or the fabrication of single-step micro-airbridges. Briefly, the micro-airbridge is directly written by e-beam lithography by using various resists which are incorporated into the fabrication process. The advantages of this technique are manifold, resulting in a high resolution and high yield process. In addition, by using the fact that electrons are backscattered by a metal pad, a self-aligned procedures can be achieved during contact writing.

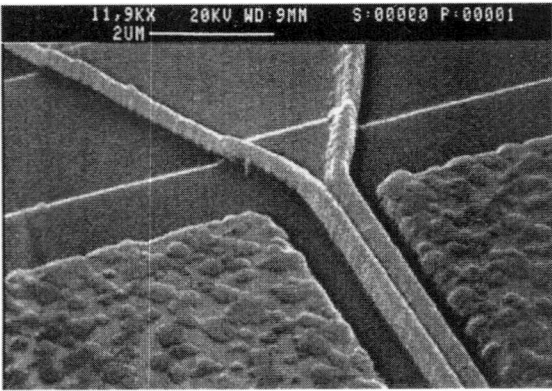

Figure 2. Y-shaped submicron gate field effect transistor. Also apparent are the ohmic contacts with a characteristic grain texture.

3.2. OHMIC AND SCHOTTKY CONTACTS

The electrical characteristics of the *ohmic contact* are now well known [4] with various current contributions viz thermionic emission (conduction over the barrier), thermionic field emission (thermally assisted process) and field emission, the latter describing pure tunnelling transmission through the barrier at the hetero-interface between metal and semiconductor. As stated in the introduction the contact resistance has to be drastically improved for operation at submillimetre wave lengths. So far, several techniques have been employed for these improvements depending on the material system. Therefore, for GaAs devices the doping concentration for a highly doped contact layer is typically 2×10^{18} cm^{-3}, with a built-in potential at the interface of ~ 0.7 eV. Under these conditions, electrons see a rather wide barrier even with local doping by the contact material. The contact resistance is about $1 \times 10^{-6} \Omega.\text{cm}^2$. To improve this contact resistance further one of the most common solution is to take benefit of low band gap material such as In$_{0.47}$Ga$_{0.53}$As lattice-matched on InP substrate. A drastic decrease in the gap and hence in the barrier height is thus obtained. This is not the sole advantage because it can also be shown that the doping concentration can be much higher ($>5 \times 10^{18}$cm^{-3}). Subsequently the contact resistance is drastically decreased with values in the 10^{-7} $\Omega.\text{cm}^2$ range and hence an order of magnitude lower than in the previous case. The same idea is applied GaAs epilayers with the growth of a low band gap capping layer which serves to fabricate small area ohmic contacts. A representative example of this fabrication technique is the n$^+$ InAs/In$_x$Ga$_{1-x}$As/GaAs layered structure [5]. The transition between InAs and GaAs has however to be optimised to permit the relaxation of strain with a good morphology of the top surface. Finally one can take advantage of planar doping in order to increase locally the dopant concentration. Such doping gives rise to a deep notch in the conduction band profile owing to local degeneracy of the material and helps greatly in the tunnelling transmission through the associated ultra thin barrier resulting in a low contact resistance. Many possibilities exist for the choice of contact material, for example by using sequential deposition of Au/Ge/Ni which when subsequently annealed gives a buried ohmic contact or PdGe used for shallow ohmic contacts.

For *Schottky contacts* the situation is also quite clear now thanks to numerous studies on this issue. The fabrication of high quality Schottky contacts on GaAs can now be realised by several techniques including evaporation in systems generally equipped with an electron gun or by anode plating. Good ideality factors under forward bias and low leakage current associated with high breakdown voltage are now obtained with GaAs epilayers Integration of the devices in anti-series or anti-parallel configurations is now employed which permits a symmetry in the C-V or I-V characteristics of Schottky diodes. This kind of symmetry is particularly interesting because the devices can operate unbiased and more importantly exhibit some special properties. In particular when using back-to-back Schottky diodes the C-V characteristic is symmetrical and only odd harmonics are generated in harmonic multiplication. Hence such an arrangement provides a great simplification in the circuit design. However, one of the main limitations in terahertz technology comes from the limited amount of power delivered by the solid state sources. In this respect, low band gap material could offer some advantages because the conduction starts at lower voltages. However, low gap structures suffer from low breakdown voltages and high leakage currents. In this context, a number of studies are now oriented towards the search for heteroepitaxies which are suitable for the fabrication of good Schottky contacts while still maintaining a good performance in terms of ideality factor or leakage current. A promising structure in this respect is the InP/InGaAs heterojunction. Recent studies have shown that the built-in potential can be continuously varied and adapted to the required application by inserting a thin InP layer in the 100Å range. By this means we can tune the voltage threshold for conduction between 0.2 and 0. 4 V[6] which is significantly lower than that of GaAs Schottky contacts.

3.3. INTERCONNECTING ELEMENTS

Turning now to the connecting elements, several solutions have been investigated for planar integrated the devices. For the diodes which are employed in a receiver to downconvert the RF signal to the lower frequency or multiplying a primary signal to the submillimetre spectrum, different planarisation techniques have been proposed including notably the dielectric cross-over [7] and the air-bridge techniques [8]. Briefly, dielectric cross-over consists in the deposition of a dielectric on the mesa walls to avoid short circuits when a beam lead is subsequently deposited interconnecting the diode to the circuit. In practice implementing such a technique requires attention to the resist edge and in particular to any undercutting effects to avoid open circuits in the metallisation. In addition, the dielectric has to be thick enough to decrease the parasitic capacitance formed by the strip connecting the diode and the highly doped regions. Despite these difficulties, this planarisation technique appears suitable for millimetre wave applications with reduced parasitic self-inductance and capacitance. In the higher part of the millimetre frequency band and at submillimetre frequencies, an airbridge technology is to be preferred with significant advantages in terms of parasitic elements notably concerning the parasitic capacitance. The decrease in the parasitic capacitance stems primarily from the use of free standing metallic structure in air. Moreover, it is possible to fabricate an air bridge with a rather large air gap whereas the dielectric thickness is often limited. Usually air bridge formation involves a two-step resist coating to define the beam lead and the pillars of the bridge. It is worth-noting that a number of difficulties stem from the mesa height and from the area difference between the connecting pad and the diode contact. For instance, a connecting pad to a 50 Ω coplanar waveguide has a dimension typically of 50 μm whereas the diode dimension is in the 1μm range. In addition if the coplanar lines are fabricated on the semi-insulating substrate the difference in height between the top and bottom contact can be several microns. Devices are often mounted in the multiplying or mixing block face down in such a way that there is no additional bonding lead. As a consequence in some cases a full planar technique is preferred with connecting pad and anode practically at the same level. For the air bridge formation a now well mastered technique involves underetching of the semiconductor below a metallic finger forming a 'surface channel etch' [9] which is the technology successfully employed by the University of Virginia. This technique which makes use of dielectric deposition for isolating the connecting pad from the semiconductor has several advantages. Notably the etch can be relatively deep (several micrometers), reducing by this means the pad-to-pad capacitance. Also at this stage, it is worth-mentioning that the fabrication of small area Schottky contacts is intrinsically easier than that of an ohmic contact of comparable dimensions [10] because this is the area of the metallic junction which defines the active region for the former while it is the mesa diameter for the latter. On the ultra small scale, it is quite difficult to etch the semiconductor to define such a mesa as seen in the following.

Figure 3. SEM photograph of a diode contacted by an air-bridge. The diode is $1\mu m^2$ and critical steps were written by e-beam lithography

3.4. ETCHING

There are two main techniques which are used for etching semiconductors in order to reach a buried highly doped region with subsequent ohmic or Schottky contact fabrication, to form a mesa or to realise metallic interconnections. The first is the very well known wet-etching process usually using acid solutions. In this respect, one innovative process concerns the use of selective etchants. Citric acid is representative of this evolution with a good selectivity. For InGaAlAs/InP heterostructures, a number of acid solutions also exhibit a high selectivity notably HCl [11]. The second technique is Reactive Ion Etching which exhibits a high directionality. For illustration, Figure 4 shows a SEM photograph of a mesa etched by this plasma etching technique.

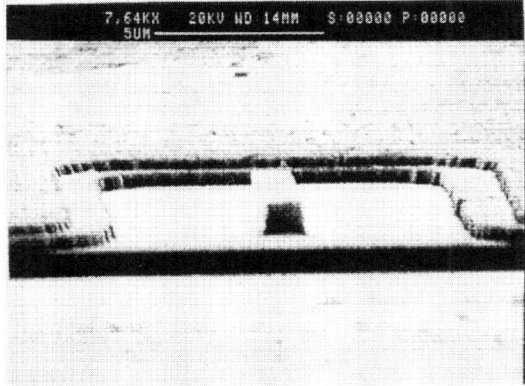

Figure 4. SEM photograph of a planar diode with a mesa defined by reactive ion etching

4. Conclusion

Planar integrated active devices intended to operate at terahertz frequencies are now fabricated in several laboratories in USA and in Europe and should compete with the whisker contacted technology. Beyond the requirement of more rugged and reliable devices for space applications, it seems that integration opens new routes towards the achievement of high performance particularly in terms of power. This integration can be made during the material growth with the series integration of several heterostructures forming a stacked layer or during the fabrication of the device itself by means of low parasitic interconnections.

Acknowledgements

I would like to thank P. Mounaix for his help in this overview

References

1.   Katz A.(1992) *Indium phosphide and related materials*: Processing, technology, and devices, Artech house, Norwood
2.   Bishop W. L. (1995) *Planar GaAs mixer diode research*, Quaterly report on millimeter and submillimeter element research, University of Virginia, 3-7
3.   Ketterson , A., Tong, M., Seao, J. W., Nummila, K., Cheng, K. Y., Morikuni, J., Kang, S., and Adesida I. (1992) Submicron modulation-doped transistor fabricated by direct-write electron beam lithography, *J. Vac. Sci. Technol.* B **10**, 2936-2940
4.   Shen T. C., Gao, G.B., and Morkoç, H. (1992) Recent developments in ohmic contacts for III-V compounds semiconductors, *J. Vac. Sci. Technol.* B **10**, 2113-2132
5.   Medhi, I., Reddy, U.K., East, J.R., and Haddad, G.I. (1989) Non-alloyed and alloyed low resistance ohmic contacts with good morphology for GaAs using graded InGaAs cap layer *J. Appl. Phys.* **65**, 867-869
6.   Kordos, P., Marso, M., Meyer, R., and Lüth, H. (1992) Schottky barrier height enhancement on n-$In_{0.53}Ga_{0.47}As$, *J. Appl. Phys.* **72**, 2347-2355
7.   Lippens, D., Barbier, E., and Mounaix, P. (1991) Fabrication of high performance AlGaAs/InGaAs/GaAs resonant tunnelling diodes using microwave compatible technology, *IEEE Electron Dev. Lett.* 12, 114-116
8.   Lheurette E., Grimbert, B., François, M. Tilmant P., Lippens, D. Nagle, J. and Vinter, B.(1992) InGaAs/GaAs/AlAs pseudomorphic resonant tunnelling diodes integrated with air bridge, *Electronics Letters*, **28**, 937-938
9.   Bishop, W.L., Crowe, T. W. and Mattauch R. J.(1993) Planar Schottky diode fabrication: progress and challenge *Proc. Fourth Int. Symp. Space Terahertz Technol.* Los Angeles, 415-429.
10.   Allen, S.T., Reddy, M., M.J.W. Rodwell, Smith R.P., Martin S.C., Liu, J. Muller E.E. (1993) Submicron Schottky collector AlAs/GaAs Resonant tunnel diodes, *Proc. IEDM*, 407-410
11.   Broekaert, T.P.E., and Fonstadt, C. G. (1992) Novel organic acid-based etchants for InGaAlAs/InP heterostructures devices with AlAs etch-stop layers, *J. Electrochem. Soc.* **139**, 2306-2309
12.   See as example Williams R. (1990) *Modern GaAs Processing methods* Artech House, Nordwood

HOT HOLE LASERS

W.TH.WENCKEBACH
Faculty of Applied Physics,
Delft University of Technology,
P.O. Box. 5046, 2600 GA Delft, The Netherlands

1. Introduction

Hot hole lasers are presently the only table top devices that produce a tuneable output at frequencies from 1 to 4 THz. Therefore, they could play an important role in the design of a scheme to characterise active and passive circuits in this range of frequencies. They were first demonstrated more than 15 years ago [1] and several groups have worked on their development since. Still, they have not yet reached the maturity needed for applications in engineering. The present contribution considers their principles of operation, some problems that still inhibit their large scale use and present trends in development. It is by no means a review of work performed in the past and references will be restricted to specifically cited results. For review articles the reader is referred to a special issue of Optical and Quantum Electronics [2] and a new review article which is expected to appear shortly [3].

Hot hole lasers are bulk devices. Their operation does not rely on a specific structure of these devices, but solely on material properties. The basic material is a p-type semiconductor. To achieve lasing, a single crystal of this material is cooled down, placed in a magnetic field **B**, while an electric field **E** is applied perpendicular to this magnetic field. It is not immediately obvious that the thus created situation yields emission of terahertz radiation. Therefore, in the following four sections we will first build up our understanding by subsequently considering four aspects:

1. the motion of holes in crossed **E** $\perp$ **B**
2. the existence of two types of holes: heavy holes with a large effective mass and light holes with a small effective mass,
3. the emission of optical phonons by hot holes,
4. the energy level scheme and radiative transitions.

2. Motion of Holes in Fields **E** $\perp$ **B**

The application of crossed fields **E** $\perp$ **B** is basic to hot hole lasers. To understand their effect we consider the simplified case of a hole with an isotropic effective mass m^* and an initial velocity **v** = 0. We take the x-axis parallel to **E** and the z-axis parallel to **B**. The time evolution of the hole's momentum $\hbar\mathbf{k}$ is solved from equations of motion,

J.M. Chamberlain and R.E. Miles (eds.), New Directions in Terahertz Technology, 163–171.
© 1997 *Kluwer Academic Publishers. Printed in the Netherlands.*

$$\hbar \dot{k}_x = m^* \dot{v}_x \quad = \quad e(E + v_y B) , \qquad (1)$$

$$\hbar \dot{k}_y = m^* \dot{v}_y \quad = \quad -e v_x B , \qquad (2)$$

$$\hbar \dot{k}_z = m^* \dot{v}_z \quad = \quad 0 , \qquad (3)$$

where $E = |\mathbf{E}|$, $B = |\mathbf{B}|$ and e is the elementary charge. The solutions are

$$v_x \quad = \quad v_0 \sin \omega_c t , \qquad (4)$$

$$v_y \quad = \quad -v_0(1 - \cos \omega_c t) , \qquad (5)$$

$$v_z \quad = \quad 0 , \qquad (6)$$

where $\omega_c = eB/m^*$ is the cyclotron resonance frequency and $v_0 = E/B$.

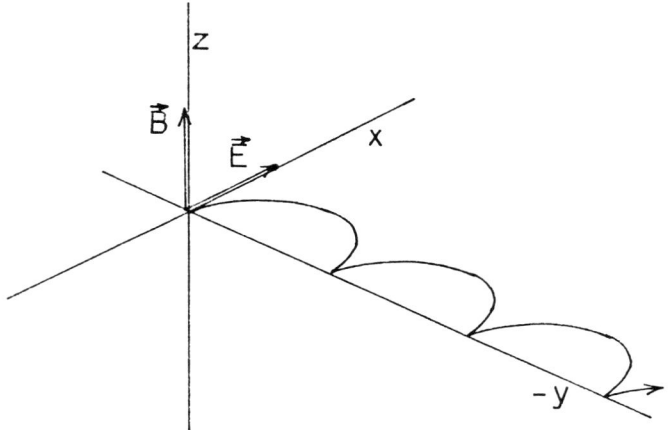

Figure 1. Orbit of a hole in crossed $\mathbf{E} \perp \mathbf{B}$ fields in normal space.

The resulting motion is depicted in Figs. 1 and 2. The hole is seen to move with an average velocity $-E/B$ in the y-direction, while in momentum space it describes a circle with a radius $m^* v_0$ and its centre at $\hbar k = (0, -v_0, 0)$. Finally, the kinetic energy of the hole varies with time as,

$$E = \frac{1}{2} m^* \left(v_x^2 + v_y^2 \right) = m^* v_0^2 (1 - \cos \omega_c t) , \qquad (7)$$

so it oscillates between 0 and $2m^* v_0^2 = 2m^* (E/B)^2$.

3. Heavy and Light Holes

Basic for the operation of hot hole lasers is the existence of two types of holes: heavy holes with a large effective mass m_{hh}^* and light holes with a smaller effective mass m_{lh}^*. Fig. 3 plots the kinetic energy of heavy and light holes as a function of momentum. According

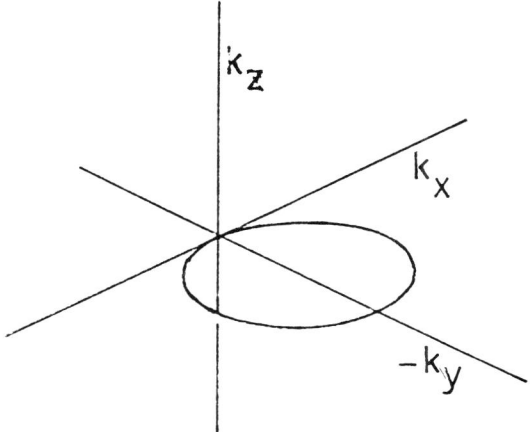

Figure 2. Orbit of a hole in crossed **E** ⊥ **B** fields in momentum space.

to the previous subsection, the momentum and the energy of both heavy holes and light holes will oscillate between zero and a maximum value. However, heavy holes reach a larger momentum $2m_{hh}^*(E/B)$ and a higher energy $2m_{hh}^*(E/B)^2$ (point A in Fig. 3) than light holes which stay below $2m_{lh}^*(E/B)$ and $2m_{lh}^*(E/B)^2$ (point B in Fig. 3).

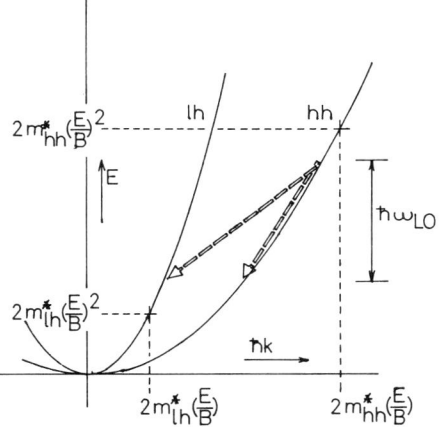

Figure 3. Energy of heavy and light holes as a function of momentum $\hbar|\mathbf{k}|$. Dashed arrows represent optical phonon emission.

4. Optical Phonon Emission

Holes would continue in the orbits shown in Figs. 1 and 2 forever, were it not that hole-phonon interaction scatters them into different orbits at irregular intervals. The most dominant process consists of emission or absorption of optical phonons that have energies varying from $\hbar\omega_{LO} = 37$ meV for germanium to 64 meV for silicon. For the moment we neglect other scattering processes. The gist of the hot hole laser is two-fold. First, we choose the temperature so low that optical phonons are virtually non-existent. Then, the only possible process consists of emission of a phonon by a hole. Next we choose E and B such that

$$2m^*_{hh}(E/B)^2 \gg \hbar\omega_{LO} \gg 2m^*_{lh}(E/B)^2 , \qquad (8)$$

so only heavy holes reach enough energy to emit optical phonons while the energy of the light holes remains too low for this process to occur.

Now, when emitting an optical phonon a heavy hole sometimes becomes light (arrow A in Fig. 3). The opposite transition will not take place, because we have chosen the conditions in such a way that light holes do not reach a high enough energy to emit optical phonons. So we have created a one way pumping process where heavy holes are transformed into light holes. As a result of this pumping process the population is inverted with respect to a thermal distribution: the higher energy light hole states are heavier populated than the lower energy heavy hole states.

5. Energy Levels and Photon Emission

Thus far we have considered the holes to behave classically. Of course, this is not exactly true and to obtain a correct picture, we need to take quantum mechanical effects into account. Here the most important quantum effect is the quantisation of the cyclotron orbits calculated above in section 2. As a result, the heavy hole energy band is split into discrete so-called Landau levels separated by a splitting $\hbar\omega_{chh} = \hbar eB/m^*_{hh}$ and the light hole energy band into discrete Landau levels split by $\hbar\omega_{clh} = \hbar eB/m^*_{lh}$. However, the picture is further complicated by the uncertainty relation, which broadens the resulting discrete levels by $\hbar/\tau$, where τ is the life time of the holes, i.e., the average time between scatterings. Now heavy holes emit optical phonons very rapidly. So for heavy holes the discrete quantum levels are smeared out into the continuous energy band shown in Fig. 3. On the other hand, light holes do not emit optical phonons. So their life time τ is long and discrete Landau levels must be taken into account. The resulting more realistic picture of the energy of light and heavy holes is given in Fig. 4. Now the motion of the heavy holes is still classical but via optical phonon emission it leads to populating the lower Landau levels of the light holes band.

Photons have negligible momentum, so radiative transitions are vertical in Fig. 4. We distinguish two types of transition, intervalence band transitions denoted by arrow A in Fig. 4 and light hole cyclotron resonance transitions denoted by arrow B in Fig. 4. Both transitions give rise to laser action as is shown in Figs. 5 and 6 giving the regions where lasing was observed in p-Ge as a function of E and B and the observed laser frequency as a function of B. The data were taken from [4].

Both types of laser have their advantages and disadvantages. The intervalence band laser has a very broad band width (Δ in Fig. 6) which makes it difficult to obtain a well defined output frequency contrary to the cyclotron resonance laser which lases at $\omega_{clh} = eB/m^*_{lh}$. On the other hand, the broad band intervalence band laser should be capable of emitting very short pulses, in theory down to a few ps. Therefore attempts to

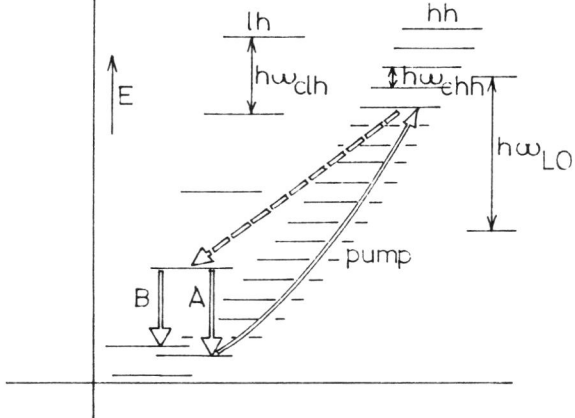

Figure 4. Energy level scheme for hot hole laser action. Arrow A represents intervalence band lasing. Arrow B represents cyclotron resonance lasing.

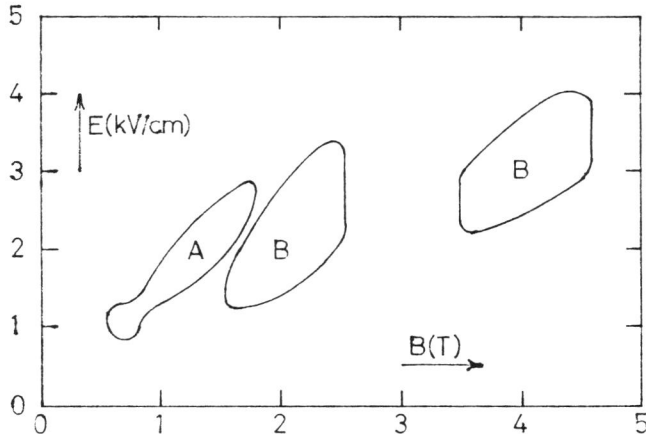

Figure 5. Regions where lasing is observed as a function of E and B. Intervalence band lasing is denoted by A, while cyclotron resonance lasing is denoted by B.

mode-lock a hot hole laser (creating ps pulses at THz frequencies) are made using an intervalence band laser.

6. Refinements of the Theory

The valence bands of real semiconductors are far more complicated than assumed above. They are strongly anisotropic and, as in the case of silicon, they are often non-parabolic. This implies in the first place that the equations of motion (1) to (3) have to be adapted to such anisotropies and non-parabolicities. The probabilities of scattering processes and the optical transitions rates also will reflect these effects and must be adjusted accordingly.

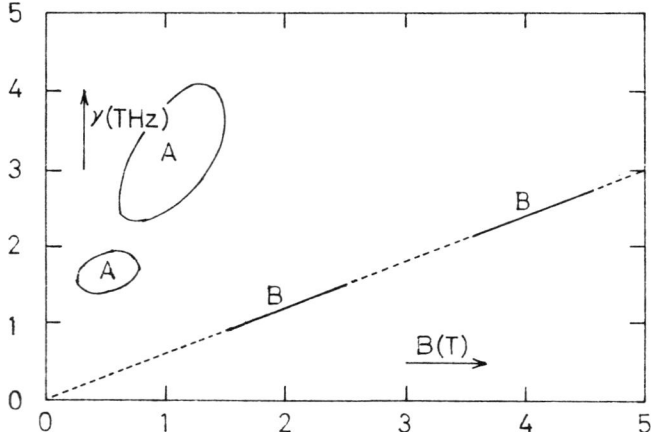

Figure 6. Lasing frequency as a function of *B*. Intervalence band lasing is denoted by A, while cyclotron resonance lasing is denoted by B.

As a result the situation becomes too complicated to be treated analytically and numerical methods must be used to simulate the operation of the laser.

In practice two approaches have been used. As intervalence band lasing can be understood classically, Monte Carlo simulations have been used successfully to describe its properties [5]. In these simulations a hole is followed as a function of time. The development of its momentum $\hbar\mathbf{k}$ is calculated classically by solving the equations of motion, while scattering processes are determined to occur stochastically using their quantum mechanical transition rates. By sampling the relevant parameters at regular intervals, their statistical average is obtained. Thus, one may obtain distribution functions, determine whether population inversion occurs and calculate the amplification in the laser medium.

A second approach is used to study cyclotron resonance lasing. This is a purely quantum mechanical effect and out of reach for quasi-classical Monte Carlo methods. Then, the energy levels and eigenstates of heavy and light holes in crossed $\mathbf{E} \perp \mathbf{B}$ fields are calculated by solving the Schrödinger equations and its solutions are used to calculate radiative transition rates and frequencies [6]. Clearly, both approaches have restrictions because the latter cannot adequately predict population inversions, while the former does yield correct transition frequencies. Unfortunately, until now a satisfactory amalgamation of both approaches is lacking.

7. Experimental Aspects

Large scale use of hot hole lasers has thus far been inhibited by some unpleasant features. To avoid optical phonon absorption destroying the population inversion between the light and heavy holes bands, one needs to operate the laser at liquid helium temperature. Here the high power dissipation in the laser medium comes into play. To obtain lasing a large electric field must be applied (typically 2 kV/cm) leading to a high current density (typically 100 A/cm^2). Furthermore, the gain of the laser is only of the order of 1.2 cm^{-1}, so the radiation must pass through several hundred cm to grow from its thermal intensity to a typical saturation density of 1 kW/cm^3. Hence, a long crystal is needed to reach enough gain to overcome reflection losses at the ends. Typical sizes are $50 \times 5 \times 5$ mm^3. The typically 200 kW/cm^3 which is dissipated in this crystal heats it much more rapidly than the liquid helium is able to cope with, restricting lasing to pulses of a few microseconds with a repetition rate of 10 Hz. For most applications this is a very unpleasant situation. Either cw operation would be preferred or really short pulses with a length of less than 100 ps.

Present work on hot hole lasers includes efforts in both directions. One line of effort concentrates on improvement of materials and advanced cavity design in order to be able to diminish the laser crystal size and thus lengthen the pulses and increase their repetition rate. Recently, using a very high quality germanium crystals with volumes down to $2.8 \times 3.2 \times 2.8$ mm^3 Bründermann et al. obtained pulses of 11 μs and a repetition rate of 6.7 kHz [7].

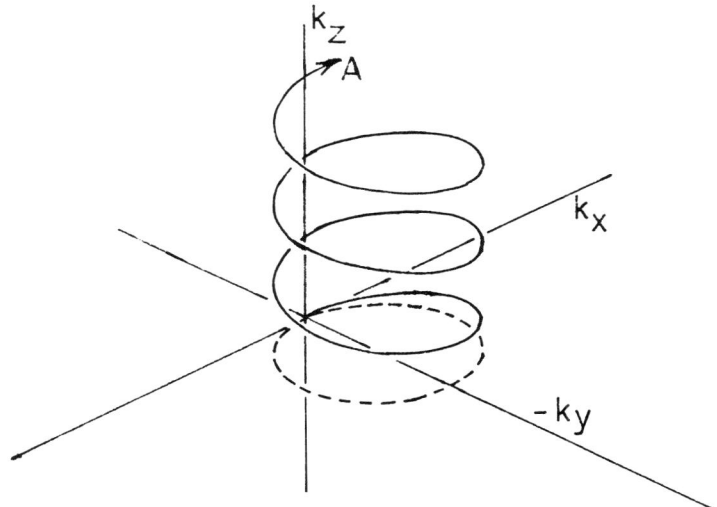

Figure 7. Motion of light holes outside the population inversion region when **E** and **B** are not perpendicular.

Another line of effort concentrates on modulating the gain of the laser as a function of time. We have experimentally explored a possibility using the fact that lasing is killed if the electric field **E** is not exactly perpendicular to the magnetic field **B** [8]. The reason is shown is Fig. 7. The component of **E** parallel to **B** simply accelerates the light holes along the arrow A until they reach the optical phonon energy and emit such a phonon, sometimes becoming heavy again. As a result an inverse process to the pumping described in section 2 is opened up and the population inversion is destroyed. We may use this effect to modulate the laser gain by applying an rf electric field parallel to **B** [9]. Then, as shown in the upper curve of Fig. 8, the laser gain is 'normal' when the electric field passes through zero, while it is reduced when it passes through its extrema. Now the intensity of the radiation grows exponentially and the moderate modulation shown in the upper curve

170

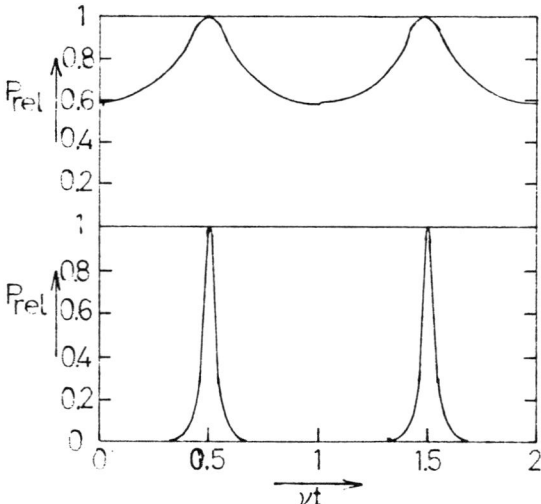

Figure 8. Upper curve: modulation of the gain per cm by an RF electric field parallel to **B**. Lower curve: the gain after 10^3 cm using the same modulation.

in Fig. 8 will eventually lead to the sharp peaks in the intensity shown in the lower curve in Fig. 8. This so-called mode-locking has not yet been proven unequivocally but first results are promising.

Thus far lasing has only been observed in germanium which combines a large ratio m^*_{hh}/m^*_{lh} of the heavy and light hole masses with low scattering rates. Unfortunately, germanium is out of favour in the electronics industry and good quality germanium for hot hole lasers is difficult to obtain. The obvious alternative is silicon. In this material m^*_{hh}/m^*_{lh} is much smaller and the resulting smaller splitting between the light and heavy holes bands should lead to lower frequencies (centred around 1 THz instead of 3 THz). Unfortunately, this smaller value of m^*_{hh}/m^*_{lh} also makes it more difficult to create a situation where the heavy holes emit optical phonons and the light holes are trapped in orbits as shown in Figs. 1 and 2. According to Monte Carlo simulations one needs to go to much higher magnetic fields (6T instead of 2T) to achieve lasing. An even more difficult problem is the greater depth of the acceptors in silicon. As a result the holes are frozen out at liquid helium temperature. Neither raising the temperature to liquid nitrogen temperature nor irradiation with a strong near infrared laser has thus far lead to promising results.

References

1. Andronov, A.A., Kozlov, V.A., Mazov, L.S. and Shastin, V.N. (1979) *Amplification of far infrared radiation in germanium during the population inversion of 'hot' holes*, Pis'ma Zh. Eksp. Teor. Fiz. **30**, 585-589, translation: JETP Lett. **30**, 551-555.
2. Gornik, E. and Andronov, A.A., editors, (1991) *Far-infrared semiconductor lasers*, special issue of Optical and Quantum Electronics **23** S111-S349.
3. Kimmitt, M. (1996), to be published.
4. Shastin, V.N. (1991) *Hot hole inter-sub-band transition p-Ge FIR laser*, Optical and Quantum

Electronics **23** S111-S131.

5. Starikov, E.V. and Shiktorov, P.N. (1991) *Numerical simulation of far infrared emission under population inversion of hole sub-bands*, Optical and Quantum Electronics **23** S177-S193.

6. Murav'jov, A.V. and Shastin, V.N. (1991) *Landau quantization and hot hole stimulated FIR emission in crossed electric and magnetic fields*, Optical and Quantum Electronics **23** S313-S321.

7. Bründermann, E., Linhart, A.M., Röser, H.P., Dubon, O.D., Hansen W.L. and Haller, E.E. (1996) *Miniaturisation of p-Ge lasers: Progress towards continuous wave operation*, Appl. Phys. Lett. **68**, 1359-1361.

8. Murav'jov, A.V., Nevedov, I.M., Pavlov, S.G. and Shastin, V.N. (1993) *Tunable narrowband laser that operates on interband transitions of hot holes in germanium*, Quantum Electronics **23**, 119-124.

9. Strijbos, R.C., Lok, J.G.S. and Wenckebach, W.Th. (1994) *A Monte Carlo simulation of mode-locked hot hole laser operation*, J. Phys. Condens. Matter **6**, 7461-7468.

10. Strijbos, R.C., Murav'jov, A.V., Blok, J.H., Hovenier, J.N., Lok, J.G.S., Pavlov, S.G., Schouten, R.N., Shastin, V.N. and Wenckebach, W.Th. (1995) *Active mode-locking of a p-Ge light-heavy hole band laser by electrically modulating its gain: theory and experiment*, IXth Int. Conf. on Hot Carriers in Semiconductors, July 31 - August 4, 1995, Proceedings to be published by Plenum Press, New York in 1996.

Theme 3

Integration

INTEGRATED WAVEGUIDES AND MIXERS

An Overview of Waveguide Mixer Design and Construction

C.M. MANN
Rutherford Appleton Laboratory
Chilton, Didcot, Oxfordshire, UK

1. Introduction

Integration of waveguide components at terahertz frequencies offers many benefits. The obvious are reduced cost, hence availability and the possibility to combine many discrete components together in much the same way that circuits are laid out in the semiconductor industry.

Benefits that are not so obvious however relate to the many types of components that are simply not possible to make even at millimetre wave frequencies because the structures required cannot be machined using the conventional techniques presently available. Such components include notch filters and diplexers, balanced mixers and power splitters and combiners. Designs for such components are readily available at microwave frequencies and could be scaled directly to terahertz frequencies if a fabrication technique became available.

It is therefore not adequate to consider the design of a terahertz device as simply an exercise in circuit theory. At these high frequencies it is vital to consider the fabrication techniques available hand in hand with the component's physical requirements. It is in this area that important lessons can be learnt from the mature techniques that have been used to realise waveguide components at frequencies as high as 2.5 THz. This chapter will discuss some of the technical developments that have allowed terahertz devices to be realised and how perceived fundamental limitations can be overcome by reconsidering the dual design problems of circuit and structure.

2. Conventional Mixer Construction

Early millimetre wave components were realised using a technique that dates back to the earliest cat's whisker devices. In such a device a sharpened wire known as a whisker was brought into contact with the surface of a semiconductor. The resulting Schottky contact could then be used to rectify a signal coupled into the whisker wire. Thus the whisker served two purposes firstly to make the contact and secondly to act as the antenna that

J.M. Chamberlain and R.E. Miles (eds.), New Directions in Terahertz Technology, 175–181.
© 1997 *Kluwer Academic Publishers. Printed in the Netherlands.*

couples the electromagnetic wave to the diode. This is an important point to remember and will be discussed in greater detail later.

Figure 1. A honeycomb dot matrix Schottky diode chip.

This technique was refined by the production of a honeycomb matrix of lithographically produced diode wells [1] (see figure 1). The area of the diode was now determined accurately by the lithographic process which allowed the designer to tailor his mount for a specific size of diode. Also the structure was far more mechanically reliable because the whisker tip found itself effectively trapped in the well rather than being free to skid around. This approach has become the mainstay of device fabrication to the present day.

Figure 2. A conventional whiskered waveguide mixer.

An example of this approach is shown in figure 2, here a bent wire whisker is soldered onto a removable post which is pushed into the waveguide block until it contacts one of the diodes on the chip. Most device manufacturers used more or less this technique though some preferred the machined coaxial filter instead of the microstrip circuit shown here. The process sounds straightforward enough until the dimensions are considered more carefully. At 300GHz the waveguide is typically 100μm high by 800μm wide. Considerable machining and assembly skills are required to fabricate such a device and hence their use has been limited to specialist applications such as astronomy and atmospheric science.

As crude as these devices sound they were sufficiently reliable to be used in space. Indeed, the vast majority of submillimetre wave devices that are produced today still rely on this early approach.

3. Early Moves to Integration

One of the main limitations of the conventional approach to device fabrication was the inability to realise circuits with more than one diode. Many applications do not require absolute sensitivity but are more driven by the need for system simplicity. This is particularly true of space based instruments where simplicity means greater reliability along with reduced power and mass requirements. A radiometer based around a sub-harmonic mixer has some highly desirable features. Most importantly the LO is at half the signal frequency and can be fed to the diodes by use of a separate waveguide. This obviates the need for a diplexer in front of the mixer. Also subharmonic mixers are inherently broadband due to the lower impedance presented to the RF and IF signals. Their main disadvantage is the need for an RF circuit that requires more than one diode.

This would be very difficult to do for a conventional waveguide block as two diodes would have to be contacted simultaneously by two separate whiskers or alternatively two posts would have to be fitted into the block which is feasible but would be very mechanically demanding bearing in mind the restricted size.

The desire to realise the subharmonic mixer lead to two major developments both described in an excellent paper by Carlson et al [2]. The first seems trivial at first but represents an important step towards device integration. Until this work the chip on which the Schottky diode array was patterned had the Ohmic contact on the face that was on the opposite side to the diodes (see figure 2). A diode chip was now fabricated with the Ohmic contact on the four faces that were orthogonal to the diode array. The consequence of this was that the diode could be mounted monolithically onto a flat substrate and have the diode array in the vertical plane. The second development took advantage of the first in that it now became possible to contact the diode with a whisker and then fix the whisker on the flat substrate. As such the substrate containing the contacted diodes and RF filters became an independent element or quasi-integrated. This allowed more than one diode to be fixed onto the substrate and hence allow a subharmonic mixer to be realised without the need for movable posts in the waveguide block.

This approach became the most popular way of realising the subharmonic mixer circuit and was extended to frequencies as high as 200GHz. An example of a substrate used in the mixer described in [3] is shown in figure 3.

Just to give an idea of the miniature nature of the circuit the quartz substrate is about 0.3mm wide, 5mm long and 75μm thick. This approach was used with great success to fabricate many mixers that had excellent performance, were indeed very broadband and as an added benefit proved to be very reliable. The circuit shown in Figure 3 has been in constant use aboard the MARSS experiment [4] (Microwave Airborne Radiometer Scanning System) since 1989. Although this technique was highly successful it had the disadvantage that the assembly of the RF circuit was complex. In a very short time a new type of diode became available that was to take the millimetre wave device to a new level of integration.

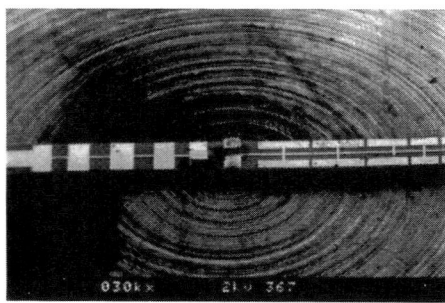

Figure 3. An example of the subharmonic double diode mixer substrate.

4. The Airbridge Planar Diode

Planar or beam lead diodes had been used at lower frequencies for many years and offered good performance up to 100GHz or so, however, as the frequency of operation was increased towards 200GHz the performance invariably degraded. The reason for this soon became clear, the diodes were made on the flat surface of GaAs. As the diodes were contacted via a flat lead the parasitic capacitance of the lead in conjunction with the solder pads effectively shorted out the diode. In 1984 a new feature was introduced into the diode structure [5]. Basically the lead that led to the diode anode itself was raised above the surface of the GaAs forming an airbridge. This acted by removing some of the parasitic capacitance associated with the lead.

The approach was further refined by Bishop *et al* in 1987 [6]. Here the air gap under the diode finger extended down into the GaAs substrate. An example of this approach is shown in Figure 4. With these diodes for the first time RF performance that closely competed with the whiskered devices was obtained at 200GHz. This development has played the most important role in directing effort away from the whiskered device and towards one that is quasi integrated.

All effort in this field has concentrated on pushing this approach to ever higher frequencies. However the best performance is still obtained using whiskered devices particularly at frequencies above 200GHz but one area where the planar airbridge diode has dominated is in the production of subharmonic mixers and balanced multipliers. The ability to incorporate more than one diode into the chip is paramount such that the submillimetre wave subharmonic mixers required by future limb sounding space missions rely on the use of packaged airbridge diodes.

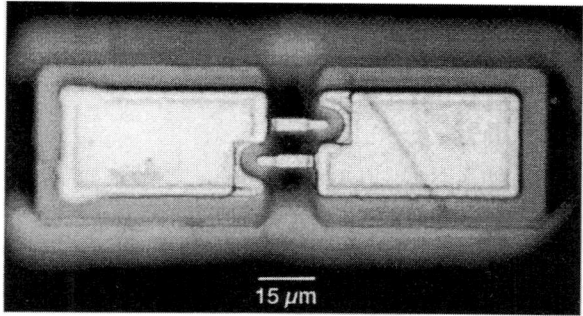

Figure 4. One of the most recent planar airbridge diodes from UVA

5. The Planar Whisker

For a period the future looked bleak for the whiskered device, planar chips proved to be very easy to mount onto microstrip circuits and because they were fabricated using lithographic techniques it was expected that they should prove easy to directly scale to submillimetre wavelengths. However, it is only recently that they are now edging towards competing with whiskered diodes at around 600GHz [7]. The reason for this is unclear but is possibly due to the complex nature of their fabrication process. Unlike the whiskered device it is not so easy to optimise the structure of the diode. A whisker may be shortened easily and this can be a useful feature. The circuit can be fine tuned by changing the diode or by moving its position within the waveguide mount or by adjusting the whisker's length. All of these adjustments vary the embedding impedance seen by the diode. For the planar airbridge diode the finger length is set by the lithographic process. A limited number of variations can be included in the photomask set but it soon becomes difficult to produce a complete suite of permutations especially when yields become important. In addition, the complete fabrication process can take a number of months so it becomes difficult to perform more than two or three iterations in a year.

A development which has put the whiskered diode back in the race is the planar whisker. This is a whisker that is made using lithography. Unlike the wire whisker which is usually made by hand the planar whisker is made via a simple process in batches of hundreds or more. Because its dimensions are set by lithography it can be made very small and can also be integrated with the microstrip filter. This removes most of the assembly problems and has resulted in its use at frequencies as high as 2.5THz. Figure 5 shows a 2.5THz mixer circuit implementing the integrated whisker/RF filter which contacts onto the 0.5µm NF1T2 diode produced at the University of Virginia. The waveguide is 26µm high by 96µm wide. This mount would be almost impossible to realise using the conventional wire whisker. Problems of reliability seem to be greatly eased which can be mainly attributed to their suppleness. When compressed they behave like a ruler and bend along their thinnest axis. This presents a very low force

180

compared with the wire whisker and allows them to be taken to high levels of compression without imparting an excessive force on the diode or whisker tip. As a result they become trapped, such that the mixer shown in Figure 5 easily passed space qualification testing.

Figure 5. An SEM of a 2.5THz mixer circuit incorporating a planar whisker

Ironically, perhaps the most important role to be played by the planar whisker is in the optimisation of planar diodes. Whilst they have their specialised uses it is hard to imagine contacting large numbers of diodes as part of a routine manufacturing process. Because they are planar the way they behave electrically in a waveguide mount is essentially the same as the metallisation (i.e, the finger and solder pads) of a planar diode. The planar whisker is able to separate the two functions played by the wire whisker into two discrete areas. Because the 'spring' that holds the whisker tip in place is provided by its natural ability to bend, the main body of the waveguide probe can be patterned to provide the ideal shape required to maximise the coupling between the diode and the waveguide circuit. Once this process has been achieved the same pattern can then be reproduced in the metalisation of the planar diode. This will increase the likelyhood of the diode performing well at the first iteration.

References

1. Young, D. T., Irvin, J. C. (1965) Millimetrewave Frequency Conversion Using Au-n-Type GaAs Schottky Barrier Epitaxial Diodes with a Novel Contacting Technique, *Proceedings of the IEEE*, 2130-2131.
2. Carlson, E. R., Schnieder, M. V., McMaster, T. F. (1978), Subharmonically Pumped Millimeterwave Mixers, *IEEE Trans on Microwave Theory and Techniques*, **MTT-26**, 706-715.
3. Mann, C. M., Matheson, D. N., Jones, M. R. B. (1989) 183 GHz Double Diode Subharmonically Pumped Mixer, *International Journal of Infrared and Millimeter Waves*, **10**, 1043-1049.
4. Jones, D. C. (1991) Met O(RSI) Branch Memorandum No.3.
5. Mills *et al* , (1986) Glass Reinforced GaAs Beam Lead Schottky Diode with Airbridge for Millimetre wavelengths, *Electronic Letters*, **20**, No. 19.

6. Bishop, W. L., McKinney, K., Mattauch, R. J., Crowe, T. W. and Green, G. (1987) A Novel whiskerless Schottky Diode for Millimeter and Submillimeter wave Applications, *IEEE MTT-S Int. Microwave Symp.Dig*, 607-610.
7. Hessler, J. L., Hall, W. R., Crowe, T. W., Weikle, R. M., Beaver, B. S., Bradley, R. F. and Pann, S. K. (1996) Submillimeter Wavelength Waveguide Mixers with Whiskerless Diodes for Spaceborne Missions, *Seventh Inernational Symposium on Space Terahertz Technology, Charlottesville*.

INTEGRATED WAVEGUIDES AND MIXERS

Future Technologies for Device Integration

C.M. MANN
Rutherford Appleton Laboratory
Chilton, Didcot, Oxfordshire, UK

1. Introduction

The waveguide/microstrip configuration where a planar diode is incorporated as an independent component has been used extensively in the millimetre wave region and is now an accepted way of realising subharmonic mixers up to 300GHz and has been extended to 600GHz for the case of a fundamental mixer [1]. A typical example of such a device is shown in Figure1. The technique is directly scaleable to around 1 THz, the limitations being the difficulty in mounting the diode onto the microstrip and the parasitic capacitance due to the diode chip/microstrip combination.

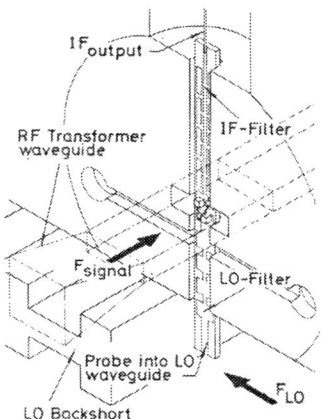

Figure 1. Typical waveguide/microstrip configuration.

The waveguide/microstrip arrangement suffers from a number of undesirable features. Firstly, a low loss wear free backshort is required that is increasingly difficult to realise above 300GHz, as is the fabrication of the waveguide itself. Also the waveguide requires a well matched antenna feed.

J.M. Chamberlain and R.E. Miles (eds.), New Directions in Terahertz Technology, 183–191.

Such feedhorns have now been realised to 3 THz [2] but the cost is high and the construction technique complicated and not yet suitable for routine manufacture.

This chapter will examine the possibility of replacing discrete assembly of the device with an integrated fabrication approach using a combination of micromachining, lithography and careful design.

2. Critical Circuit Components

In order to achieve integration it is necessary to investigate the possibility of replacing certain critical parts of the device. There is no doubt that the intrinsic Schottky diode is capable of heterodyne operation at terahertz frequencies, early work using the corner cube mount and recently with a more sophisticated waveguide circuit [3] has produced mixers with excellent performance. In order for a planar or integrated approach to succeed it is likely that the parasitics associated with the planar airbridge diode, for example, must be considered. There are two obvious routes for tackling this problem:

- A technique where the accurate removal of the diode package parasitics via an integral tuning circuit is developed; or,
- The planar diode package must be modified in order to reduce the parasitics to similar levels to those obtained for a whiskered device.

In reality it is likely that a combination of the two will provide the desired result but for the purpose of this discussion it is easier if we consider them as independent tasks.

2.1. TUNING

Tuning structures have been developed for SIS devices [4], but the capacitance of the SIS junction is much higher than that for a Schottky diode and the junction is generally fabricated on a substrate that has a much lower dielectric constant i.e. quartz. The lack of series resistance in the SIS device allows in theory, the reactive parasitics to be tuned out perfectly. For a planar diode chip this is not the case. The planar diode package capacitance is distributed throughout the chip, and is, in general of the same order or greater than the diode capacitance. A tuning circuit would therefore be more difficult to determine. However, if the additional parasitic capacitance is caused by a feature that is in the direct vicinity of the diode, it can then be considered as part of the embedding circuit. In this instance it is possible to design the waveguide probe (hence embedding impedance) to remove the effect of this capacitance provided it is linear.

2.2. REMOVAL OF THE PARASITICS

The realisation of a waveguide device at terahertz frequencies has been made possible with the use of a planar whisker. The main reason is that it allows the photolithographic integration of the RF filter circuit, waveguide probe and contacting tip whilst providing parasitics identical to the conventionally whiskered device. A schematic and SEM of the 2.5THz waveguide mixer described in [3] is shown below in Figure2.

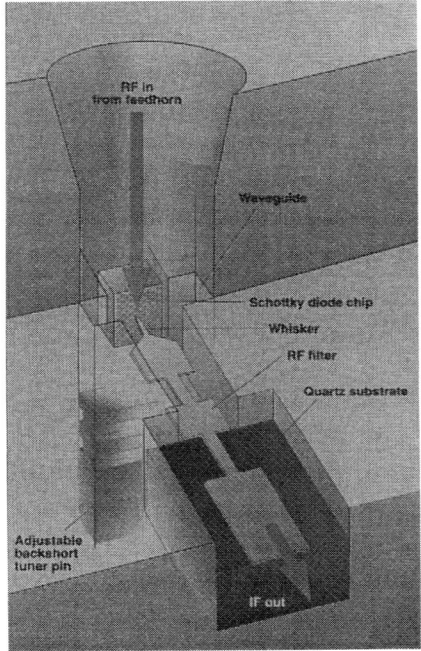

Figure 2. Schematic diagram and SEM image of the 2.5THz waveguide mixer

By comparison, the airbridge planar diode does not mimic the whiskered diode very accurately. Use of the airbridge does remove some of the finger overlay capacitance, but the capacitance between the solder pads of the diode chip is still present.

Environmental testing of the mixer shown above has shown that the flat free-standing gold structure (filter/whisker) has a very tolerant mechanical behaviour. Space qualification was successful on the first attempt. With this in mind it is now interesting to consider how far back can the finger of a planar diode be supported before it becomes mechanically unstable? For the 2.5THz mixer shown above the total length of the RF filter is only 120µm. Planar varactor diodes fabricated at the University of Virginia (UVA) have unsupported diode fingers that are comfortably in excess of this length, albeit with much larger anodes than those that would be required for terahertz operation (see Figure3).

With respect to the diode's operation GaAs is only required on one side of the chip for the diode and Ohmic contact formation. The GaAs on the finger side of the chip only fulfils the role of acting as a mechanical support. The GaAs of the chip itself plays no role in the RF circuit other than to introduce an unwanted parasitic capacitance.

Therefore it would be highly desirable to remove the GaAs from the area of the mixer that handles the RF signal. This could be achieved by fabricating the RF circuit as part of the finger and leaving it free-standing in air. It could be supported by a thick GaAs support outside the region of the RF circuit.

By taking this approach the package parasitic capacitance could then be expected to reduce to similar levels as that obtained using a whiskered diode. This concept is shown below in Figure4 and has been has been dubbed the window diode. Of course this approach is not entirely integrated as the device would still require the mating of the window diode chip to a machined waveguide mount. Extending this concept further to full integration is discussed in a later section.

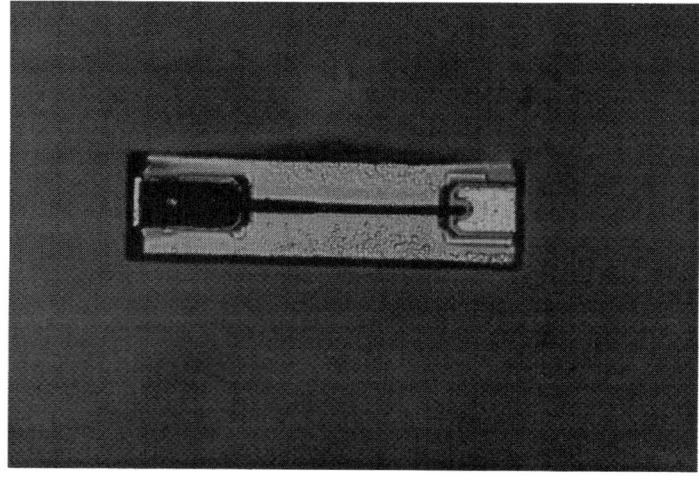

_____150µm_____

Figure 3. UVA planar varactor diode chip showing an unsupported finger ≈150µm long

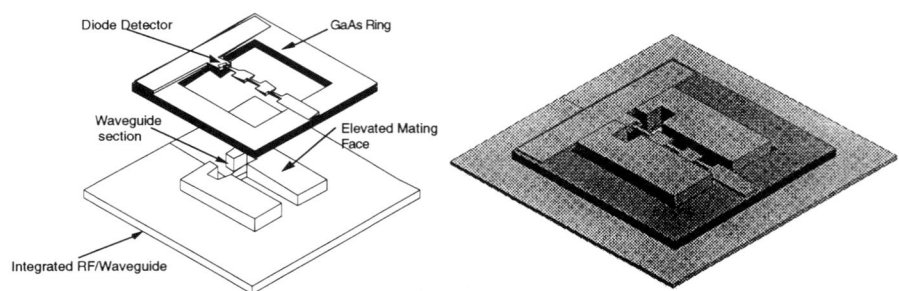

Figure 4. A schematic diagram of the window diode and waveguide mount

3. Device Design Techniques

One of the drawbacks of device integration was mentioned in an earlier chapter, that is, once the structure has been fabricated it is difficult to perform an empirical optimisation other than by having a series of devices that include systematic variations in the RF design and trying each one in turn to see which, if any, performs the best. Bearing in mind the number of possible variations needed to ensure that all permutations are covered this can become a time consuming and prohibitively expensive exercise. Therefore the importance in having a high degree of confidence in the design procedure cannot be overestimated for an integrated approach. This section examines how this can be achieved.

It was mentioned in a previous chapter that the role of the conventional whisker was to perform two basic tasks. Firstly it formed a low parasitic electrical contact to the diode and its spring ensured that the contact was mechanically stable. Secondly it acted as an antenna within the waveguide and coupled the signal into the diode. The first task it performed adequately but the second task left much to be desired. The reason for this is that because the whisker had to have some spring it would invariably require some kind of bend. This made the mount difficult to analyse in order to determine the embedding impedance presented to the diode. Also the whisker's size was set by the wire that was available and there is a very limited choice.

If we now consider the planar whisker in the 2.5THz mixer or the flat finger in the window diode as they are effectively equivalent, it is clear that they now perform three tasks. The additional task is in the provision of an RF filter. The design of such a filter is very straightforward and the task is made even easier by keeping the structure in air.

It should, however, now be apparent that the action of the whisker/finger as an antenna is no longer restricted to a bent wire as in the case for the conventional whisker. Because it can be patterned lithographically it can take any two dimensional form. It is now possible to fit the shape of the whisker/finger to that required by an accurate analysis.

Freedom to pattern the RF coupling circuit using lithography allows the engineer to tailor it to a proven analytical model. A well known analysis [5] offering accurate prediction of embedding impedance can then be used to describe the behaviour of the waveguide mount.

The analysis is valid for any waveguide size and any position of the diode within the waveguide. In addition, the waveguide mount can be terminated by any two complex reflection coefficients. It is therefore possible to include the effects of idler circuits, movable shorts or even E-plane tuners. The analysis has been verified using low frequency scale models, some results of which can be seen below in Figure5.

The power of this novel approach has been put to the test by using it to investigate the electrical behaviour of a classically designed millimetre wave multiplier [6]. It has been re-assembled to include an integrated planar waveguide launching probe. An SEM image of the new RF circuit and launching probe is shown in Figure6.

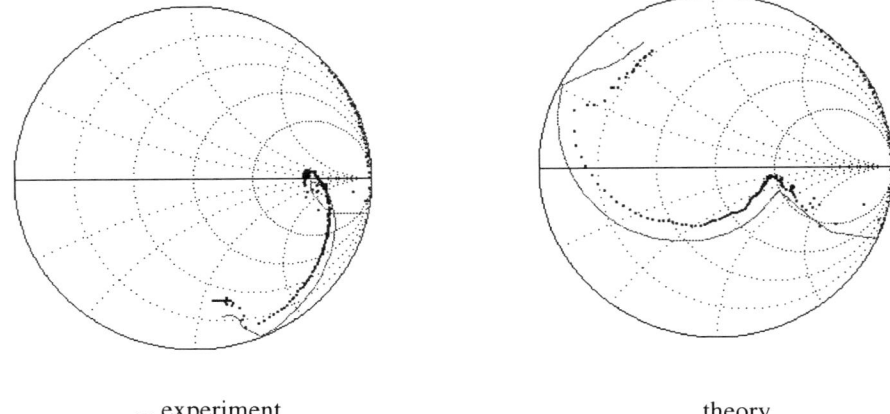

... experiment   ___theory

Figure 5.   Comparison between predicted and measured embedding impedances for two different waveguide/ launching probe configurations.

The physical dimensions of the actual tripler mount (including those of the idler circuit) were input into the analysis and a 'virtual' movable short was retracted in steps as carried out in the real device. A list of embedding impedances is calculated for the first four harmonics each set relating to a specific position of the backshort. This list along with the diode's physical parameters (size, capacitance, series resistance etc.,) was used as the input data for a modified version of the Seigel and Kerr harmonic balance analysis and a list of predicted output powers versus backshort positions was obtained.

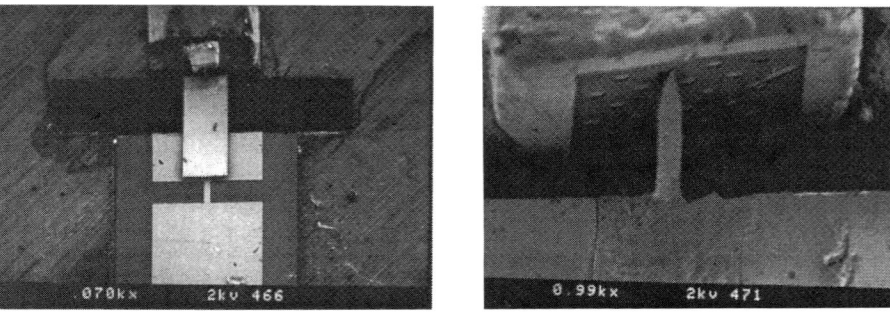

Figure 6. An SEM image of the planar launching probe used.

Next the real multiplier was tested at the same frequency used in the analysis in order to retrieve the actual output power as a function of backshort position. The results at 278GHz are shown in Figure7. The high level of agreement obtained is extremely encouraging and could only exist if both the circuit analysis and the diode model accurately predict the electrical behaviour of the waveguide circuit.

4. Complete Device Integration

It should now be apparent that the basic route for device integration is taking shape but one final hurdle exists. Whilst the window diode eases most of the assembly tasks, if a large number of devices are required in a system, as part of an array for example, then assembly could be prohibitively difficult. In addition, it would still be necessary to machine each waveguide mount. What is required is a means of performing the assembly and fabrication of the diode, RF circuitry and the waveguide mount as part of the fabrication process. Fortunately there are now some techniques becoming common place that may offer a solution.

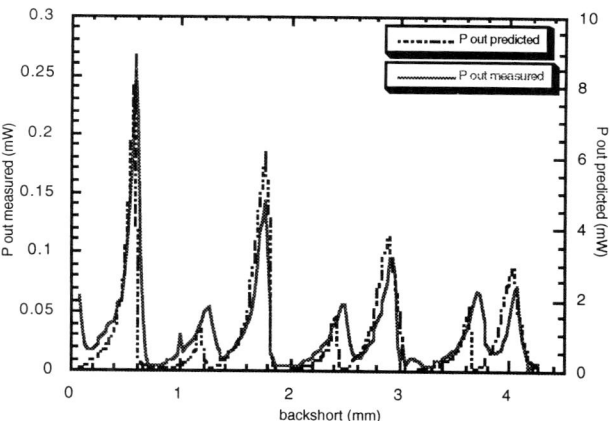

Figure 7. A plot of predicted and measured output power from the 278GHz tripler

At terahertz frequencies the waveguide cavities dimensions are very small. At 1THz half height waveguide is approximately 250μm wide by 65μm high. There are now lithographic processes available that can be used to fabricate structures with such dimensions. The most well know is the LIGA [Lithographie Galvanoformung Abformung] process. Here a photoresist structure having dimensions up to 1mm high is produced using deep x-ray lithography. The structure is then electroplated in order to turn the resist structure into one that is metal. The main disadvantage of this technique is that the wavelength of the x-rays required to penetrate to such depths in the photoresist needs to be 1Å. Such x-rays are produced by a synchrotron source and hence make the process expensive. Because such depths are not really required workers in this area are now using UV lithography to replace this requirement [7]. Other methods used to produce such structures include reactive ion etching, anisotropic etching of silicon and also soft x-ray lithography. All of these techniques have their advantages and disadvantages but are capable of producing the resist formers required for terahertz waveguide.

An example of how the structures can be realised using lithographic techniques is shown in Figure8. A resist former for the window diode approach has been fabricated using soft x-ray lithography. The waveguide is approximately 50μm high which is about right for a 2.5THz mixer. This is then electroformed to make the waveguide cavity. This work is in the early stages but the results are very encouraging.

For the window diode the waveguide mount consists of a waveguide placed in the vertical plane. This is not desirable from a micromachining perspective. Also it may become difficult to link mixers with local oscillator sources if the waveguide is in this orientation. Ideally therefore the waveguide should be formed on the horizontal plane. As to whether the broad wall of the waveguide should also lie in the horizontal plane is open to question but the approach put forward here will have the broad wall of the waveguide lying in the vertical plane relative to the horizontal.

 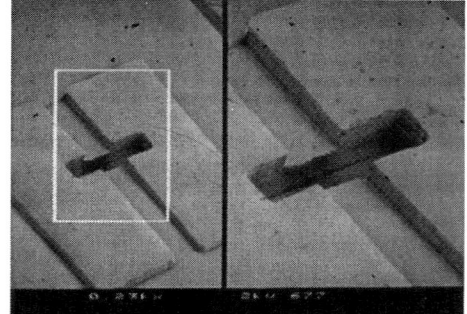

Figure 8. A 2.5THz waveguide cavity formed by soft x-ray lithography and electroforming

Upon examination of the window diode it is apparent that the role of the GaAs could be further reduced. For the device to work at RF the only GaAs required is that necessary to form the Ohmic contact and diode itself. The ring of GaAs performs the simple function of mechanical support. For an integrated approach the GaAs ring could be replaced by an alternative support such as a dielectric film linking the RF filters to the waveguide itself. Putting all these features together we might come up with the arrangement shown below in Figure9 which relies on the whole of the RF circuit being suspended within the waveguide cavity and microstrip channels via dielectric supporting pillars. This configuration has been dubbed the suspended filter approach.

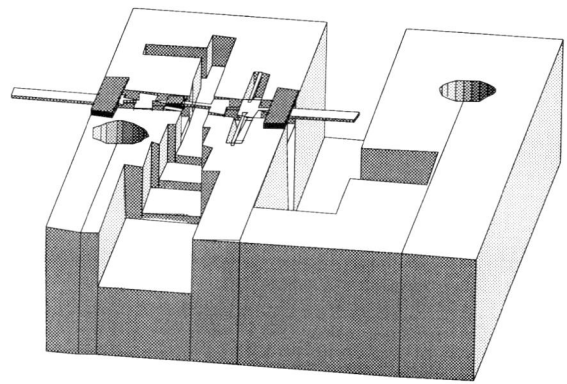

Figure 9. A schematic diagram of the suspended filter approach

References

1. J. L. Hessler, W. R. Hall, T. W. Crowe, R. M. Weikle, B. S. Beaver, R. F. Bradley, S. K. Pann, *Submillimeter Wavelength Waveguide Mixers with Whiskerless Diodes for Spaceborne Missions,* , Seventh International Symposium on Space Terahertz Technology, Charlottesville, March 1996.

2. B N Ellison, M L Oldfield, D N Matheson, B J Maddison, C M Mann and A F Smith, *Corrugated Feedhorns at Terahertz Frequencies - Preliminary Results,* The Fifth International Symposium on Space Terahertz Technology, Ann Arbor, May 1994.

3. B. N. Ellison, M.L. Oldfield, D.N. Matheson, B.J. Maddison, C.M. Mann, S. Marazita, T. W. Crowe, P. Maaskant, W. M. Kelly, *First Results for a 2.5 THz Schottky Diode Waveguide Mixer,* 7th International Symposium on Space Terahertz Technology, University of Virginia, March 1996.

4. Kerr A R, (1995), *Some fundamental and practical limits on broadband matching to capacitive devices and the implications for SIS mixer design,* IEEE MTT, 43, No. 1,.

5. R.L.Eisenhart, P.J.Khan, (1971*) Theoretical and Experimental Analysis of a Waveguide Mounting Structure,* IEEE Transactions on Microwave Theory and Techniques VOL. MTT-19 no.8.

6. J. Thornton, C. M. Mann, *A Design Approach for Planar Waveguide Launching Structures* 7th International Symposium on Space Terahertz Technology, University of Virginia, March 1996.

7. D.A. Brown, A. S. Treen, N. J. Cronin*, Micromachining of Terahertz Waveguide Components with Integrated Active Devices,* 19th International Conference on Infrared and Millimetre Waves, Sendai, Japan, 1994.

MATERIAL ISSUES FOR NEW DEVICES

D. LIPPENS
Institut d'Electronique et de Microélectronique du Nord
U.M.R.S. C.N.R.S. 9929, Université des Sciences de Lille 1
Avenue Poincaré, BP 69, 59652 Villeneuve d'Ascq Cedex, France

Abstract

This paper is an overview of material issues in connection with the design and the fabrication of new heterostructure devices operating at terahertz frequency. Notably, I will discuss in terms of frequency and power capabilities the advantages afforded by the shrinking in dimension and by the possibility to temporarily confine the charge carriers and to transfer them by tunnelling. In this context, special attention will be devoted to recent progress in strained layers pseudomorphically grown on GaAs or InP substrate. Finally, selected recent achievements dealing with the Single Barrier Varactor and the resonant tunnelling diode will be reported.

1. Introduction

The remarkable advance of the last decade in the techniques of crystal growth has stimulated active research area in III-V semiconductor heterostructures. In fact, the advantage of using semiconductor heterojunctions are numerous from the physics but also from the technological point of view by alleviating or overcoming most of the trade-offs which have to be satisfied in the design and in the fabrication of high frequency devices. Therefore, it is well known that making a low resistance ohmic contact requires a low gap material. This is in contrast to the fabrication of good metal semiconductor junction for which it is often imperative to deposit the metal on wide gap semiconductor. On the other hand, using a low material is often synonym of high mobility semiconductor material with the associated benefit of short transit time, high depletion layer modulation or low access resistance. Here again the use of heterojunction gives some flexibility in the design which could not be achieved otherwise using homojunctions. These are not the sole advantages because, on short scale, new physical effects are involved with notably the occurrence of quantum effects when the electron wavelength compares to the structure dimensions. In this paper, I will try to give an overview of the main issues related to the design and also to the growth of these semiconductor heterostructures. In the first section the electronic transport will be addressed. A second section will be devoted to the heterojunction schemes starting from the basic quantum well and barrier heterostructure. Some representative device

193

J.M. Chamberlain and R.E. Miles (eds.), New Directions in Terahertz Technology, 193–201.

194

structures which are now operating at millimetre and submillimetre wave lengths will be reported in a third section.

2. Material Systems

2.1. ELECTRONIC PROPERTIES

To have an overall view of how the energy band structure of these semiconductors looks like and to illustrate the complexity of the band structure, Figure 1 gives the isoenergy lines in the first conduction band in the ΓXK cross section for the ternary alloy $Al_xGa_{1-x}As$ with an aluminium content x of 45 %. Seen in the foreground is the central valley at Γ point. At X points are located the minima of energy corresponding to the satellite X valley with an ellipsoidal shape. Higher energy sites can also be seen appearing as red regions. The present case corresponds to the cross-over between the variations of the X and Γ energy gaps as a function of Aluminium content and hence to the transition between a direct to an indirect band gap [1]-[2].

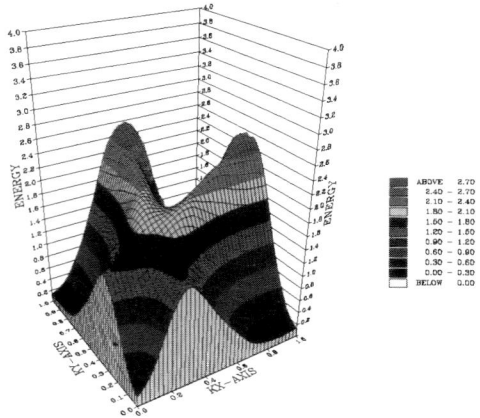

Figure 1. Isoenergy lines in the first conduction band in ΓXK cross section ($Al_{0.45}Ga_{0.35}As$) pseudopotential calculations

The first issue which can be addressed on the basis of band structure calculation is the maximum velocity which can be reached by electrons. In fact, it can be demonstrated that the band structure carrier velocity is given by the simple equation $v = \hbar^{-1} \partial E/\partial k$. This means that the velocity is maximum when the slope of the E(k) function is maximum and hence at the inflexion point in energy variation versus wave number. This consideration motivates most of the developments on *hot electron injection* so that the carriers take benefit of a high velocity prior to relaxation mechanisms.

At last, a first idea of impact ionisation mechanisms can be obtained by studying the energy dispersion relationship including the high energy states. To this aim, it is useful to remember that at high energy the scattering rates are extremely high [3].

This means that the carriers are very often scattered and subsequently able to reach the impact ionisation energy threshold by the combined action of the deterministic force due to the electric field and of the random action of these scatterings. On short scale, these impact ionisation mechanisms are of prime importance because any applied voltage even of moderate magnitude can lead to relatively high magnitude of electric field, let us say in excess of 100 kV/cm. In a first approach, the ionisation energy threshold is Ei = 1.5 Eg. Also, the magnitude of electric field where ionisation will take place also depends on the doping concentration (N_d). Taking these two facts into account a first estimate of the breakdown voltage and hence of the voltage handling can be achieved using the universal relationship proposed by Sze and Gibbons [4]. For $In_{0.53}Ga_{0.47}As$ which appears as a key material for advanced high speed devices, the breakdown voltage corresponding to doping concentration Nd=1×10^{17} cm^{-3} is ~ 6V. It is only 2V for InAs which consequently appears with a limited voltage handling.

2.2. TRANSPORT PROPERTIES

In order to give some insight into the transport properties, let us consider in a first approach only the vicinity of the conduction band edge which can be in general described by a parabolic approximation $E = \hbar^2 k^2 / 2m^*$. Under this assumption, quite elementary design rules can be proposed for the dependence of the mobility versus energy band gap. Therefore, is can be demonstrated that the mobility is high for narrow gap and decreases at increasing forbidden gap. For instance, the electron mobility for InAs is as high as 20,000 cm^2/Vs whereas for AlAs it is around 300 cm^2/Vs for undoped material at room temperature. With respect to the maximum velocity, it will depend strongly on the critical field for scattering towards the lateral valleys. This field magnitude is ~3.5 kV for GaAs with a maximum peak velocity ~ 2 10^7cm/s and around 2 kV/cm for InAs corresponding to a maximum velocity twice higher. In contrast the saturation velocity, which is reached in general for field exceeding 100 kV /cm, is below 1×10^7cm/s [5].

For most devices, room temperature operation is preferred. In some specific cases however, it is imperative to cool down the samples either at liquid nitrogen or helium temperatures. It can be shown that the mobility is limited by optical phonon scattering at room temperature and by impurity scattering at low temperature with a maximum in the variation of the mobility as a function of temperature around 77K. For modulation doped heterostructure, the low temperature mobility deviates from this typical behaviour with a monotonous increase in the mobility values at decreasing temperature typically up to 10^6cm^2/Vs. Such a very high mobility is particularly promising for *ballistic* devices and in particular for the so-called electron waveguides [6].

At this stage, it is worth-noting that most of the arguments given above depict a stationary situation in the sense that the electrons are always in equilibrium with the electric field. In others words, the carrier energy corresponds to the average energy of an electron gas without unbalance between the heating effect due to the electric field, assumed uniform, and the energy loss due to energy relaxation. In fact in most real devices the electric field is inhomogeneous and more importantly the spatial and temporal gradients are very strong. Under these conditions, nonstationary effects become very important and the fact to consider the transport as a steady state regime is no longer valid. One of the most representative effect which was very popular in the theoretical studies of ultra short gate field effect transistors is the so-called *overshoot* effect. As seen previously, the mobility of carriers in semiconductor is a function of their energy. If the electron gas is cold its average mobility is high. In contrast for hot electrons this mobility is low. Because carriers cannot adapt instantaneously their energy to the electric field, their velocity will exceed the steady state velocity (*overshoot*) when cold electrons are injected in a high field region. Conversely, when hot carriers experience a low field region *undershoot* effects can be observed. To summarise, various transport regimes can be achieved according to the time and distance scales involved ie: (i)collision dominated transport (ii) velocity overshoot, (iii) ballistic transport, (iv) hot electron injection and (v) the band structure limit which is about 1×10^8 cm/s in the [100] direction for GaAs.

3. Heterostructures

3.1. EPITAXIAL GROWTH

Figure 2 gives the variation of band gap as a function of lattice constant for III-V binary and their alloys. The forbidden gap magnitude varies typically between ~0.4 eV (InAs) and 2.4 eV (AlP) whereas the lattice parameter ranges between 5.5 Å to 6.5 Å. Therefore basically a large variety of band gap can be achieved by using ternary alloys. Nevertheless, some rules have to be respected to maintain a good crystal quality. Schematically, this requirement can be satisfied using three different ways. The first one is to use *lattice-matched* semiconductor compounds. In Figure 2 some arrangements can be deduced according to this selection rule and to the availability of a substrate. GaAs-based system has been extensively used over the past because there is no significant mismatch between GaAs and AlAs. However we have seen previously that the cross-over gives some drawbacks not only in the carrier transport but also in the conduction band offset which can be expected using this system. This often means that the aluminium content is limited to ~0.4. For this Al concentration, the conduction band offset at hetero-interface between AlGaAs and GaAs is typically 300 meV. The other lattice-matched system concerns InGaAs/InAlAs layers grown on InP substrate. The conduction offset achieved in that particular case is 500 meV and hence relatively higher than ΔEc for AlGaAs/GaAs. Moreover, this scheme does not suffer of indirect arrangement.

In order to further increase ΔEc we can also use strained layers *pseudomorphically* grown on InP substrate. At this stage two solutions can be distinguished. The first one consists to grow an AlAs layer sandwiched between two InGaAs layers. For this case the discontinuity in the conduction band is quite high > 1 eV. However the epilayer thickness has to be chosen less that the critical length in order to avoid dislocation formation. Also both lattice-matched and strained layer can be combined. For instance, a stair-like InGaAs/InAlAs/AlAs scheme may satisfy the trade-off between thick barrier and dislocation-free material.

Lastly we can make use of the *metamorphic growth* of InAs/AlSb epilayers on GaAs substrate with the potential of high discontinuity in the conduction band and low effective mass in the adjacent layers. In that case the transition between the substrate and the mismatched layer is made by a buffer layer.

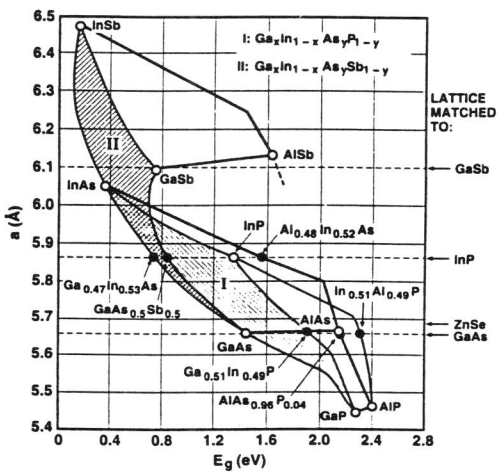

Figure 2. Variation of band gap as a function of lattice constant for III-V binary and alloy semiconductors

3.2. HETEROJUNCTION TYPE AND SUPERLATTICES

Figure 3 illustrates different types of band structure arrangements by considering n-type, undoped and p-type materials and also type I and type II heterojunctions respectively [7]. Square quantum wells can be obtained by means of two heterojunctions which induce two potential walls. By alternating narrow gap and wide gap materials it is also possible to create a barrier potential. The thickness of this barrier with respect to the evanescence length of the electronic wave will have a direct implication about the transparency of these barriers. Indeed for thickness of about 10 Å (~3 monolayers) the quantum probability for tunnelling through the barrier can be very high even with barrier height as high as 1 eV. If we now form a double barrier-well structure with thin barriers the tunnelling effect is resonant because of the dramatic increase in the quantum probabilities when the electron energy matches the quantum states of the well. This resonant tunnelling gives rise to a negative differential resistance effect in the current voltage characteristics. At last, superlattices can be considered here again with two schemes. On one hand a multiple quantum well heterostructures (MQW) can be fabricated with rather thick barriers in order to minimise the inter-well coupling. In contrast, if the barriers are thin in such a way that the coupling between wells is tight the structure is comparable to a superlattice [8].

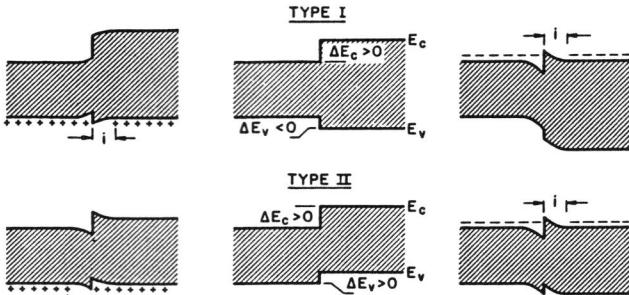

Figure 3. Heterostructure band alignments

4. Quantum Well and Tunnelling Structures

I will distinguish two classes of devices. The first one concerns a heterostucture barrier which shows a varactor effect. It is now used at very high frequencies for harmonic multiplication. The second class concerns the Resonant Tunnelling Diodes which exhibit a negative differential resistance and are used as oscillators.

4.1. HETEROSTRUCTURE BARRIER VARACTORS

An Heterostructure Barrier Varactor also called Single Barrier Varactor (SBV) or Quantum Barrier Varactor (QBV) consists of a large band gap sandwiched between two n-type narrow gap cladding layers. The potential barrier which is thus formed in the conduction band prevents the conduction through the structure so that each cladding layer can be depleted according to bias conditions. In practice this modulation of the capacitance of the devices as a function of applied voltage can be used for harmonic multiplication at very high frequencies by using the non linear capacitance to transform power at a frequency fp to power n.fp [9].

The requirements for successful operation of such a device are: (i) a low leakage current and (ii) a rapid modulation of the depleted zone. Moreover the ohmic contact and access resistance have to be optimised. Basically we can expect that a thick barrier of let us say 500 meV (InAlAs/InGaAs) is sufficient for blocking any parasitic current contribution. Indeed it can be expected that the thermionic current even at room temperature (activation energy ~25 meV above the Fermi level) and that the tunnelling current is also negligible. In practice however, this is not the case for conventional heterostructure such as AlGaAs/GaAs because the conduction discontinuity is limited and also because under bias the barrier exhibits a triangular shape with a high level of conduction via Fowler-Nordheim tunnelling mechanisms. One way to overcome this drawback is to use AlAs strained layer epitaxy sandwiched by InGaAs cladding layers in single or stack configuration (Fig. 4). With such a scheme it can be shown that the leakage current can be maintained to a very low level (<10 A/cm^2) up to 6 V [10]. For the capacitance voltage modulation the use of InGaAs layer is also beneficial because it permits a rapid modulation of the depletion layer edge and hence an operation at very high frequency.

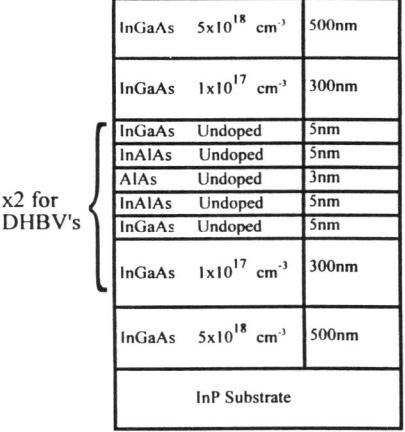

InGaAs	5×10^{18} cm^{-3}	500nm
InGaAs	1×10^{17} cm^{-3}	300nm
InGaAs	Undoped	5nm
InAlAs	Undoped	5nm
AlAs	Undoped	3nm
InAlAs	Undoped	5nm
InGaAs	Undoped	5nm
InGaAs	1×10^{17} cm^{-3}	300nm
InGaAs	5×10^{18} cm^{-3}	500nm
InP Substrate		

x2 for DHBV's

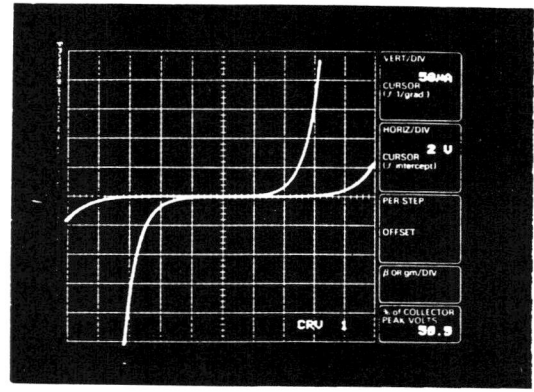

Figure 4. Illustration of the current voltage characteristics for a pseudomorphic InGaAs/InAlAs/AlAs heterostructure barrier varactor

4.2. RESONANT TUNNELLING HETEROSTRUCTURES

The resonant transmission of the Double Barrier Heterostructure (DBH) can be understood by considering Figure 6 which shows the time evolution of a Gaussian wavepacket interacting with a DBH [11]. The wave packet was initiated with an average energy near resonance. Moreover for clarity, the time was reversed. It can be seen that the DBH behaves as an energy filter and that only the carriers with energy close to the quantum level tunnel through the barrier into and out of the quantum well. In practice it can be shown that the temporal evolution of the probability for electron to be trapped within the well is governed by the lifetime in the well. In fact the lifetime of any resonant state is given by $\tau = \hbar/\Delta E$ where ΔE is the full width at half maximum of the transmission probability. Therefore, with a proper choice of material parameter ΔE can be increased to yield a subpicosecond intrinsic response time.

From this view point the comparison of a DBH with a resonator is very fruitful with a resonance sharpness related to the coupling of the cavity with the outer regions. In others words, the resonant tunnelling effect is an intrinsically fast process provided the barriers are chosen thin enough to achieve a strong coupling with the access regions. Under this condition, the lifetime in the well is very short and hence the structures exhibit a very high cut-off frequency $f_c = 1/2\pi\tau$. On the other hand, the potential barrier has to be high to improve the peak-to-valley current ratio in the negative differential resistance current-voltage characteristics. The best results so far with simultaneous high PVR and current peak were obtained with pseudomorphic InGaAs/AlAs structures and metamorphic InAs/AlSb heterostructures [12].

200

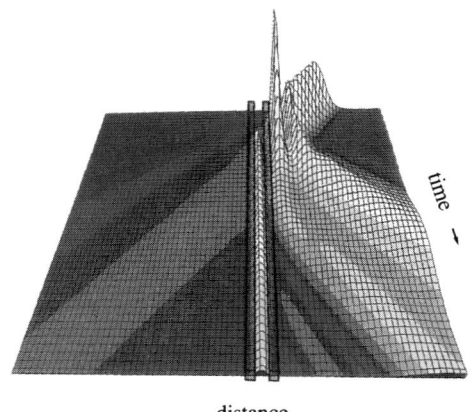

distance

Figure 5. Time series for a Gaussian wave packet interacting with a Double Barrier Heterostructure.

5. Conclusion

In summary, we have seen that there are numerous ideas for optimising the material and/or the structure for devices intended to operate at terahertz frequency. Indium-based heterostructures with tensile or compressive strains offer a great flexibility in the design and permit to alleviate most of the trade-offs which have to be satisfied. Generally speaking the quantum devices seem less mature than conventional devices such as Schottky varactors. Nevertheless they appear promising for the long term.

ACKNOWLEDGEMENTS.
    I would like to thank O.Vanbésien for his help in this overview.

References

1.   See as example O. Madelung (1991) *Semiconductors Group IV and II-V Compounds*, Data in Science and Technology, Springer Verlag
2.   Adachi S.(1985) GaAs, AlAs, and $Al_xGa_{1-x}As$: materials parameters for use in research and device applications, *J. Appl. Phys.* **58**, R1-R27
3.   Capasso F. (1985) *Physics of avalanche photodiodes* Semiconductors and semimetals **22**, 2-168, Academic Press
4.   Sze, S.M., and gibbons, G. (1966) Avalanche breakdown voltages of abrupt and linearly graded p-n junctions in Ge, Si, GaAs, and GaP, Appl. Phys. Lett. 8, 111-113
5.   See as example Sze S.M. (1985) Semiconductors devices: Physics and Technology John Wiley & Sons
6.   Vanbésien O. and Lippens D. (1995) *Directional coupling in dual-branch electron waveguide junctions,* Phys. Rev. B, **52**, 5144-5153
7.   lnes and Feucht, *Heterojunction sand metal semiconductor junctions* Academic press.
8.   stard G. (1988) *Wave mechanics applied to semiconductor heterostructures*, Les éditions de physique.
9.   llberg E. L., and Rydberg, A. (1989) Quantum barrier varactor diode for high efficiency millimeter-wave multipliers , Electronics Lett. **25**, 1696-1697

10. Lheurette, E., Mounaix, P., Salzenstein, P., Mollot, F., and Lippens, D. (1996) *High performance heterostructures barrier varactors in single and stack configuration* to appear in Electronics Letters

11. de Saint Pol, L., Lippens D., Clérot, F., Lambert, B.,. Deveaud, B.and Sermage B.(1990) Time domain analysis of resonant tunnelling in double barrier heterostructure, *Institute of Phys. conference proceedings*, 106, 801-806

12. Brown E. R., Söderström, J.R., Parker, C. D., Mahoney, L.J., Molvar, K.M., and McGill, T.C. (1991) Oscillations up to 712 GHz in InAs/AlSb resonant tunneling diodes *Appl. Phys. Lett.* **58**, 2291-2293

ACTIVE ANTENNA POWER COMBINING, BEAM CONTROL AND 2-DIMENSIONAL COMBINING

S.T.CHEW and T.ITOH
Department of Electrical Engineering
University of California, Los Angeles
405 Hilgard Avenue,
Los Angeles, CA 90095,
USA

Abstract

A review of the 2- and 3-dimensional active power combiners, and beam-control arrays was presented. Emphasis was placed on the arrays, instead of single-elements. 3-dimensional arrays include the active grid and that of the array-type. 2-dimensional arrays use the hybrid dielectric slab beam waveguide as the transmission medium. Beam-control, including beam-scanning, beam-switching and retrodirective, will be discussed. In all these arrays, both amplifier and oscillator arrays will be presented.

1. Introduction

As the microwave frequency spectrum gets overcrowded by increasing demands and applications in the commercial and military sectors, the operating frequencies of wireless systems are pushing towards that of the millimeter-wave. Also, the advantageous abilities of millimeter-wave signals to penetrate fog and smoke, and to provide pencil-beam accuracy of the target position makes millimeter-wave systems highly attractive. To date, most of the high power devices in the millimeter-wave frequencies are tubes. These devices need high voltage power supply. With progress in solid-state technology, solid-state devices are now providing more RF power at millimeter-wave frequencies. To provide reasonable RF power, many active devices must be combined. However, conventional power combiners are inefficient at these frequencies as the transmission lines are lossy. Thus, solid-state active antenna pioneered by J. W. Mink [1] provide a viable alternative. By integrating each active device with an antenna, each device radiates its RF power into free-space. The radiated power is combined in free-space, hence the term 'quasi-optical'.

Research in active antennas is extensive and can be broadly categorized into two major groups, as shown in Figure 1. One group is the power-combining arrays and the other is the active integrated antennas. Power combining arrays are designed with

J.M. Chamberlain and R.E. Miles (eds.), New Directions in Terahertz Technology, 203–220.

the purpose of delivering more RF power while active integrated arrays emphasize compact integration of more RF functions at the antenna front-end. In both areas, amplifier and oscillator arrays have been investigated. Power combining arrays include the active grid, array-type, and 2-dimensional power combiner. Active integrated antennas can be further grouped according to their applications. They are transponder, transceiver, beam-control arrays and others. In this paper, a review will be made with emphasis on power combining and beam-control arrays. The design approach of the circuits and their significant contributions to solving critical active antenna problems will be discussed and highlighted.

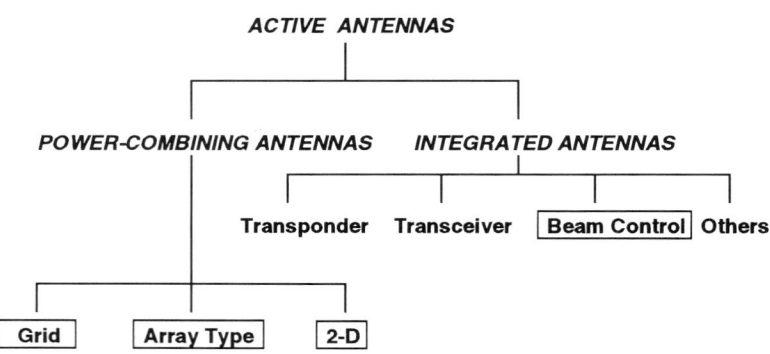

Figure 1. Categorization of active antennas.

2. Power Combining Arrays

As mentioned above, power combining arrays can be broadly divided into the active grid, array-type, and 2-dimensional power combiner. The grid approach models the ideal optical combiner. The grid has to be tested as an entity, as high density of the cells is needed to impose symmetry in the circuit. However, the size of the active device limits the packing density. The array-type active antenna is easier to implement as the conventional antenna array theory is well established. Testing of the individual element is possible. The 2-dimensional power combiners use hybrid dielectric slab beam waveguide as the transmission medium for power combining. This approach allows direct monolithic fabrication.

2.1. ACTIVE GRID

Shown in Figure 2 is an active grid. Due to symmetry of each cell, electric and magnetic walls can be imposed at the symmetry axes when excited by a plane wave. As such, the design and analysis are simplified to that of a single element. The interspacing of the cell is small compared to the wavelength of the signal. Because of the boundary conditions imposed, the cell can be treated as a parallel plate waveguide with only the dominant TEM mode excited. Evanescent modes present reactances at the device terminals and are determined using the induced EMF method. The leads of the active devices are modeled as posts in a waveguide. To date, amplifier [2-3] and oscillator [4-7] grids have been implemented.

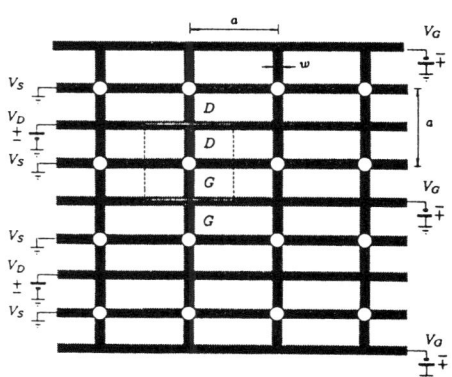

Figure 2. An active grid. (© 1991 IEEE. Reprinted with permission from Reference [4])

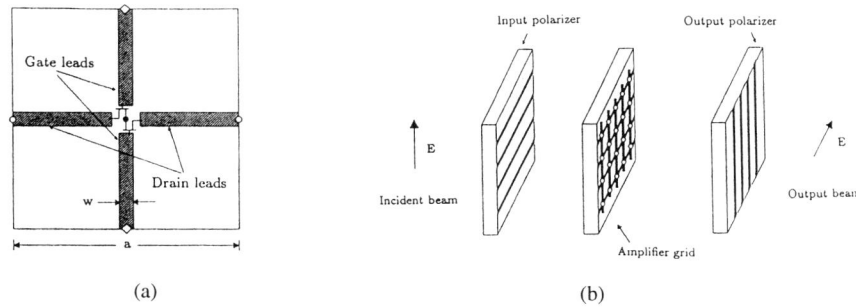

(a)   (b)

Figure 3. (a) An amplifier grid cell; (b) The active grid system.
(© 1991 IEEE. Reprinted with permission from Reference [2])

2.1.1. *Amplifier Grids*

In the amplifier grid, each cell has a differential pair active device, as shown in Figure 3a. FET [2] and HBT [3] active devices have been used. Polarizers are needed to provide isolation between the input and output of the system, as shown in Figure 3b.

The MESFET grid consists of 25 differential-pairs while the HBT grid consists of 100 elements. The measured amplifier gain of the MESFET grid is about 11 dB at 3.3 GHz with a 3-dB bandwidth of 90 MHz. The HBT grid exhibits 10 dB gain at 10 GHz with a 3-dB bandwidth of 1 GHz.

2.1.2. *Oscillator Grids*

Here, the grid is placed in front of a mirror (Perfect Electric Conductor) located about one-wavelength away to form a Fabry-Perot cavity [4]. When the grid is powered up, all cavity modes compete for steady-state oscillation. Ultimately, the grid locks to the mode that suffers the lowest diffraction loss per round-trip. The oscillation frequency is 5.2 GHz with an Effective Radiated Power (ERP) of 22 W.

To increase the output power, several grids were connected in parallel with dielectric spacers placed between them [5-6]. This configuration is favorable as the dielectric spacers can also be a heat-sink for the grids. Copper tape is placed on the top and bottom of the spacers to enforce the TEM mode and prevent slab radiation. The mirror provides passive feedback while the grids provide active feedback. In [5], a double grid shows an increase of 2 dB in the output power and 3% increase in the efficiency at 5 GHz, as compared to that of a single grid. In [6], a 100-transistor quadruple oscillator grid was also built. Measurements show an increase of 6.5 dB in the ERP, as compared to a single grid.

To provide some form of frequency tuning, a diode grid, consisting of varactors, is placed behind a MESFET grid, forming a Voltage-Controlled Oscillator (VCO) [7], as shown in Figure 4. To preserve the symmetry of the cell boundary conditions, the diode grid also has the same periodicity as that of the MESFET. As the diode grid was placed in the near field of the MESFET grid, the transmission-line model is no longer applicable. Several grids consisting of dipole and bow-tie antennas were investigated to determine the frequency tuning range. The dipole grid operates at 2.6 GHz while the bow-tie grid is at 3.7 GHz. A 10% tuning bandwidth with less than 2 dB power change was measured for the bow-tie grid.

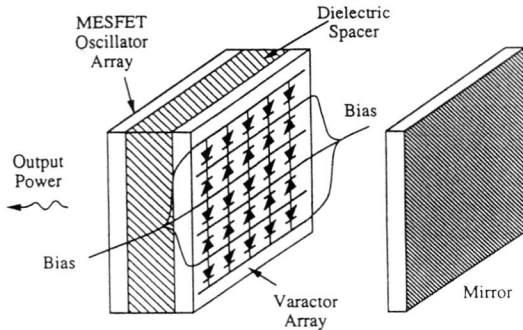

Figure 4. A VCO active grid. (© 1993 IEEE. Reprinted with permission from Reference [7])

Theoretical analysis of the active grid was established using the full-wave method of moments for the driving-point impedance of the cell [8]. Once the impedance is derived, MDS™ is used to simulate the nonlinear mechanism.

2.2. ARRAY-TYPE

In array-type combiners, each active device is integrated directly to one or more antennas. As such, conventional antenna theory can be applied to the design of this class of active antenna. Due to grating lobe criteria, the elements are normally spaced less than one-half wavelength apart. As frequency increases, the wavelength gets smaller. But, the size of active device cell does not reduce in the same proportion. This can lead to overcrowding at very high frequencies. Amplifier and oscillator arrays have been extensively reported.

2.2.1. *Amplifier array*

In the amplifier array, an antenna each is connected to the input and output of the amplifier. The antennas are of different polarization to reduce coupling and feedback that can cause oscillation. Input signal is received quasi-optically and then amplified by each element. The amplified signal is then radiated out and combined quasi-optically at the output.

As mentioned above, real-estate is a concern in the array-type active antenna. To solve this problem, folded slot antenna was proposed [9]. Folded slot also exhibits broader bandwidth as compared to the rectangular patch antenna. In [9], use of CPW feed allows the entire circuit to be fabricated on a single layer, as shown in Figure 5. A 4X4 array was measured with an EIPG of 32 dB at 4.24 GHz. The 3-dB bandwidth is 8%.

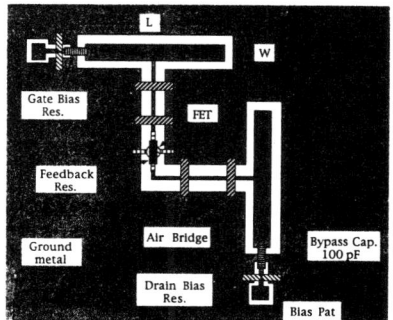

Figure 5. A folded slot amplifier cell. (© 1994 IEEE. Reprinted with permission from Reference [9])

Another circuit uses stacked patch antenna to provide a wider bandwidth [10]. Here, the orthogonal modes of the stacked patch antenna are used to provide the isolation between the input and output. The array operates in the reflection mode. With the ground plane, heat sinking was made possible.

One of the problems of the amplifier array is that it is difficult to excite all the elements uniformly. To improve the aperture efficiency, hard horns were used [11]. Slabs with conducting strips on their surfaces are placed on the E wall of the horn to improve the uniformity of the aperture fields. The array is placed in the near field of the horns to couple energy in and out of the array. Measurement shows a small signal gain of 16.2 dB at 9.98 GHz with 90 mW of RF power.

To increase RF power generation, one alternative is to increase the active device packing density per antenna. This is demonstrated in [12] with patch antenna. The input and output antennas are on different substrates. Energy is coupled between substrates through broadband slotlines. The power received by the input antenna is divided four-ways. Each divided signal is amplified separately. The signals are then combined at the output antenna. Measured results show a gain of 10 dB at 10 GHz.

To increase the frequency bandwidth and optimize the real-estate of the circuit, slot antennas are used instead [13]. The circuit is shown in Figure 6. The horizontal slots are the receiving antenna while the vertical slot is the transmitting antenna. The microstrip-to-slotline transition serves not only as the coupling network but also as the power-dividing network. This results in efficient use of the circuit real-estate. As the power divider is a balun, a 180° delay line is needed to provide the necessary phase. Measurement in the transmission mode shows an EIPG of 18 dB with a 3 dB bandwidth of 17% at 4 GHz. It must be stressed that the circuit in Figure 6 represents a unit cell. When the array size increases, the ports currently terminated by 50 ohm resistors can be used as input to other amplifiers, which are to be fed to other vertical slot antennas.

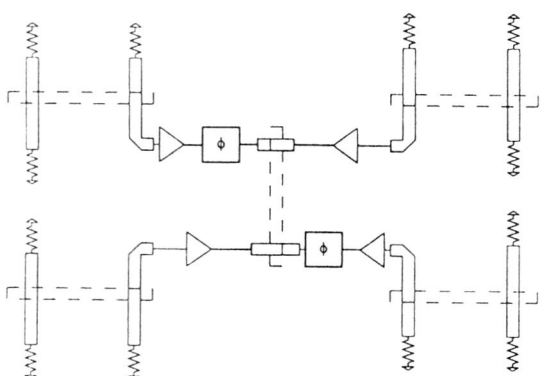

Figure 6. Multi-amplifier slot active antenna.

2.2.2. *Oscillator Arrays*

In oscillator arrays, emphasis is placed on how to synchronize the oscillator elements to oscillate in-phase, resulting in broadside radiation. To date, several methods have been reported. They can be broadly grouped into weakly or strongly coupled oscillator arrays. This coupling results in interinjection-locking [14].

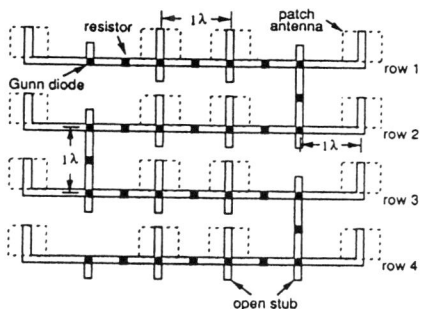

Figure 7. Weakly-coupled oscillator array. (© 1994 IEEE. Reprinted with permission from Reference [16])

In weakly coupled oscillator array, oscillators are mutually coupled weakly to achieve synchronization. One approach uses the free-space coupling of the antenna [15]. By controlling the interspacing between elements, the coupling phase can be varied. Using the amplitude and phase dynamics, differential equations are setup to determine the relationship between the coupling phase and synchronization. A four-element array has been demonstrated with broadside radiation at X-band. Another form of weak coupling is lossy transmission line, as shown in Figure 7 [16]. A 10X1 array was implemented at 8.4 GHz with an ERP of 10.5 W.

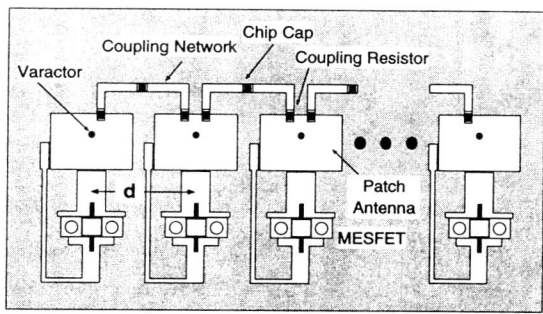

Figure 8. Strongly-coupled oscillator array. (© 1994 IEEE. Reprinted with permission from Reference [17])

210

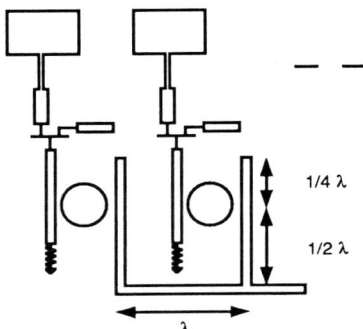

Figure 9. A dielectric resonator oscillator array.

In strongly coupled oscillator array, mode analysis is used as the analytical tool [17]. By inter-connecting each individual element with a transmission line, modes are being set up. By using average potential theory, these modes can be predicted. Also, the most stable and dominant mode can be identified. It has been shown that by inserting a resistor in the transmission line, the dominant and stable mode is the in-phase mode. A 4X4 array, as shown in Figure 8, shows an ERP of 21.3 dBm at 11.75 GHz. To extend the above concept, dielectric resonator is used as the coupling network between elements, as shown in Figure 9 [18]. By doing so, the phase noise of the system is improved and oscillation at the desired frequency is ensured. Measurement with a 4X1 linear array operating at 4.88 GHz shows a phase noise of -108 dBc/Hz at 100 KHz offset. The measured ERP is about 30.9 dBm. This design concept can be implemented in the receiver system. The dielectric resonator serves also as a high Q bandpass filter, allowing the frequency mixing to be treated individually for each element. By spacing the elements one-half fundamental wavelength apart, the second harmonic signal of the oscillator can be made to radiate effectively [19].

Another approach, also linked to the strongly coupled oscillators, is the extended resonance method [20]. Here, the coupling network is still the transmission line. However, the system is treated as an entity with one resultant resonator. The coupling transmission line transforms the reactance at one node to another for cancellation of reactance at the other node. The oscillation frequency will be that which can cancel the reactance effectively. A 3X3 array was designed at 7.33 GHz with an ERP of 6.8 W.

Lastly, injection-locking with an external source can be used to synchronize the array elements, as shown in Figure 10 [21]. Through Kurokawa's theory [22], there is a phase relationship between the injected and locked signals, depending on their frequencies and power levels. As such, by tuning the individual element to a common frequency, in-phase oscillation results when locked to an external source. A 4X4 array shows an ERP of 28.2 W at 6 GHz. Advantage of this implementation allows the phase noise of the system to be improved by using a low phase noise injected signal.

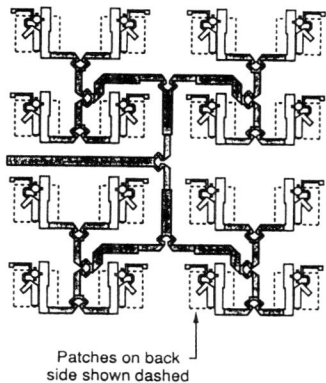

Patches on back
side shown dashed

Figure 10. Injection-locked oscillator array. (© 1992 IEEE. Reprinted with permission from Reference [21])

2.3. 2-DIMENSIONAL POWER COMBINERS

As opposed to the above discussed arrays, 2-dimensional power combiners are planar and appropriate for monolithic fabrication. J. W. Mink *et al* suggested that a hybrid dielectric slab beam waveguide can be used for transmission of millimeter-wave signals [23]. 2-dimensional arrays are not restricted in real-estate and DC bias circuitry can be easily incorporated. An important advantage is that being planar, the lenses and array do not require mounting structures. Also, alignment of the array and lenses is easier than that of the 3-dimensional arrays. The slab can also serve as the heat-sinking material and mechanical support for the array. Measurement is easier than the 3-dimensional arrays as the 2-dimensional arrays can be treated as a two-port connectorized component.

2.3.1. *Oscillator Combiner*
Here, a 4-element oscillator combiner was demonstrated at 8 GHz, as shown in Figure 11 [24]. The oscillator elements are mutually coupled through a curved reflector. Vivaldi antenna is used as the radiating element. By its inherent property, the Vivaldi antenna also decouples the forward and backward waves. The oscillator is designed so as not to excite the surface-of-slab to ground-plane resonance. TE mode, that is the E field is parallel to the ground plane, is excited. Through experiments, the oscillators are placed as such that repeatable oscillation at the desired frequency is ensured. The spacing between elements is also experimentally determined to reduce mutual coupling,

212

other than that from the reflector. By using injection-locking, the output signal is locked to the injected frequency. FM modulation is also demonstrated.

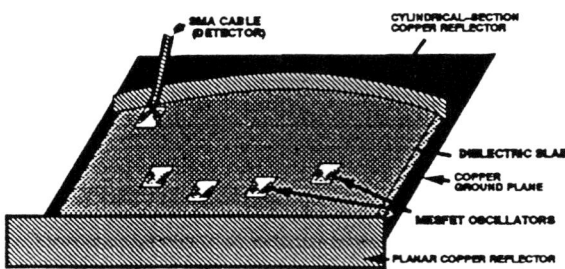

Figure 11. A 2-dimensional oscillator combiner.
(© 1995 IEEE. Reprinted with permission from Reference [24])

2.3.2. *Amplifier Combiner*

In the amplifier combiner, a lens each is placed at the input and output of the array, as shown in Figure 12 [25]. Several configurations of the amplifier element placement were investigated. The placement was experimentally determined to give the highest gain and prevent oscillation. Again, Vivaldi antenna is used as the input and output antenna of the amplifier element. Amplifier gain is defined here as the ratio of the output power with and without bias. Measurement shows an amplifier gain of 10 dB at 7.4 GHz. To reduce beammode perturbation, scattering losses and reflection from the amplifier's impedance mismatch, amplifier cells are located on the ground plane [26].

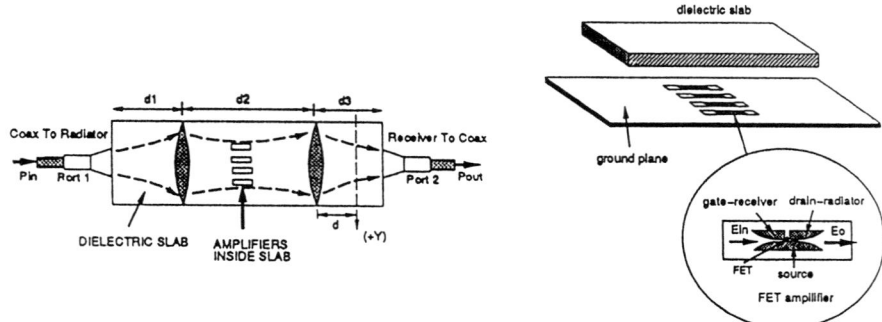

Figure 12. A 2-dimensional amplifier combiner using Vivaldi antennas.
(© 1995 IEEE. Reprinted with permission from Reference [26])

Another amplifier combiner uses the Yagi-Uda antenna as the antenna element, as shown in Figure 13 [27]. The Yagi-Uda antenna excites the dominant E field that is perpendicular to the ground plane. This mode is more lossy than the TE mode with E field parallel to the ground plane, but is easier to excite cleanly. Microstrip lines are used as delay lines to focus the beam. The thickness of the slab is chosen as such that the center frequency is at 90% of the cut-off frequency of the second order TM mode. This maximizes the coupling to the slab waveguide. A ten-element amplifier was designed at 8.25 GHz with a gain of 11 dB and 3-dB bandwidth of 0.65 GHz.

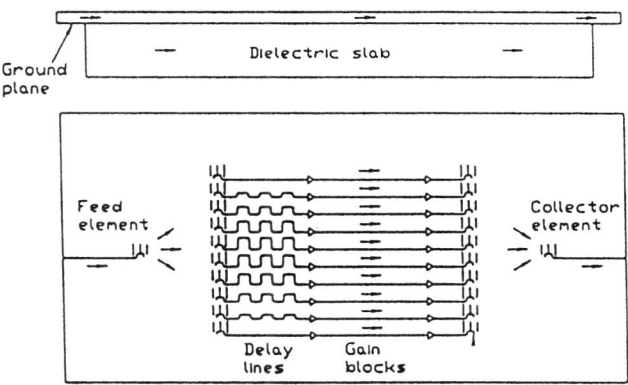

Figure 13. A 2-dimensional amplifier combiner using Yagi-Uda antennas.
(© 1996 IEEE. Reprinted with permission from Reference [27])

2.4. OTHERS

It must be stressed that active antenna can also be extended to waveguide structure [28]. By using short length of waveguides, power is guided to an amplifier. With microstrip probes, signal is tapped from and injected into the waveguides. The input and output waveguides are orthogonal to achieve the isolation.

3. Beam Control Arrays

In this section, emphasis is placed on the active integrated antennas. As mentioned earlier, the purpose of this active antennas is to integrate as many RF functions as possible at the antenna front-end. Based on the applications, this class of active antennas can be further categorized as follows : transponder, transceiver, beam-control arrays and others. Here, only beam-control arrays will be discussed. Beam-control includes beam-scanning, beam-switching and agile beam polarization.

3.1. BEAM-SCANNING ARRAYS

Beam-scanning arrays are designed to scan the main beam of the antenna. To scan the main beam, a consistent phase difference between adjacent elements must be established. Conventional arrays use a ferrite-based phase shifter, which is not suitable for monolithic integration. Therefore, electronic phase tuning is required. There are several design approaches to the beam-scanning arrays.

3.1.1. *Strongly-Coupled Oscillator Array*
Here, strongly-coupled oscillator array is used to implement the beam-scanning array, as shown in Figure 14 [29]. From the amplitude and phase dynamics, it was found that a progressive phase shift exists when the end-elements of a linear array are detuned in frequency. As the elements are mutually locked, the array is still operating at a common frequency. A phase tuning of more than 180° can be achieved using this method. A 4X1 linear array was tested with a scanning range of 27.5^O at X-band. To increase the phase tuning range, a frequency multiplier is connected at the output of each oscillator [30]. Intuitively, a frequency multiplier scales not only the frequency but also the phase, hence increasing the scanning range.

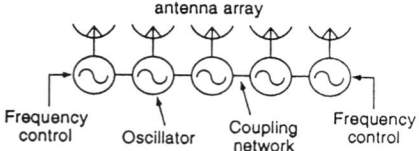

Figure 14. A strongly-coupled oscillator array with end oscillators as beam-scanning tuning elements.
(© 1993 IEEE. Reprinted with permission from Reference [29])

3.1.2. *Unilateral Injection-locking*
In this design approach, injection-locking is used to control the phase of each oscillator, as shown in Figure 15 [31]. From Kurokawa's theory, the phase of the oscillator can be controlled by injecting a reference signal to lock the oscillator. Once locked, the phase of the oscillator can be tuned about 180° by varying its free-running frequency. A progressive phase shift is achieved by coupling part of the signal of a neighboring element to lock the next element. By tuning the free-running frequency of the locked element, the phase can be tuned. The oscillator is locked to the common frequency. Then, the signal is tapped for subsequent element locking. The first element is locked to a stable external source. One concern is that of reverse injection-locking. Thus, an amplifier is used as an active isolator. The amplifier also boosts the signal level of the injecting signal to maintain the injection-locking power. A 4X1 array shows a scan angle of 24°.

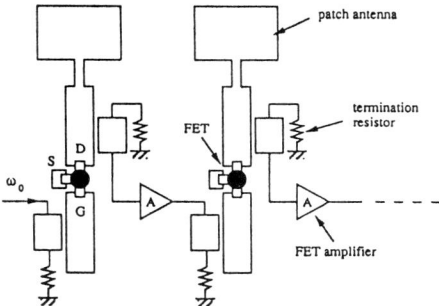

3.1.3. *Phase-locked loop*

In a phase-locked loop, a fraction of signal from the neighboring element is tapped as a reference signal for the next element. The reference signal is used to lock the next element through the phase-locked loop. Once locked, the reference and oscillating signals suffer a phase difference. The amount of phase difference can be tuned by adding a DC offset in the phase-locked loop. A locking range of 180° can be achieved using this approach. A two-element array was designed and tested with a scanning range of 15° at 10 GHz [32].

3.2. BEAM-SWITCHING ARRAYS

In target tracking systems, a monopulse technique is used to track the position of the target. This technique requires the sum and difference channels. These correspond to the sum and difference patterns of the array. By multiplying the IF signals from these channels, a DC error voltage is generated. This DC voltage is then used to determine the position of the target. As described, the system is a receiver. Here, different transmitting circuits capable of generating the sum and difference patterns will be discussed.

3.2.1. *Strongly-Coupled Oscillator Array*

Using average potential theory, it was found that there exists different modes when the elements of the array are interconnected by transmission line [33]. With a through transmission line of one-wavelength, the most stable mode is a out-of-phase mode for a two-element array. Therefore, a difference pattern is generated. When the transmission line has a resistor connected at the center in series, the stable mode is the in-phase mode. Here, a sum pattern is generated. One observation is that the operating frequency of these modes are different, as inherent in their distinct mode characteristics. Also, the switch between the patterns are mechanical in nature.

216

To solve this problem, electronic and optical switching were introduced [34]. The electronic switching uses a MESFET at the center of the transmission line, as shown in Figure 16. The MESFET is connected in parallel. By tuning the gate voltage, R_{ds} of the MESFET can be tuned from low to high resistance. By doing so, a short or an open circuit can be enforced at the MESFET. As a result, the patterns are switched. Optical switching exploits the photoconductive and photovoltaic effects of a FET. The R_{ds} can be tuned optically by varying the intensity of the light source.

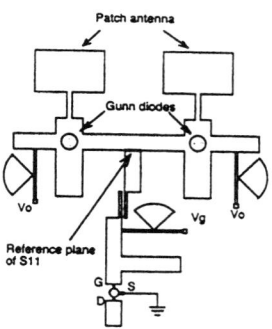

Figure 16. A beam-switching array with MESFET electronic tuning.
(© 1995 IEEE. Reprinted with permission from Reference [34])

3.2.2. *Injection-locking*

Through injection-locking, a $\pm 90^\circ$ phase shift can be generated. By injecting a reference signal and allowing one element of the array to have a 90° lag while the other a 90° lead, a difference pattern is generated. This is achieved by tuning the frequencies of one element to one end of the locking range while the other is tuned to the opposite end. By tuning the free-running frequencies to a common one, a sum pattern results. It is worth noting that the patterns here operate at a common frequency as a result of injection-locking. This is important as the system needs only to deal with one frequency. A 2X2 planar array, as shown in Figure 17, is designed to provide beam-switching in the azimuth and elevation planes [35].

3.2.3. *Active Grid*

Here, active grid has demonstrated a fixed beam diffraction [36]. By using varying gap size in the grid, different capacitances are implemented on the grid. As such, the reflection phase of the grid can be controlled.

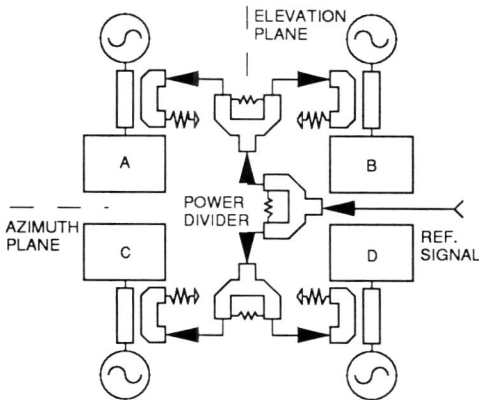

Figure 17. A 2X2 switching array using injection-locking.
(© 1995 IEEE. Reprinted with permission from Reference [35])

3.3. RETRODIRECTIVE ARRAYS

Retrodirective arrays are designed to reflect the in-coming signal back to the source without prior knowledge of the location of the source. This is particularly suitable for communication and the RF identification transponder. Advantage of this circuit is that retrodirectivity is not destroyed by conformal mounting or partial blocking of the array.

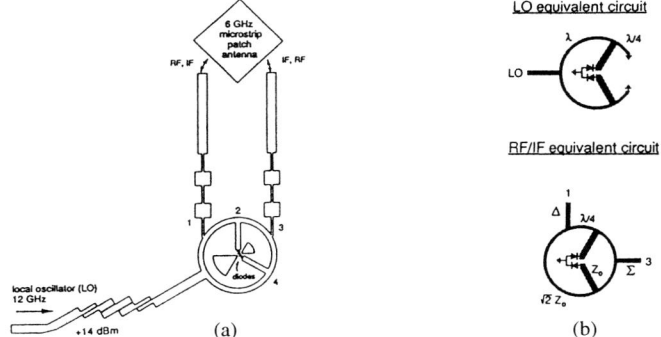

Figure 18. (a) A retrodirective antenna (© 1995 IEEE. Reprinted with permission from Reference [37]);
(b) Equivalent circuit of the element.

Here, a 8X1 linear array operating at 6 GHz was designed [37]. To redirect the signal back to the source, the array must provide phase conjugation of the in-coming signal. This is achieved using heterodyne method. The schematic diagram of an element is shown in Figure 18a. A LO of 12 GHz is used for the mixing. The mixer is designed as such that the RF/IF ports are open-circuit at the LO frequency. The LO voltage is in-phase at the anti-parallel diode pair. This results in anti-phase diode conductance waveforms. The IF, which is also 6 GHz, combines constructively at the other RF port. Thus, RF/IF isolation is achieved. The IF signal is then fed to the orthogonal mode of the patch antenna. The equivalent circuit at the LO and RF frequencies are shown in Figure 18b. A planar 4X4 array is also implemented using slot ring antenna on a ground plane [38].

3.4. AGILE BEAM POLARIZATION

Besides beam-scanning and beam-switching, another form of beam-control is beam polarization control [39]. By integrating two devices per antenna element and an external source for injection-locking, the polarization of the antenna can be linear or circular. Different polarization is controlled by turning 'on' the appropriate active device. To date, only single element antenna has been developed.

4. Conclusion

A review of the 2-dimensional and 3-dimensional power combiners, and beam-control arrays has been given. It is important to note that this review is restricted to arrays. However, there are extensive reports on single-element active antenna. When properly designed, these approaches can be used in active arrays. Another area that is lacking in active antenna is the rigorous full-wave electromagnetic analysis with consideration for nonlinear circuits. To date, only preliminary results are reported [40].

5. References

1. Mink, J. W. (1986) Quasi-Optical Power Combining of Solid-State Millimeter-Wave Sources, *IEEE Trans. Microwave Theory Tech.* **34**, 273-279.
2. Kim, M., Roseberg, J. J., Smith, R. P., Weikle II, R. M., Hacker, J. B., DeLisio, M. P., and Rutledge, D. B. (1991) A Grid Amplifier, *IEEE Microwave and Guided Wave Letters* **1**, 322-324.
3. Kim, M., Sovero, E. A., Hacker, J. B., DeLisio, M. P., Chiao, J.-C., Li, S.-J., Gagnon, D. R., Roseberg, J. J., and Rutledge, D. B. (1993) A 100-Element HBT Grid Amplifier, *IEEE Trans. Microwave Theory Tech.* **41**, 1762-1771.
4. Popovic, Z. B., Weikle II, R. M., Kim, M., and Rutledge, D. B. (1991) A 100-MESFET Planar Grid Oscillator, *IEEE Trans. Microwave Theory Tech.* **39**, 193-200.
5. Shiroma, W. A., Shaw, B. L., and Popovic, Z. B. (1994) Three-Dimensional Power Combiners, *1994 IEEE Int. MTT-S Digest*, 831-834.
6. Shiroma, W. A., Shaw, B. L., and Popovic, Z. B. (1994) A 100-Transistor Quadruple Grid Oscillator, *IEEE Microwave and Guided Wave Letters* **4**, 350-351.

7. Mader, T., Bundy, S., and Popovic, Z. B. (1993) Quasi-Optical VCOs, *IEEE Trans. Microwave Theory Tech.* **41**, 1775-1781.

8. Bundy, S. C. and Popovic, Z. B. (1994) Analysis of Planar Grid Oscillators, *1994 IEEE Int. MTT-S Digest*, 827-830.

9. Tsai, H. S., Rodwell, M. J. W. and York, R. A. (1994) Planar Amplifier Array With Improved Bandwidth Using Folded-Slots, *IEEE Microwave and Guided Wave Letters* **4**, 112-114.

10. Benet, J. A., Perkons, A. R., Wong, S. H. and Zaman, A. (1993) Spatial Power Combining for Millimeterwave Solid State Amplifiers, *1993 IEEE Int. MTT-S Digest*, 619-622.

11. Ivanov, T. and Mortazawi, A. (1996) A Two-Stage Spatial Amplifier with Hard Horn Feeds, *IEEE Microwave and Guided Wave Letters* **6**, 88-90.

12. Ivanov, T. and Mortazawi, A. (1995) A Double Layer Microstrip Spatial Amplifier with Increased Active Device Density, *25th European Microwave Conf.*, 320-323.

13. Chew, S. T. and Itoh, T. (1996) High Active Device Density Quasi-Optical Amplifier, *to be presented at 26th European Microwave Conf.*.

14. Stephan, K. D. and Morgan, W. A. (1987) Analysis of Interinjection-Locked Oscillators for Integrated Phased Arrays, *IEEE Trans. Antennas Propagation* **35**, 771-781.

15. York, R. A. (1993) Nonlinear Analysis of Phase relationships in Quasi-Optical Oscillator Arrays, *IEEE Trans. Microwave Theory Tech.* **41**, 1799-1809.

16. Liao, P. and York, R. A. (1994) A 1 Watt X-Band Power Combining Array Using Coupled VCOs, *1994 IEEE Int. MTT-S Digest*, 1235-1238.

17. Lin, J. and Itoh, T. (1994) Two-Dimensional Quasi-Optical Power-Combining Arrays Using Strongly Coupled Oscillators, *IEEE Trans. Microwave Theory Tech.* **42**, 734-741.

18. Chew, S. T. and Itoh, T. (1996) Improved Phase Noise Active Antenna Arrays Using DR-Coupled Oscillators, *to be presented at PIERS 1996, Austria.*

19. Mortazawi, A., Foltz, H. D. and Itoh, T. (1992) A Periodic Second Harmonic Spatial Power Combining Oscillator, *IEEE Trans. Microwave Theory Tech.* **40**, 851-856.

20. Mortazawi, A. and De Loach, B. C. Jr. (1993) A Nine-MESFET Two-Dimensional Power-Combining Array Employing an Extended Resonance Technique, *IEEE Microwave and Guided Wave Letters* **3**, 214-216.

21. Birkeland, J. and Itoh, T. (1992) A 16 Element Quasi-Optical FET Oscillator Power Combining Array with External Injection Locking, *IEEE Trans. Microwave Theory Tech.* **40**, 475-481.

22. Kurokawa, K. (1973) Injection Locking of Microwave Solid-State Oscillators, *Proceedings of IEEE* **61**, 1386-1410.

23. Mink, J. W. and Schwering, F. K. (1993) A Hybrid Dielectric Slab-Beam Waveguide for the Sub-Millimeter Wave Region, *IEEE Trans. Microwave Theory Tech.* **41**, 1720-1729.

24. Poegel, F., Irrgang, S., Zeisberg, S., Schuenemann, A., Monahan, G. P., Hwang, H., Steer, M. B., Mink, J. W., Schwering, F. K., Paollela, A. and Harvey, J. (1995) Demonstration of an Oscillating Quasi-Optical Slab Power Combiner, *1995 IEEE Int. MTT-S Digest*, 917-920.

25. Hwang, H., Monahan, G. P., Steer, M. B., Mink, J. W., Harvey, J., Paollea, A. and Schwering, F. K. (1995) A Dielectric Slab Waveguide With Four Planar Power Amplifiers, *1995 IEEE Int. MTT-S Digest*, 921-924.

26. Hwang, H., Nuteson, T. W., Steer, M. B., Mink, J. W., Harvey, J. and Paollea, A. (1995) Quasi-Optical Power Combining in a Dielectric Substrate, *URSI Int. Symp. on Signals, Systems, and Electronics Dig.*, 89-92.

27. Perkons, A. R. and Itoh, T. (1996) A 10-Element Active Lens Amplifier on a Dielectric Slab, *1996 IEEE Int. MTT-S.*, 1119-1122.

28. Kolias, N. J. and Compton, R. C. (1993) A Microstrip-Based Unit Cell for Quasi-Optical Amplifier Arrays, *IEEE Microwave and Guided Wave Letters* **3**, 330-332.

29. Liao, P. and York, R. A. (1993) A New Phase-Shifterless Beam-Scanning Technique Using Arrays of Coupled Oscillators, *IEEE Trans. Microwave Theory Tech.* **41**, 1810-1815.

30. Alexanian, A., Chang, H.-C. and York, R. A. (1995) Enhanced Scanning Range of Coupled Oscillator Arrays Utilizing Frequency Multipliers, *1995 IEEE Int. AP-S Digest*, 1308-1310.

31. Lin, J., Chew, S. T. and Itoh, T. (1994) A Unilateral Injection-Locking Type Active Phased Array for Beam Scanning, *1994 IEEE Int. MTT-S Digest*, 1231-1234.
32. Martinez, R. D. and Compton, R. C. (1994) Electronic Beamsteering of Active Arrays With Phase-Locked Loops, *IEEE Microwave and Guided Wave Letters* **4**, 166-168.
33. Lin, J., Itoh, T. and Nogi, S. (1993) Mode Switch in a Two-Element Active Array, *1993 IEEE Int. AP-S Digest*, 664-667.
34. Minegishi, M., Lin, J., Itoh, T. and Kawasaki, S. (1995) Control of Mode-Switching in an Active Antenna Using MESFET, *IEEE Trans. Microwave Theory Tech.* **43**, 1869-1874.
35. Chew, S. T. and Itoh, T. (1995) A 2X2 Beam-Switching Active Antenna Array, *1995 IEEE Int. MTT-S Digest*, 925-928.
36. Kim, M., Weikle II, R. M., Hacker, J. B. and Rutledge, D. B. (1991) Beam Diffraction by a Planar Grid Structure at 93 GHz, *1991 IEEE Int. AP-S Digest,* 150-153.
37. Pobanz, C. and Itoh, T. (1995) A Conformal Retrodirective Array for Radar Applications Using a Heterodyne Phased Scattering Element, *1995 IEEE Int. MTT-S*, 905-908.
38. Pobanz, C. and Itoh, T. (1996) A Two-Dimensional Retrodirective Array using Slot Ring FET Mixers, *to be presented at 26th European Microwave Conference.*
39. Haskins, P. M., Hall, P. S. and Dahele, J. S. (1994) Polarization-Agile Active Patch Antenna, *Electronic Letters* **30**, 98-99.
40. Toland, B., Lin, J., Houshmand, B. and Itoh, T. (1994) Electromagnetic Simulation of Mode Control of a Two Element Active Antenna, *1994 IEEE Int. MTT-S Digest*, 883-886.

GRID AMPLIFIERS

D.B. RUTLEDGE
California Institute of Technology,
Pasadena, CA 91125, USA

1. Introduction

Active grids are periodic structures loaded with transistors or diodes that interact with electromagnetic beams. These new quasi-optical components may make possible a new generation of low-cost, high-power solid-state millimeter-wave communications, broadcast, and radar systems. Quasi-optical power combining allows the output powers from large numbers of individual transistors and diodes to be combined in free space without transmission-line losses. These components can often be made as planar structures that are suitable for large-scale monolithic integration, particularly at millimeter and submillimeter wavelengths. This should allow these systems to have low cost. Often these devices are multi-mode devices that work with beams at different angles, or even with several beams simultaneously. This should make it possible to use these components in electronicallyscanned systems. The devices can also tolerate high failure rates—a 10% transistor failure rate may only reduces the gain by 1 dB. This could give advantages in increasing the fabrication yield to lower the cost, or enabling the construction of ultra-reliable systems for space applications, or allowing a device to survive partial destruction in military use. These circuits are analyzed by equivalent waveguide circuits. The diameter of the equivalent waveguide is determined by the period of the structure, which is fixed by photolithography, rather than by fabrication of a metal waveguide. This makes it possible to make grids for the terahertz frequency range. A variety of grids have been demonstrated [1], including phase shifters, multipliers, oscillators [2], and switches. In these grids, the power is proportional to the area, while the circuit impedances are determined by the dimensions of the unit cell. This allows great design flexibility. One can achieve high power and high efficiency simultaneously, or large dynamic range and low noise at the same time. An oscillator with an output power as high as 10 W have been demonstrated at 10 GHz [3], and a doubler has shown an output of 330 μW output at 1 THz [4].

2. Hybrid Grid Amplifiers

In 1991, Moonil Kim *et al.* demonstrated the first grid amplifier [5]. The grid amplifier is an active quasi-optical device that amplifies a beam as it passes through it (Figure 1). The grid had a gain of 11 dB at 3.3 GHz. Our experience has been that it helps in several ways to have the input and output beams cross-polarized. It is easier to stabilize the grid against oscillations. In addition, the polarizers can tune the input and output circuits independently.

J.M. Chamberlain and R.E. Miles (eds.), New Directions in Terahertz Technology, 221–226.

We have also found that gain measurements are easier if the input and output beams are cross-polarized, because interference is reduced. This first grid was small, with only 25 differential-pair elements, and the wiring arrangement was complicated, with radiating elements on one side of the board, bias lines on the other, and connecting feedthroughs. The fabrication used hybrid techniques, with 50 discrete packaged transistors mounted on a Duroid board.

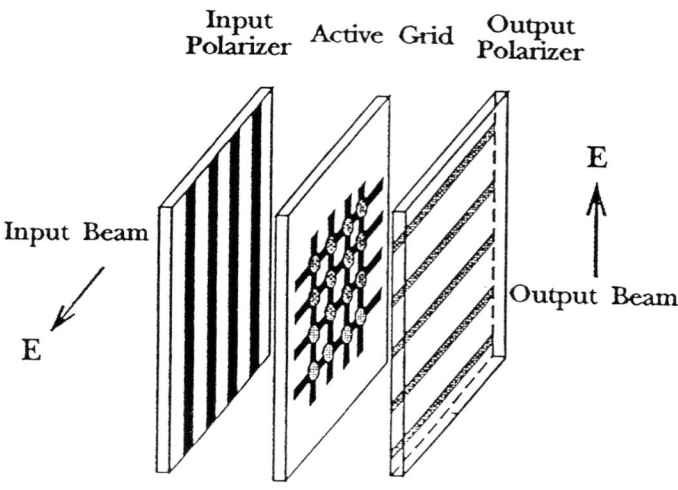

Figure 1. A grid amplifier [5]. In the figure, horizontally polarized power is incident from the left, and passes through the polarizer. The grid amplifier amplifies the beam and radiates it as vertically polarized power, which passes through the output polarizer on the right.

In 1993, Kim extended this idea to make a 100-element, 10-GHz grid amplifier [6]. The active devices in the grid were chips with heterojunction-bipolar transistor (HBT) differential pairs (Figure 2). Kim measured a 10 dB gain at 10 GHz with a 3-dB bandwidth of 1 GHz. The input and output return loss were better than 15 dB. The maximum output power was 450 mW, and the minimum noise figure was 7 dB.

HBT's typically have less gain and poorer noise figures than pseudomorphic high-electron mobility transistors (pHEMT's), and this means that it was important to be able to demonstrate a pHEMT grid amplifier. DeLisio *et al.* recently extended the grid-amplifier idea to a 100-element X-band HEMT amplifier [7]. They reported 11 dB of gain at 9 GHz, 3-dB noise figure, and 3.7 W of output power, considerably better than the previous HBT grid. Moreover, it was found that a grid amplifier can amplify beams at angles up to $\pm 30°$, so that it could be used in an electronic beam-steering system. In addition, DeLisio

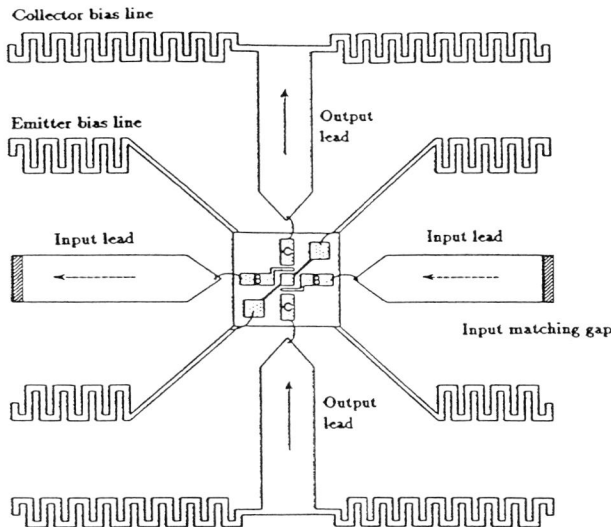

Collector bias line

Emitter bias line

Output lead

Input lead

Input lead

Input matching gap

Output lead

Figure 2. Grid-amplifier unit-cell design for a 100-element 10-GHz HBT grid amplifier [6]. The arrows indicate the direction of current flow. The unit-cell size is 8 mm.

demonstrated a transmission-line model for the grid, together with extensive comparisons of theory and experiment to validate the model.

3. Monolithic Grid Amplifiers

At millimeter-wavelengths, it is attractive to build monolithic grid amplifiers to avoid the extensive fabrication required in hybrid grid amplifiers. Liu *et al.* reported a 36-element monolithic HBT grid amplifier [8, 10]. Liu employed a phase-shifting capacitor in the base leads to stabilize the grid against oscillations. He measured a gain of 5 dB at 40 GHz (Figure 3), with a maximum power of 670 mW. However, the efficiency is poor, with a peak efficiency of only 4% at 500-mW output power, and the grids could only be operated for short periods of time, of the order of a second.

For higher-frequency millimeter-wave operation, it is important to demonstrate a pHEMT amplifier. DeLisio *et al.* reported a 36-element pHEMT Grid Amplifier that could be tuned to operate from 44 GHz to 60 GHz by changing the positions of external polarizers and tuning slabs [9, 10] (Figure 4). The gain at 44 GHz was 6.5 dB, while at 60 GHz, the grid has a peak gain of 2.5 dB. Again, the transmission-line model predicted gain and tuning curves that were consistent with the measurements.

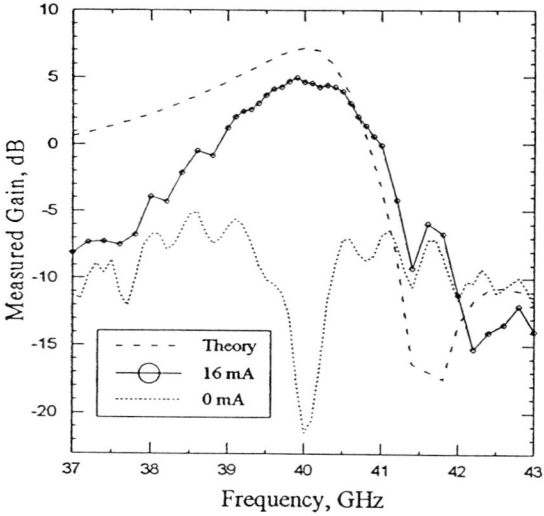

Figure 3. Gain measurements for the monolithic HBT amplifier [8, 10]. The peak gain is 5 dB at 40 GHz. The theory is calculated from the transmission-line model.

4. The Future

Research in monolithic quasi-optical amplifiers is just beginning, and the field is wide open, with several competing groups. Heat dissipation, high-power, and efficiency are major challenges. Grid amplifiers have demonstrated gain at millimeter wavelengths in monolithic circuits. However, the efficiency is quite poor, only 4%. It is important to improve this efficiency to make high-power transmitters. Currently two-thirds of the power is dissipated in bias resistors. This suggests reducing, or eliminating these resistors. Our theory indicates that the grid should be stable if we do this. For this we will need to mount the grids on substrates with high thermal conductivity like diamond and aluminum nitride. This should make the goal of battery-operated, watt-level millimeter-wave transmitters practical. High-power quasi-optical millimeter-wave amplifiers are the subject of a major new U.S. Defense Department Inititave. This will allow us to see whether high powers can actually be achieved.

5. Acknowledgements

We appreciate the support of the U.S. Army Research Office and the U.S. Air Force Material Command/Rome Laboratory.

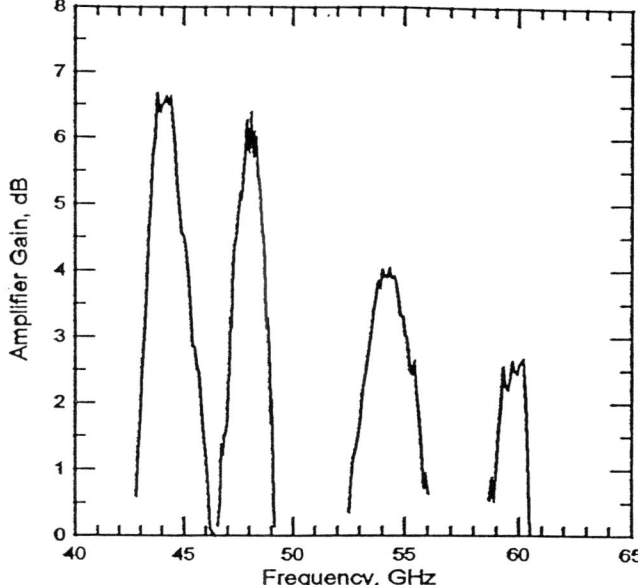

Figure 4. Tuning range of the 36-element pHEMT grid amplifier [9, 10]. The polarizers were adjusted to give peaks at various frequencies in the range from 44 GHz to 60 GHz. The 60-GHz measurement includes an output tuner.

References

1. 1. Hacker, J.B., and Weikle, R.M., (1995) "Quasi-Optical Grid Arrays," in T.K. Wu (ed.), *Frequency Selective Surface and Grid Array*, Wiley, chapter 8. This is an excellent review article on quasi-optical grids.
2. Popović, Z.B., Weikle R.M., Kim, M., and Rutledge, D.B., (1991) "A 100-MESFET Planar Grid Oscillator," *IEEE Trans. on Microwave Theory and Tech.*, MTT-39, pp. 193–200.
3. Hacker, J.B., DeLisio, M.P., Kim, M., Liu, C.M., Li, S.J., Wedge, S, and Rutledge, D.B., (1994)"A 10-Watt X-band Grid Oscillator," International Microwave Symposium, San Diego CA, pp. 823–826.
4. Chiao, J.C., Markelz, A., Li, Y.J., Hacker, J., Crowe, T., Allen, J., and Rutledge, D., (1995) "Terahertz Grid Frequency Doublers," 6th International Symposium on Space Terahertz Technology, Pasadena CA.
5. Kim, M., Rosenberg, J.J., Smith, R.P., Weikle, R.M., Hacker, J., DeLisio, M.P., and Rutledge, D.B., (1991)"A Grid Amplifier," *IEEE Microwave and Guided Wave Letters, MGWL-1* pp. 322–324.
6. Kim, M., Sovero, E., Hacker, J., DeLisio, M.P., Chiao, J.C., Li, S., Gagnon, D., Rosenberg, J., and Rutledge, D.B., (1993) "A 100-Element Grid Amplifier ," *IEEE Transactions on Microwave Theory and Techniques*, pp. 1762–1771.
7. DeLisio, M.P., Duncan, S., Tu D.W., Liu C.M., Moussessian, A., Rosenberg, J., and Rutledge, D., (to be published) "Modelling and Performance of a 100-Element pHEMT Grid Amplifier," *IEEE Transactions on Microwave Theory and Techniques*.
8. Liu, C.M., Sovero, E., Ho, W.J., DeLisio, M.P., and Rutledge, D.B., (1966) "Monolithic 40-GHz 670-mW HBT Grid Amplifier," The International Microwave Symposium, San Francisco, CA., pp. 1123–1126.

9. DeLisio, M.P., Duncan, S., Weinreb, S., Liu, C.M., and Rutledge, D.B., (1996) "A 44–60 GHz Monolithic pHEMT Grid Amplifier," The International Microwave Symposium, San Francisco, CA., pp. 1127–1130.

10. DeLisio, M.P. and Liu, C.M., (1996) "Grid Amplifiers," in R. York and Z. Popović (eds.), *Active and Quasi-Optical Arrays*, Wiley, chapter 9. This chapter is a comprehensive review of recent hybrid and monolithic grid amplifiers.

SYSTEM CHARACTERISATION ISSUES FOR INTEGRATION

J.W. BOWEN
Department of Cybernetics, The University of Reading
Whiteknights, PO Box 225, Reading, RG6 6AY, UK

1. Introduction

Vital to the successful development of terahertz systems and devices is the establishment of measurement systems and techniques to quantify their performance characteristics reliably and probe the underlying physical mechanisms. Many of the new systems we will want to characterise will be integrated units, the form of which is largely dictated by the fabrication processes involved. This poses a particular problem in their characterisation. Before addressing this and explaining the approach we have adopted in the UK Terahertz Integrated Technology Initiative (TINTIN) programme, we will take a look at the parameters of interest and how we may go about measuring them.

1.1. PARAMETERS OF INTEREST

Apart from the overall system performance there are a number of other parameters of interest. S-parameter measurement can yield important information about impedance and losses. Mixers and detectors are characterised in terms of parameters such as conversion loss, responsivity, noise temperature and noise equivalent power. Sources are characterised in terms of power spectra, phase noise and stability. Antenna amplitude and phase patterns need to be measured. The determination of the complex optical constants of materials is important for systems design.

1.2. MEASUREMENT TECHNIQUES

The measuring instruments may be waveguide-based, quasi-optical, or a hybrid of the two. The techniques involved may be broadly categorised in terms of the type of source used (in active systems).

1.2.1. *Techniques Using Coherent Sources*
These utilise the interaction of coherent radiation with the device under test and, as such, have an instantaneous bandwidth equal to the line width of the source. The technique affords high frequency resolution but the source must be swept to carry out wide bandwidth measurements.

J.M. Chamberlain and R.E. Miles (eds.), New Directions in Terahertz Technology, 227–234.

1.2.2. *Techniques Using Wide-band Noise Sources*
These analyse the interaction of radiation from a wide-band noise source with the device under test using a spectrometer and wide-band detector. This approach yields measurements over a multi-octave instantaneous bandwidth at the expense of a reduced frequency resolution.

1.2.3. *Time Domain Techniques*
These employ the interaction of short pulses of radiation, the frequency spectra of which extend into the terahertz region, with the device under test. Small photoconductive probes allow high-spatial resolution mapping of electric fields and direct measurements on small-scale devices.

2. The Interface Problem

In order to make any measurements we need some means of coupling power between the system to be characterised and the measuring instrument. In other words we need some sort of interface. A higher coupling efficiency leads to an increased signal-to-noise ratio and thus an improved measurement precision. The effects of non-optimal coupling will appear as a power loss which it will be necessary to calibrate out to make accurate measurements of S-parameters or power.

It is particularly difficult to ensure good coupling to integrated systems. The waveguiding in the measuring instrument and the system under test will usually be based on different technologies and fabrication techniques. Moreover, it is often impossible to fabricate a flange on the integrated system that allows direct connection to the measuring instrument.

In the absence of a flange, waveguides of equal size could be butted together end-to-end. However, this would require the waveguides to have very accurately machined end surfaces and their alignment would be non-trivial.

An alternative that has been explored is dielectric coupling. Taking the example of an integrated system based around micro-machined rectangular waveguide, one end of a bar of dielectric material may be inserted into the micro-machined waveguide so that it protrudes clear of it. If the inserted end of the dielectric bar is tapered, the bar will act as a dielectric waveguide, the energy being coupled into the bar with low back reflection. The other tapered end of the bar may then be inserted into the conventional waveguide of the measuring instrument so that the energy is coupled from the dielectric guide into the measuring instrument. This has been found to work well for full-height rectangular waveguide but does not work with reduced height waveguides. There may also be problems, particularly as the frequency is increased, with losses in the dielectric guide and from the potential for damaging the fragile integrated waveguides with the dielectric guide.

The best solution appears to be to use quasi-optical coupling. A beam is launched from the measuring instrument and this is coupled, with the aid of a system of lenses or focusing mirrors, into an antenna monolithically integrated on the system under test. A returned beam would be coupled back reciprocally to the measuring instrument. This has the major advantage that physical contact to the delicate integrated structure is not necessary. In any case, an antenna is needed for most applications as there usually has to be some interaction between the radiation in the system and the outside world if the system is to perform some useful function. This is the approach that has been adopted by TINTIN and I will describe the antenna we have developed before explaining, in general terms, how the degree of coupling can be quantified and optimised.

3. The TINTIN Antenna

The TINTIN micro-machined waveguides are produced by building up a thick layer of photo-resist on top of a metallised surface on a semiconductor wafer. Areas of photo-resist are etched away leaving ridges of photo-resist intact which define the locations of the waveguides on the wafer. A layer of gold is then deposited over the raised areas of photo-resist. The remaining photo-resist is subsequently removed to leave a free-standing rectangular waveguide circuit.

While it is straightforward to taper the width of a waveguide or, to a limited degree near 1 THz, produce a step change in its height, the fabrication process does not lend itself to the production of tapers in the vertical plane. This imposes severe constraints on the type of antenna that can be fabricated monolithically with the waveguide. Thus, while it is impossible to fabricate anything approaching a pyramidal horn, H-plane sectoral horns can be made. However, while they have well-directed radiation patterns in the H-plane they produce very broad patterns in the E-plane, especially if they are less than the full waveguide height (as is the case for the lower frequency TINTIN waveguides). The reason for their poor E-plane patterns is that the width of the E-field distribution in the E-plane is small in the plane from which the beam is radiated (the source plane): the horn aperture here. The far-field radiation pattern is related to the E-field distribution in the source plane by a Fourier transform relationship and a broad beam-width is the result. In order to reduce the E-plane beam-width we need to expand the E-plane E-field distribution before radiation is allowed to occur. We have done this by cutting a tapered slot in the upper broadwall of the horn, Figure 1.

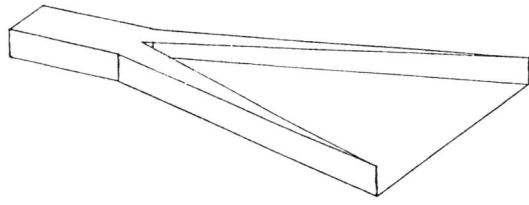

Figure 1. TINTIN antenna.

Where the slot is sufficiently narrow the E-field distribution extends outside the confines of the waveguide but elements of it in close proximity are in anti-phase and radiation is negligible. Radiation occurs when the slot is large enough for the in-phase component to become dominant over any region in the E-field distribution. Correct choice of the slot

and horn tapers produces symmetrical beams with a 3 dB width of 27° angled at 22.5°
up out of the plane of the substrate. Versions with exponential tapers have been found to
work best for reduced height (eighth-height) waveguide. A complication is that the
launched beam is astigmatic.

4. Beam-mode Analysis

In order to determine the degree of coupling between a free-space beam and an arbitrary
antenna it is helpful to decompose the beam launched by the antenna into Hermite-
Gaussian or Laguerre-Gaussian beam-modes [1].

If we consider an arbitrary paraxial beam propagating in the z direction we can write
the field in an arbitrary near-field plane $z = z_S$, which we term the source plane, as a
superposition of orthonormal basis functions

$$u(x, y; z_S) = \sum_{m,n} C_{mn} u_{mn}(x, y; z_S)$$
(1)

where the coefficients C_{mn} are, in general, complex. In equation (1) and those following,
I have omitted the plane-wave phase term $\exp{-i(kz - \omega t)}$ which, when multiplied by the

modulating function $u(x, y, z)$, provides a complete description of the beam-field. It

proves advantageous to choose Hermite-Gaussian functions, which form a complete
orthonormal set, as the basis functions.   Laguerre-Gaussian functions would be an
alternative. The (non-astigmatic) Hermite-Gaussian function of order m,n is

$$u_{mn}(x, y, z) = \left(2^{m+n-1} \pi m! n!\right)^{-1/2} \frac{1}{w} \cdot H_m\left(\sqrt{2} \frac{x}{w}\right) \cdot H_n\left(\sqrt{2} \frac{y}{w}\right) \cdot \exp{-\frac{\left(x^2 + y^2\right)}{w^2}}$$
$$\cdot \exp{-ik \frac{\left(x^2 + y^2\right)}{2R}} \cdot \exp{i(m+n+1)\Theta}$$
(2)

where the $H_m(X)$ and $H_n(Y)$ denote Hermite polynomials of order m,n respectively
$(m,n = 0,1,2,...)$. The arbitrary constants w, R and Θ can be chosen freely to give a good
fit to the field with, for example, the minimum number of Hermite-Gaussian functions.

It can be shown that a beam with a field in the source plane that can be described by
a single Hermite-Gaussian function will have a field in any arbitrary down-beam
constant-z plane that is a Hermite-Gaussian function of the same order with a scaled
argument and added spherical wavefront curvature. It follows that the propagating field
can be considered to be a superposition of beam-modes

$$u(x, y, z) = \sum_{m,n} C_{mn} u_{mn}(x, y, z)$$
(3)

The beam-modes $u_{mn}(x,y,z)$ take the form of the Hermite-Gaussian function above but w,
R and Θ vary with z as

$$w^2 = w_0^2 + \left\{2(z - z_0)/kw_0\right\}^2$$
(4)

$$R = (z - z_0) + \left\{kw_0^2/2\right\}^2 / (z - z_0)$$
(5)

$$\Theta = tan^{-1}\left\{kw^2/2R\right\} + \Theta_0$$

$$= sin^{-1}\left(1 + \left\{kw^2/2R\right\}^{-2}\right)^{-1/2} + \Theta_0 \tag{6}$$

w describes the width of the beam, which takes a minimum value w_0 at the beam-waist located at $z = z_0$. R describes the radius of curvature of the spherical wave-fronts of each beam-mode. Θ governs the phase-slippage between successive modes and relative to an on-axis plane-wave. The values of w_0, z_0 and Θ_0 are determined by the values assigned to w, R and Θ when fitting the field in a constant-z cross-section with a superposition of Hermite-Gaussian functions.

It is a property of complete orthonormal sets of functions that the coefficients C_{mn} in the superposition that describes the arbitrary field $u(x,y;z_S)$ in the plane $z = z_S$ can be determined from the integral

$$C_{mn} = \iint_{z_S} u^*_{mn}(x, y; z_S) \cdot u(x, y; z_S) dxdy \tag{7}$$

Thus, for an antenna with a theoretically calculable field in a near-field plane, usually an aperture, this field can be decomposed into Hermite-Gaussian functions by evaluation of the above integral, the values of w, R and Θ being chosen for maximum simplicity and computational economy. The launched beam will be a superposition of beam-modes described by the same C_{mn}.

This approach has been applied to the beam launched by a corrugated feed-horn in [2]. The field at the horn aperture has a spherical wave-front with a radius of curvature equal to the length of the horn. Thus simplicity dictates that we should set R equal to its length and Θ equal to zero. The value of w chosen is that which maximises the coefficient of, and thus power carried by, the fundamental beam-mode. Once R and w at any constant-z plane through the beam are known it is possible to determine the location and size of the beam-waist, parameters which will be the same for all modes in a superposition.

For some antennas a theoretical description of the near-field will not be available and then the measured far-field amplitude and phase patterns will have to be taken as a starting point. As the far-field of an antenna is related to its near-field by Fourier transformation and the Fourier transform of a Hermite-Gaussian function is itself a Hermite-Gaussian function of the same order, the far-field can be described as a superposition of Hermite-Gaussian functions. The process outlined above can thus be used to determine the C_{mn} by fitting the far-field with Hermite-Gaussian functions in polar co-ordinates.

5. Beam-mode Coupling

Once the modal composition of the beam formed by an antenna is known, the degree of coupling between an incident beam and that received by the antenna may be determined.

The situation is illustrated in Figure 2. The relative amplitude and phase of the signal coupled from beam S to beam A is given by the coupling or overlap integral between the fields of the two beams, $u_S(x,y,z)$ and $u_A(x,y,z)$, over any constant z plane through the beams, c.

$$\langle u_A | u_S \rangle = \iint_c u_A^*(x, y, z) \cdot u_S(x, y, z) dx dy \tag{8}$$

The result is independent of the plane of integration.

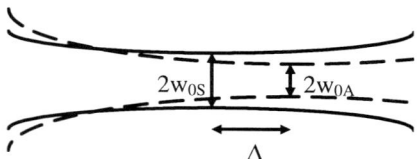

Figure 2. Coupling between two beams.

If u_A and u_S represent two co-axial fundamental beam-modes the amplitude coupling integral becomes

$$|\langle u_A | u_S \rangle| = \left(k \overline{\overline{w_0}} \right)^2 \left\{ \left(k^2 \overline{w_0^2} \right)^2 + (k\Delta)^2 \right\}^{-1/2} \tag{9}$$

where $\overline{\overline{w_0}}$ is the geometrical mean of w_{0A} and w_{0S}, and $\overline{w_0^2}$ is the mean of w_{0A}^2 and w_{0S}^2. Similar expressions exist for laterally and rotationally displaced fundamental beam-modes. More general expressions which quantify the coupling between beam-modes of different orders are given in [3].

The ideal situation would be to have perfect coupling between the test beam and that launched by the antenna on the integrated system under test. Even if this is achieved, it is likely that there will be some reflection at the antenna and calibration procedures [4] will have to be undertaken to de-embed the system and device parameters.

6. Astigmatism Correction

If the beam launched by the antenna is astigmatic it will be necessary to correct the astigmatism to achieve good coupling to the stigmatic beams produced by the measurement system. The quasi-optical element to provide this correction can also be used to provide the desired beam-waist size and location for optimal coupling. Three methods of achieving this are:
a) to use a train of two plano-cylindrical lenses acting orthogonally on the antennas E and H-planes. If space permits, the focal length of each lens may be chosen so that it may be positioned at a distance equal to its focal length from its respective input beam-waist thereby forming a co-located output beam-waist. Such a beam-waist would have a frequency invariant location at a distance from each lens equal to its respective focal length.
b) to use a single non-axially symmetric lens.
c) to use an off-axis spherical mirror.

Unfortunately a) and c) give an elliptical rather than circular output beam-waist for all but a special case ratio of astigmatic difference and desired output beam-waist size. On the other hand lens b), is considerably more difficult to manufacture. The mirror c) gives lower throughput losses and freedom from interference effects. The design of an off-axis spherical mirror for astigmatism correction is covered in detail in [5].

7. Zoom Optics

A zoom system which allows adjustment of the output beam-waist size and location of the beam launched by the measuring instrument can provide maximum versatility for coupling to different antennas. Figure 3 schematically illustrates the addition of a two lens (focal lengths f_1 and f_2) zoom system to the beam thrown by a single lens. The dependence of d_e and w_{0e} on the distances d_1 and d_2 for $f_1 = 250$ mm, $f_2 = 50$ mm, $w_{0i} = 25.2$ mm at 700 GHz are shown in the contour plots of Figures 4 and 5. The negative values taken by d_1 indicate a virtual input beam-waist. The spacing to give a desired combination of d_e and w_{0e} can be found by overlaying the two contour plots.

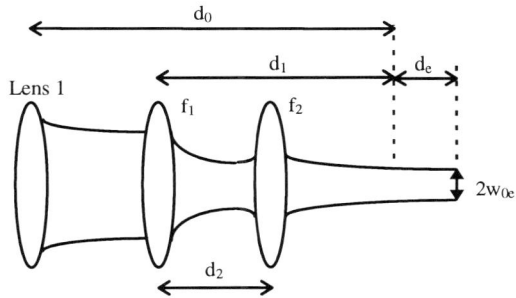

Figure 3. Zoom system. Lens 1 alone throws a beam-waist w_{0i} at d_0

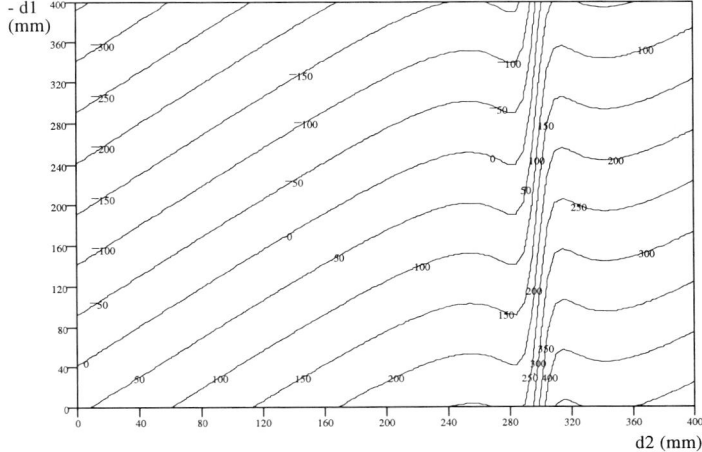

Figure 4. Output beam-waist location d_e (mm) for a typical zoom system.

234

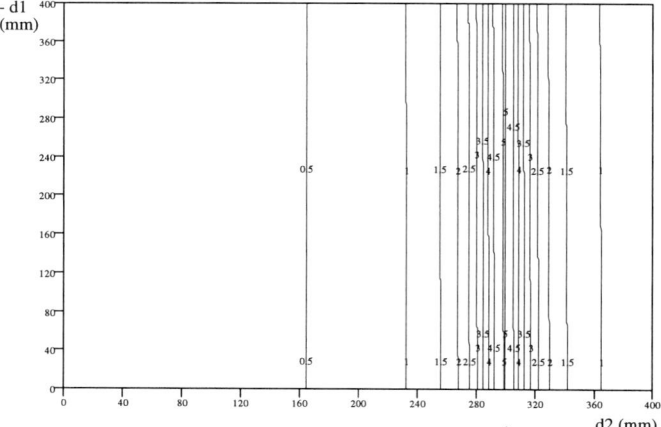

Figure 5. Output beam-waist size w_{0e} (mm) for a typical zoom system.

8. TINTIN Measurement System

The measurement system designed for the characterisation of the TINTIN components is a quasi-optical system based on a hybrid Mach-Zehnder/Martin-Puplett interferometer. The same system can be used over a frequency range of 90 GHz to 2.5 THz and may be configured as either a single-pass transmission or reflection dispersive Fourier transform spectrometer, a null-transmissometer or null-reflectometer. In wide-band mode it uses a mercury arc lamp source and InSb hot electron bolometers achieving a maximum resolution of 0.1875 GHz. The narrow-band mode uses an optically-pumped FIR laser as a source. A detailed description is given in [6].

The techniques that I have described here are currently being employed to optimise the coupling of power between the measurement system and the integrated components under test.

References

1. Martin, D.H. and Bowen, J.W. (1993) Long wave optics, *IEEE Trans. Microwave Theory Tech.* **41**, 1676-1690.
2. Wylde, R.J. (1984) Millimetre-wave Gaussian beam-mode optics and corrugated feed horns, *Proc. Inst. Elec. Eng.,* **131**, pt. H, 258-262.
3. Kogelnik, H. (1964) Coupling and conversion coefficients for optical modes, *Proceedings of the Symposium on Quasi-optics, Microwave Research Institute Symposia Series,* **14**, Polytechnic Press, New York, 333-349.
4. Pollard, R.D. (1996) Millimetre wave and terahertz waveguides and measurements, *this volume.*
5. Bowen, J.W. (1995) Astigmatism in tapered slot antennas, *Int. J. Infrared and Millimeter Waves,* **16**, 1733-1756.
6. Bowen, J.W. et al. (1995) A quasi-optical measurement system for the characterisation of terahertz integrated components, *20th International Conference on Infrared and Millimeter Waves, Orlando, Florida,* 93-94.

Theme 4

Applications

THz ASTRONOMY FROM SPACE

Th. de GRAAUW
SRON-Groningen,
PO Box 800, 9700 AV Groningen,
the Netherlands.

Abstract

The THz frequency range is the last section of the electromagnetic spectrum to be fully explored in astronomy. The astrophysical questions that can be addressed by space-borne facilities operating in this frequency range with state of the art detectors and heterodyne mixers will be summarised. The derived requirements for the technical capabilities of presently planned observatories and their instrumentation will be discussed and an overview will be given of ongoing and future THz space missions.

1. Introduction

From the ground, observations of the Universe in the THz frequency range (or at Submillimeter (Submm) and Far InfraRed (FIR) wavelengths) are only possible through a few atmospheric windows. Although air- and balloonborne telescopes allow the avoidance of a fair quantity of the hindering absorption by watervapour and other atmospheric constituents, full exploration, with the highest sensitivity, of the cold component of the universe requires space facilities and observatories. There are several good reasons to go to space for Submm and FIR observations: to avoid the absorption and emission by the earth's atmosphere and to be able to observe with cryogenic telescopes and /or instrumentation directly into space, i.e. without a vacuum window in the cryostat. An impression of the brightness temperature of the earth's atmosphere from a high altitude site can be obtained from Figure 1 in the contribution by B. Carli, this volume.

Until today there have been two all-sky survey/explorer space missions for this frequency/wavelength range: the InfraRed Astronomical Satellite (IRAS) and the Cosmic Background Explorer (COBE). See Figure 1 for an overview of (FIR)/Submm/THz missions in the 1980-2020 period. Both satellites made full maps of the entire sky. IRAS mapped in four bands, at 12, 25, 60 and 100 microns. COBE extended observations into the millimetre wave range. IRAS and an angular resolution of a few arcminutes; COBE had only a resolution of seven degrees. COBE's main goal was to measure the black body spectrum and the (an)isotropy of the cosmic background. The third (F)IR mission was the Japanese IRTS (InfraRed Telescope in Space), with a 20 day lifetime. Its scientific goal was to observe the diffuse infrared emission and survey about 10% of the sky at

J.M. Chamberlain and R.E. Miles (eds.), New Directions in Terahertz Technology, 237–244.

wavelengths from 1 to 1000 microns. ESA's Infrared Space Observatory (ISO) is the most sophistocated of all these missions with a near/mid IR camera, a photometer and two spectrometers covering the 2-240 micron region. Its spatial resolution is given by the diffraction limit of the 60 cm telescope. ISO, in orbit since November 1995, produces detailed maps and spectra of selected regions, previously surveyed by IRAS. As it is the first observatory that can make detailed spectra over the entire IR wavelength range it explores a new dimension of the (F)IR universe. The NASA equivalent of ISO, SIRTF, which will use larger arrays with extended wavelength coverage, is expected to be launched in 2002.

There are two small satellites, to be launched soon (1996,1998), which will be operating in the submm range. These are SWAS and ODIN; both are dedicated to the study of star formation and interstellar chemistry and therefore use heterodyne receivers. Both have 1m class telescopes and their instruments are optimised for detection of water and molecular Oxygen lines in a variety of galactic sources. ODIN will also be used for chemistry studies of the Earth's atmosphere, see Carli in this volume.

There are two new major ESA missions planned for the THz range (FIR to mm). These are COBRAS/SAMBA and FIRST. The first mission is an integrated programme of the Cosmic Background Radiation Anisotropy Satellite (COBRAS) and the Satellite for Measurementof Background Anisotropies (SAMBA). With its 1.5m telescope, it will make a full sky survey in 9 bands between 900 and 30 Ghz, with an angular resolution of about 9 arcminutes and with a very high precision which is needed for the assessment of anisotropy of the cosmic background.

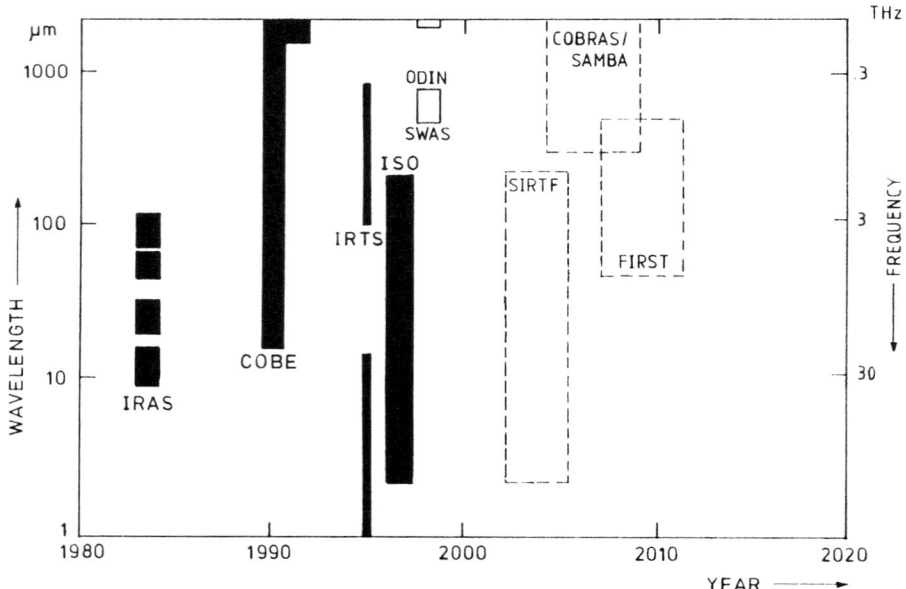

Figure 1. Summary of (F)IR/Submm/THz missions in the period 1980-2020 and their wavelength coverage. Missions in black have been carried out or are in operation, dotted missions are planned, SWAS and Odin are under construction.

FIRST, with a 3m telescope, will have an instrument complement consisting of heterodyne spectrometers, photometers and Fabry-Perot spectrometers, covering the 0.5-4 THz range (75-600 micron range). It will allow very sensitive studies of condensations in molecular clouds, signatures of dust cocoons of protostars in their very early stages of formation. The most important and strongest FIR cooling lines in the heavily obscured star-formation regions around nuclei of galaxies can also be well studied in the early, distant universe. The heterodyne receivers will allow very sensitive and very high resolution spectroscopy of important light molecules that have not been detected in interstellar space. The two missions, COBRAS/SAMBA and FIRST, to be launched in 2004 and 2007 respectively, are very complementary with respect to the scientific objectives. More details can be found in the Phase A reports on FIRST (D/SCI(93)6) and on COBRAS/SAMBA (D/SCI(96)3), which are written by two teams of European scientists [1, 2].

With the rapid progress in the development of technology for the THz range one can expect from the above-mentioned missions major breakthroughs in several scientific areas in astronomy which cannot be accomplished from high-altitude observatories or from balloon-borne (PRONAOS) and air-borne (SOFIA) platforms.

2. THz Radiation as Diagnostic Tool

The THz band is the region for continuum and line emission from cool objects and diffuse media such as circumstellar and interstellar gas and dust. Black bodies with temperatures of ten to a few hundred Kelvin emit here, and gases have their brightest ionic, atomic and molecular lines in this band.

The thermal emission from dust grains is the most common origin of the continuum. It arises from absorption by the dust of uv, visible and near-ir photons which are then re-emitted in the FIR. It requires only a small amount of dust to absorb sufficiently the short-wavelength photons. Thus even galaxies and Quasi-Stellar Objects (QSO's) emit a substantial or dominant fraction of their energy in the (F)IR. Also other radiation-emitting processes like thermal bremstrahlung, synchrotron radiation and inverse Compton scattering can be quite intensive in compact and dense sources such as Active Galactic Nuclei (AGN) and QSO's. The observed spectral energy distribution of the continuum tells us about the radiation emission mechanism and the temperature distribution and is also a measure of the mass of the emitting dust. So, continuum measurements supply us with unique information about regions with dust concentrations such as in star formation, in proto-planetary objects and in circumstellar disks and also in the central regions of AGN's and QSO's.

The THz band is also very rich in atomic, ionic and molecular lines. It contains fine-structure emission lines of abundant elements that provide efficient cooling in neutral (CI, OI, SiI) and ionised (CII, OIII, NII) gas. It also covers the ground-state rotational lines of molecules containing one heavy element and one or more hydrogen atoms (hydrides, e.g. OH, CH, NH, HCl, H2O,) which are chemically most important in the evolution and composition of the interstellar medium. The higher rotational lines of the heavy molecules (CO, HCN, CS, CN, H2CO, HCO) are also located in this band and observations will provide information on the physical conditions of the gas. In some cases, e.g. for CO and HCN, the emitted energy in the rotation lines represent a large fraction of the radiation cooling of the object. Also, the lowest rotational lines of the deuterated molecular hydrogen (HD) can be found here as well, and the detection of this species will contribute to determine the cosmologically important deuterium abundance. There are also hydrogen recombination lines in this band which are probes of the emission measure of the ionised gas and the flux of the ionising photons.

In sources with large redshifts, several important near- and mid-infrared lines of interesting ions (SiII, SIII, NIII) will be shifted into the FIR band, as might also be the case for the vibrational transitions of very large molecules such as polycyclic aromatic hydrocarbons(PAH's).

Detailed observations and analysis of the relative and absolute intensities of all these lines provides accurate determination of the physical conditions, the processes that play a role in the energy balance and the chemical evolution and composition of the gas component of circumstellar and interstellar regions. In this fashion, line emission will provide unique templates for the various excitation processes involved, and once these have been established, they can be used in further studies.

3. Astrophysical Questions to be Solved with THz Observations

The THz frequency band is primarily sensitive to cool and cold matter which is a substantial fraction of the observable Universe. This includes embedded protostellar and protoplanetary condensations, dense interstellar clouds, the diffuse interstellar dust, circumstellar shells and nuclear regions of active galaxies. However, Solar System objects such as planets, comets, asteroids and Kuiper Belt objects are part of the cool Universe. It is therefore not surprising that THz observations will be able to address succesfully fundamental questions in almost all areas in modern astronomy. However there are two main areas where these observations are essential to make progress:

(a). The formation and evolution of stars, and of disks and planets around protostars.

(b). The evolution of the very early Universe; how galaxies were formed and evolve.

3.1. STELLAR AND PLANETARY EVOLUTION

Understanding the formation of star and planetary systems is one of the most important questions in modern astrophysics. In the present theoretical and computational models it starts with the gravitational collapse of a dense molecular core and the formation of a star in the center of the surrounding infalling material within the parent molecular cloud. The early collapse phase is expected to occur with gas and dust temperatures in the 10-25 K range. Thus, young protostellar objects are very cold, heavily obscured and low luminosity sources which will radiate most of their energy in the mm and submm range. Detection will require very sensitive continuum imaging capabilities. During the earliest phases of star formation the chemical composition is evolving as well. This will have a profound effect on the energy balance of the collapsing cloud core which depends on the presence of the main cooling lines. The chemistry in the quiescent molecular cloud is dominated by ion-molecule reactions; in the collapse phase the molecules condense onto grains and the moleclar abundances can be modified by grain surface-reactions. After the formation, the young star heats up its environment and molecules are being released, back into the gasphase, most likely in an order depending on their sublimation temperature.

An example of the different chemistry for different stages of evolution is given by Helmich *et al.* [3]. They present spectra in the 242 GHz range showing the differences in molecular abundance for three regions in the so-called W3 HII/molecular cloud region. Towards W3-IRS4 a quiescent chemistry with simple molecules was found. Towards W3-IRS5 there is an indication that large amounts of molecular material are still depleted onto grains, and towards W3(H2O) a large abundance of methanol and formaldehyde seem to have been evaporated recently from the interstellar dust grains and a complex organic chemistry is taking place. W3-IRS5 seems to be the youngest in evolution, W3-IRS4 the oldest and W3(H2O) is somewhere in between.

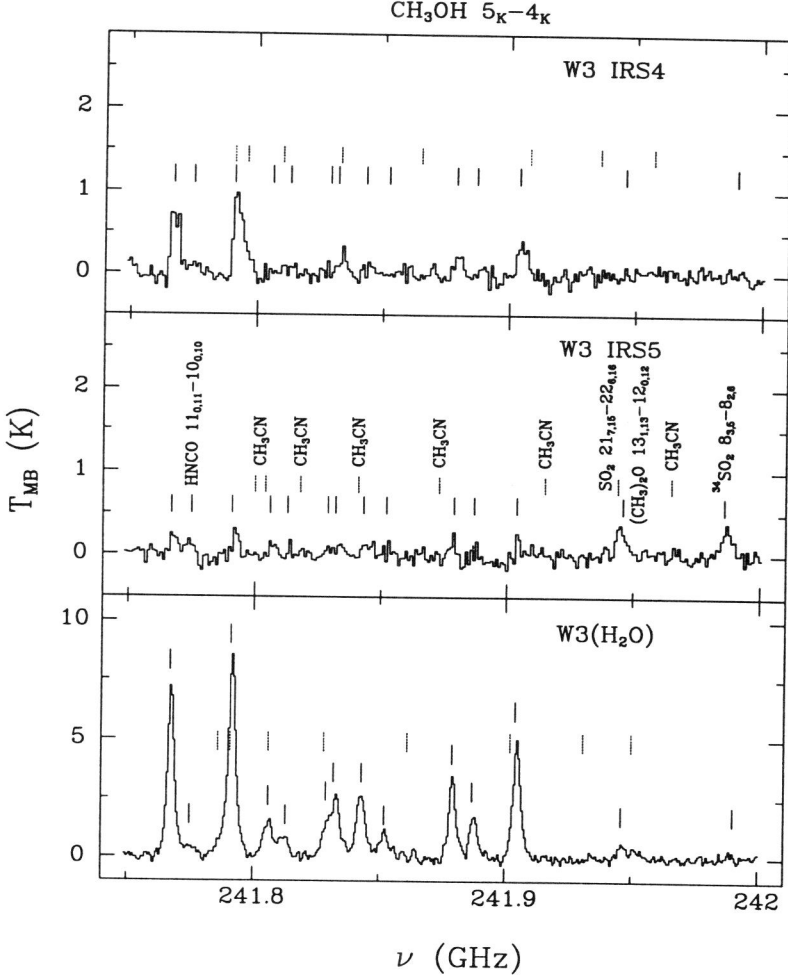

Figure 2. An example of 242 GHz spectra towards three IR sources in the W3 cloud representing different phases of evolution of gas and dust in high-mass star-forming regions.

Line emission observations with heterodyne spectrometers provide unique diagnostic information on the physical and chemical status of starforming regions. The high resolution of the spectrometers gives velocity information to a few tenths of km/sec. The variety of available transitions of molecules, atoms and ions allows to probe a wide range of gas densities, density radial profiles, excitation conditions and molecular abundances.

3.2. EVOLUTION OF EARLY UNIVERSE

With the accurate measurement of the spectrum of the cosmic microwave background (CMB) and the detection of the existence of temperature irregularities in the CMB by the satellite COBE [4] an exciting phase for observational cosmology was started. These temperature irregularities (usually named anisotropies) were imprinted on the CMB by primordial perturbations within 10^{-35} s of the Big Bang. Observations of CMB anisotropies provide us with a means of probing the ultra-high energy conditions in the very early Universe. Although the first detection of CMB anisotropy was an exciting discovery, the low spatial resolution did not allow definitive answers to be given to a number of fundamental questions, of which the origin of the primordial irregularities required to form galaxies and other structures in the Universe is one of the most intriguing. Detailed computations of the characteristics of CMB anisotropies predict that most information on the CMB will become available with observations at an angular scale of the order of 10 arcminutes. Therefore the spatial resolution to be provided by COBRAS/SAMBA is set at this level or better. The sensitivity limit for the 10 arcminute pixels is to be given by astrophysical background sources. An important issue in the determination of the anisotropy is the discrimination of the detected microwave signal from other various astrophysical foreground sources. These are weaker at 100 GHz than the CMB, and with the measurement of the spectral profile to about 1THz, where these sources are much stronger, a proper separation can be achieved. The measurements with COBRAS/SAMBA, carried out with the satellite in a spinning mode, will therefore also provide an all-sky survey at FIR, Submm and mm wavelengths of galactic and extra-galactic objects, very similar to what the IRAS survey did in the (F)IR range.

As galaxies in the early Universe seem to have a larger fraction of interstellar matter than present (nearby) galaxies, one can expect these objects to be strong FIR and Submm sources. This makes a statistical study of numerous, luminous starburst galaxies at large distances a feasible project with the bolometer array of FIRST.

4. Instrument Requirements

From the above scientific cases one can simply extract the requirements for the scientific instrumentation for FIRST and COBRAS/SAMBA, i.e. the composition of the instrument package and the specifications of the various components.

One can start with a general statement that, in all conditions, sensitivity has the top priority. The faintness of the objects, the relatively small collecting areas of the space-borne telescopes and the short lifetime of the satellites, all point into the direction of using state-of-the-art detectors and mixers.

This requirement also plays a role in the selection of the orbit of the satellite. ISO experience has shown that a daily passing through the particle belts has decreased the sensitivity of the Ge and Si photoconductors a factor two to four. This number might increase when larger detectors, necessary for longer wavelengths, are being used. An orbit like the one used for ISO is therefore not very suitable for astronomical satellites and also not for FIRST.

For FIRST's instrument complement one would like to have sensitive broadband photometers over its entire operating range. Medium resolution spectrometers are needed from 80-300 microns, for observations of redshifted lines. In both instruments one would like to have large detector arrays to map in detail, and in frequency, large areas on the sky with long integration times.

For the heterodyne instruments there are, besides sensitivity, basically three further, major characteristics that play an important role: operating frequency, instantanous bandwidth, and resolution. Table 1 gives an overview of the requirements for two types of objects: cold clouds in the Galaxy and nuclei of galaxies. In the cold cloud case the velocity resolution needed is extremely high (0.1 km/sec) and can be provided only by heterodyne instrumentation. For extragalactic observations the large instantaneous bandwidth is given by the wide range of velocity components that are present in circumnuclear regions and accretion disks. The limited spatial resolution ensures that all velocity components are within one beam.

With the present technological development it is very likely that sensitive SIS receivers up to 1.2 THz can be built for satellite applications and extension to 3 THz, if hot-electron bolometer mixers become available, may also be possible (see articles by Beaudin and Kollberg this volume) The main limitation at 3 THz may come from availability of tunable Local Oscillators. It is doubtful whether Gunn oscillators with a multiplier chain can provide sufficient power at 3 THz over a 5% tuning range. Another area that requires more development is that of backend spectrometers. Considerable study is needed to cover the large bandwidths for extragalactic observations while staying within the power-, mass- and volume budget of the satellite. It is very likely that both acousto-optical and digital spectrometers will become part of the instrumentation package for FIRST. SWAS's spectrometer consists of an acousto-optical spectrometer; ODIN is using both technologies.

Because of the frequency limitations of the heterodyne receivers, FIR spectrometers will be needed to cover the entire FIR range in order to intercept all redshifted lines from distant galaxies. Here the developments for airborne astronomy can be used for FIRST. The remaining concern here is again about the detector behaviour under particle radiation. It is for that reason that the development of Blocked Impurity Band (BIB) detectors for the FIR region will be very beneficial for these space programmes.

For COBRAS/SAMBA radio techniques will also be applied; the present plan foresees 4 HEMT radio receivers, operating from 30-125 GHz, in 4 bands. Above 125 GHz, bolometer arrays will be used, in 5 bands. The spinning satellite is expected to achieve a stability adequate to measure a temperature anisotropy of the order of one part in a million.

Both missions are planning to use similar bolometer arrays. The required detector Noise- Equivalent Power (NEP) is of the order of 10^{-17}. The development of the bolometer arrays, in particular the hot-electron bolometer design by Nahum and Richards [5], is well advanced and the above quoted NEP values seem to be within reach.

TABLE 1. Bandwidth and resolution requirements for heterodyne receivers in the THz range.

Specification	Galactic Sources				Extra Galactic Sources			
	Frequency	400 GHz	1500 GHz	2000 GHz	Frequency	400 GHz	1500 GHz	2000 GHz
Instantaneous bandwidth	100 km/s	130 MHz	500 MHz	650 MHz	1000 km/s	1.3 GHz	5 GHz	6.5 GHz
Spectral Resolution	0.1 km/s	130 KHz	500 KHz	650 KHz	10 km/s	13 MHz	50 MHz	65 MHz
Number of channels		1000	1000	1000		100	100	100

Both FIRST and COBRAS/SAMBA missions use ambient temperature telescopes. With adequate solar shields the operating temperatures will go down to about 100K. Both satellites might use mechanical closed cycle cryo-coolers although for FIRST the cryostat option is also being studied. In the cryostat case the mission will be limited to the hold time of the liquid He. The expected lifetime for the mechanical coolers is 3 years or more.

5. Conclusion

The development of THz instrumentation for space astronomy is well under way. Although some technologies are still not qualified for space, which will still require a considerable effort, there are no major obstacles for the planned THz space missions. These missions will enable astronomers to make enormous progress in a number of very fundamental questions in astrophyics.

References

1. FIRST Phase A Report (1993), European Space Agency Report Number D/SCI (93)6.
2. COBRAS/SAMBA Phase A Report (1996), European Space Agency Report Number D/SCI (96)3.
3. Helmich, F.P., Jansen, D.J., de Graauw, Th., Groesbeck, T.D., and van Dishoeck, E.F. (1994) Physical and Chemical Variations within the W3 Star-Forming Region - I : SO_2, CH_3OH and H_2CO, Astron. Astrophys. **283**, 626-634.
4. Meinhold, P., Clapp, A., Devlin, M., Fischer, M., Gunderson, J., Holmes, W., Lange, A., Lubin, P./, Richards, P., and Smoot, G. (1993) Measurements of the Cosmic Background Radiation at 0.5 Scale near the Star Mu Pegasi, Astrophysical Journal, L1-L4.
5. Nahum, M. and Richards, P.L. (1991) Design Analysis of a Novel Low-Termperature Bolometer, IEEE Trans. Magn.**27**, 2484-2487.

SUBMILLEMETRE MEASUREMENTS FROM SATELLITES

B. CARLI
Istituto di Ricerca sulle Onde Elettromagnetiche,
via Panciatichi, 64 - 50127 Firenze, ITALY.

Abstract

Submillimetre heterodyne spectroscopy finds an important application in the study of the Earth's atmosphere from space-borne platforms. The scientific problems that are present in this field and the possibilities offered by heterodyne techniques are summarised providing a review and a short description of deployed and planned instruments.

1. Introduction

Measurements of the composition of the Earth's atmosphere are necessary for the study of atmospheric chemistry and for the development of models that can predict its possible evolution. Spectroscopy provides an important tool for these measurements.

The spectrum of the atmosphere can either be measured in absorption (observing the radiation from an external source) or in emission (observing the atmospheric black-body radiance) and contains features which are characteristic of its molecular constituents. The intensity of the observed features provides quantitative information on the constituent amount. Spectroscopy is a passive technique that does not modify the sample under observation, as well as a remote sensing technique suitable for satellite observations.

The early observations made in the millimetre/submillimetre regions are guiding the development of further observations, while recent extension of heterodyne techniques to the submillimetre and terahertz spectral regions make accessible to this technique a larger fraction of the rotational spectra increasing the number of possible applications in atmospheric physics. Because of the present central role of submillimetre observations (300 - 1000 GHz), this spectral region will be the main subject of this paper.

The scientific studies that motivate the observations of the Earth's atmosphere are summarised in Section 2. An overview of the spectrum of the Earth's atmosphere in the submillimetre regions is presented in Section 3. The advantages of submillimetre observations are reviewed in Section 4. The different geometries of observation and

J.M. Chamberlain and R.E. Miles (eds.), New Directions in Terahertz Technology, 245–255.

their properties are recalled in Section 5 and a overview of existing and planned space observations is given in Section 6.

In the following discussion and figures the frequency unit will usually be GHz and THz, but occasionally when referring to far infrared photometric observations also wavenumbers (cm^{-1}) will also be used. 1 Thz is equal to about 30 cm^{-1}.

2. The scientific problem

The Earth's atmosphere is a complicated system in which numerous chemical, physical and radiative process occur. It is characterised by a large variability with latitude, altitude and season; the weather being the most evident manifestation of this variability. The complexity makes, however, its average local behaviour (the climate) rather constant and life on the Earth has adjusted to and depends on the climate.

The atmosphere is directly affected by the increased capacity of human activity modifying the surrounding matter. The resulting changes in climate that have occurred up to now are difficult to quantify and in any case are considered to be small and slow, but projections to the future indicate the risk of large and rapid changes.

The three main species that characterise the status of the atmosphere are carbon dioxide, ozone and water vapour.

Carbon dioxide is responsible of a greenhouse effect in the atmosphere. Its presence in the atmosphere reduces the radiative cooling of the Earth's surface causing an increase of its temperature. Carbon dioxide is produced in all burning processes and its concentration in the atmosphere has been constantly increasing in the last century.

Ozone controls the absorption of solar UV radiation and in so doing heats the stratosphere and prevents dangerous radiation from reaching the ground. Ozone can be depleted by four catalytic cycles involving the radicals OH, ClO, NO and BrO (hydrogen cycle, chlorine cycle, nitrogen cycle, and bromine cycle) [1]. While the first radical is present in the stratosphere in natural concentration, the other three can be increased by human activity. The molecules which lead to the formation of these radicals are called "source molecules". The molecules that temporarily subtract these radicals from the depletion cycle are called "reservoir molecules". The molecules that can permanently subtract these radicals from the stratosphere are called "sink molecules". The radical responsible of each cycle together with its sources, sinks and reservoirs form a family of species that determine the efficiency of the cycle.

Water vapour affects the radiative exchange of the Earth; in gas form this causes a warming (by way of the greenhouse effect) and in the form of clouds, causes a cooling by reflecting the solar radiation in the visible. Water vapour concentration is not varied by human activity, but climate changes directly affect the water cycle so that water vapour is part of a feedback mechanism. Existing uncertainties make it difficult to decide whether this feedback is either negative (adding stability to the system) or positive, increasing the risk of rapid climatic changes.

The local composition of the atmosphere and the global distribution of its constituents are the measurements necessary for characterisation and modelling of the processes and of the dynamics that determine the status and the evolution of the atmospheric system.

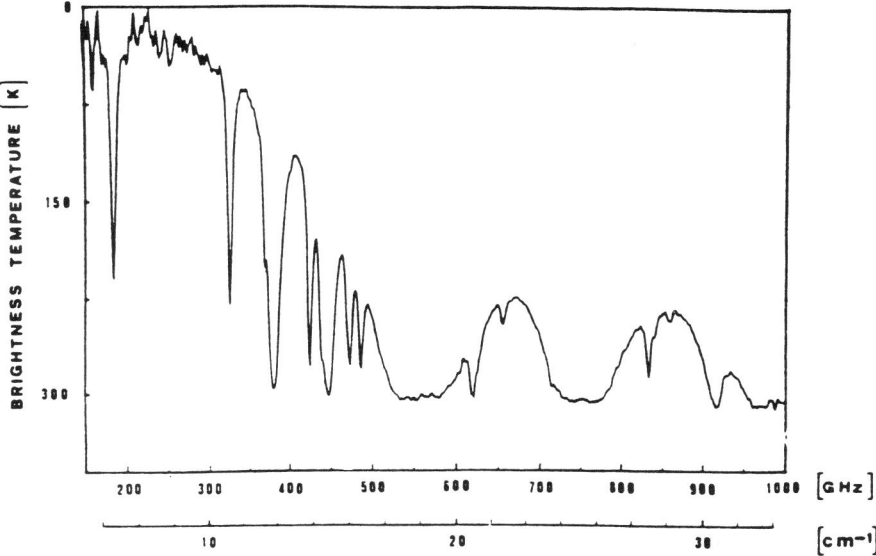

Figure 1. The brightness temperature of the atmosphere measured from the Jungfraujoch observatory. The 0-300 k temperature scale corresponds approximately to the 1 - 0 transparency scale.

3. The atmospheric spectrum in the submillimetre region

In the submillimetre regions the spectrum of the Earth's atmosphere has two characteristics. The first is the presence of signals from highly-concentrated species, such as water vapour, with a strong rotational spectrum that can, in the case of high density (at low altitudes) and of long optical paths, make the atmosphere opaque. The second is that the widths of the atmospheric lines are usually smaller than the spacing between the lines themselves so that, in the case of transparent atmosphere, its spectrum is characterised by individual lines.

An example of the transparency of the atmosphere from the ground is given in Figure 1, which shows the emission spectrum of the atmosphere at a zenith angle of 45° in the 6 - 32 cm^{-1} region measured from a mountain site with a resolution of 0.1 cm^{-1}. On an average, at high frequencies the brightness temperature is high, indicating an opaque atmosphere, while at low frequencies the brightness temperature is low, indicating a transparent atmosphere. The transparency further decreases at frequencies higher than those shown in the figure. Most of the observed brightness, with the sole exception of a few small features, is due to water. The localised spectral intervals in which some transparency is obtained (atmospheric "windows") have a width and height that depend on the water-vapour content, and vary with altitude and meteorological conditions.

In order to have some useful transparency at sea level we must use the millimetre region. Figure 2 shows the calculated atmospheric brightness-temperature in the millimetre region at the zenith from sea level. Spectrum (a), which is obtained by including water vapour, oxygen and ozone in the computations, is dominated by the water lines at about 22 and 183 GHz, and by the oxygen lines at 60 and 118 GHz. Spectrum (b) shows the result obtained when water is not included in the computations. When comparing Figure 2 with Figure 1, note that different scales of brightness temperature are used and that the line at 183 GHz appears in both figures.

Figure 3 shows an example of the stratospheric spectrum observed in the 45 - 65 cm^{-1} spectral region from a 38-km altitude in a limb sounding observation (see Section 5) that penetrates the atmosphere down to a tangent altitude of about 32 km. The water vapour transitions, the ozone Q-type branches, and the molecular oxygen transitions are the main features of the spectrum and are indicated in the figure.

Because of their dominant effect, the three molecular species H_2O, O_2, and O_3 can be considered to be the "main spectroscopic constituents" of the millimetre/submillimetre stratospheric spectrum, in contrast to the other molecular species which can be considered as "minor spectroscopic constituents". This classification is based only on the consistency of the spectral features, and leaves out of consideration the actual quantity of each constituent in the atmosphere.

Sounding of the atmosphere in the submillimetre region becomes possible only at high altitudes, preferably above 15 km in the stratosphere where, because of the reduced water concentration, many windows can be observed also for very long optical paths.

The minor spectroscopic constituents that can be observed in this spectral region increase in number and intensity as we move to higher frequencies, with a peak in the region between 2 and 3 THz and a slow decrease up to 6 THz and above. Minor spectroscopic constituents that have been observed include:

- isotopes of the three main spectroscopic constituents ($O^{16}O^{18}O^{16}$, $O^{18}O^{16}O^{16}$, $O^{17}O^{16}O^{16}$, HDO, $H_2^{18}O$, $H_2^{17}O$, $^{16}O^{18}O$)
- species of the four family involved in the catalytic ozone destruction:
  - ♦ OH , H_2O, and H_2O_2, of the hydrogen family,
  - ♦ ClO, HOCl, CH_3Cl, and HCl, of the chlorine family,
  - ♦ NO, N_2O, HNO_3, of the nitrogen family,
  - ♦ HOBr, and HBr, of the bromine family.
- and other species such as HF, CO, HCN, atomic oxygen, and SO_2.

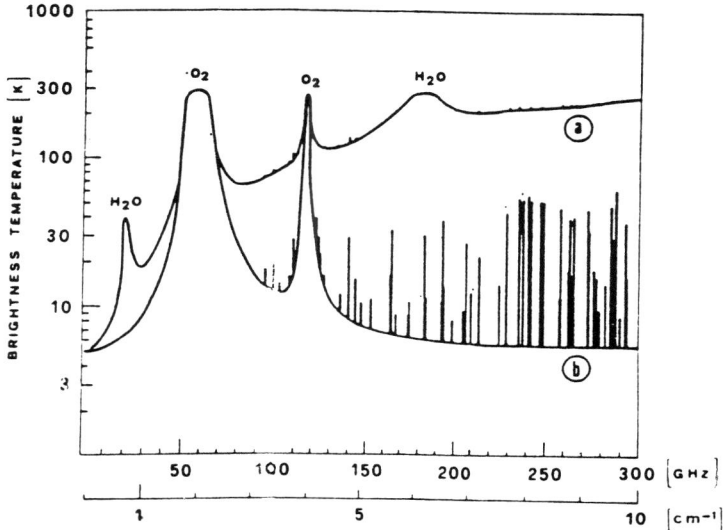

Figure 2. Calculated brightness temperature for a zenith observation from sea level. Water vapour, ozone and molecular oxygen are considered in the spectrum (a), and only ozone and molecular oxygen are considered in the spectrum (b).

Furthermore, low-lying vibrationally-excited states produce in some cases detectable pure rotational transitions.

Many transitions contribute to the observed spectra, and the resolution of the spectrometer is important for limiting the effect of spectral blending. With a spectral resolution of 0.01 - 0.0025 cm^{-1} (300 - 75 MHz), as it is typically attained with radiometric measurements [2,3], the individual transitions of the different atmospheric constituents are usually resolved. An example is given in Figure 4, which shows an expansion in the 61 - 63 cm^{-1} spectral interval of the spectrum of Figure 3 measured with a resolution of 0.0033 cm^{-1}.

The line width of the atmospheric features depends on Doppler broadening, equal to about 1 MHz at 1 THz, and on pressure broadening, proportional to pressure and equal to about 10 GHz at one atmosphere. Therefore, a resolution better than 1 MHz is necessary in order to fully resolve the atmospheric line shape at high altitudes (above 60 km) and a bandwidth greater than 10 GHz is necessary to measure the whole feature at ground level.

Present broadband knowledge of the submillimetre atmospheric spectrum results from radiometric measurements, which have been used to compile an atlas of the submillimetre atmospheric spectrum. Two sections of the atlas have been published so far [4, 5], covering in total the 7-40 cm^{-1} interval. A few narrow band observations have been performed with very high resolution using heterodyne techniques.

250

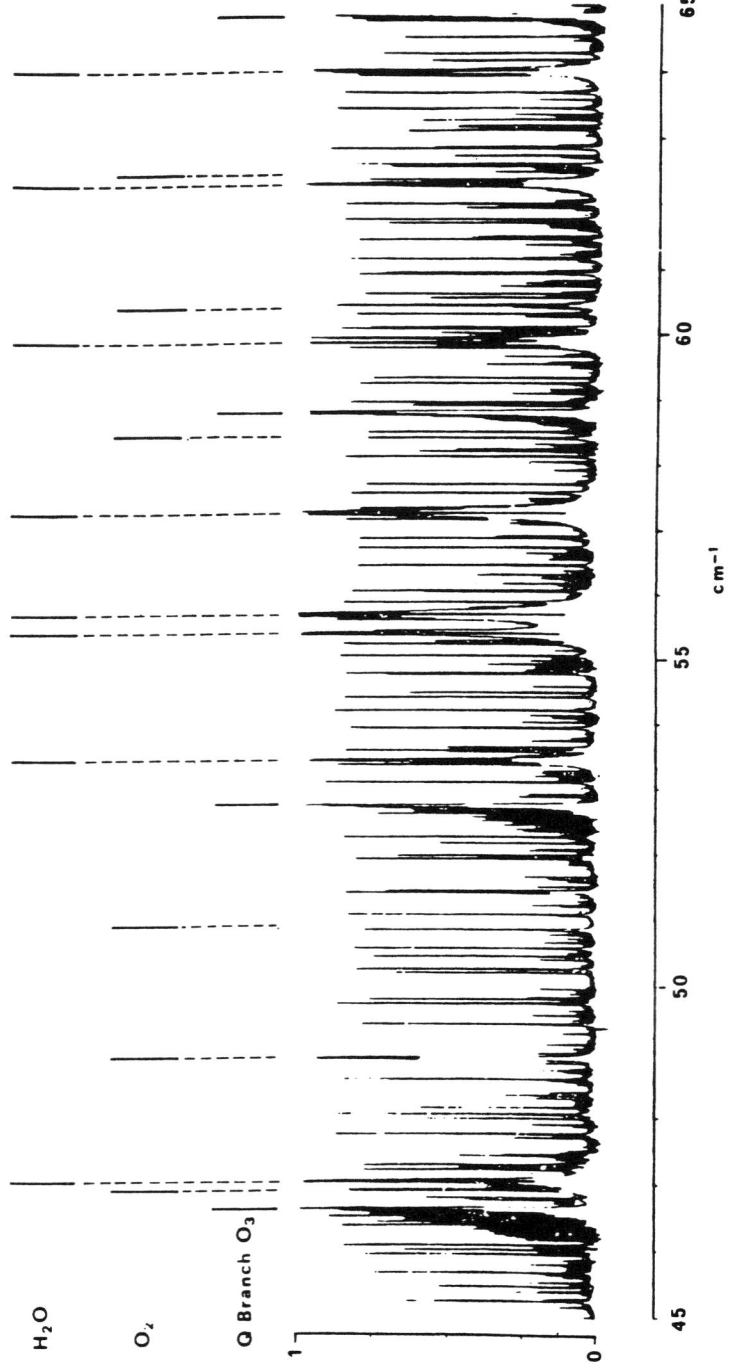

Figure 3. A sample of the atmospheric emission spectrum in the 45-65 cm⁻¹ spectral region. The main transitions of water vapour, molecular oxygen and ozone Q-type branches are indicated.

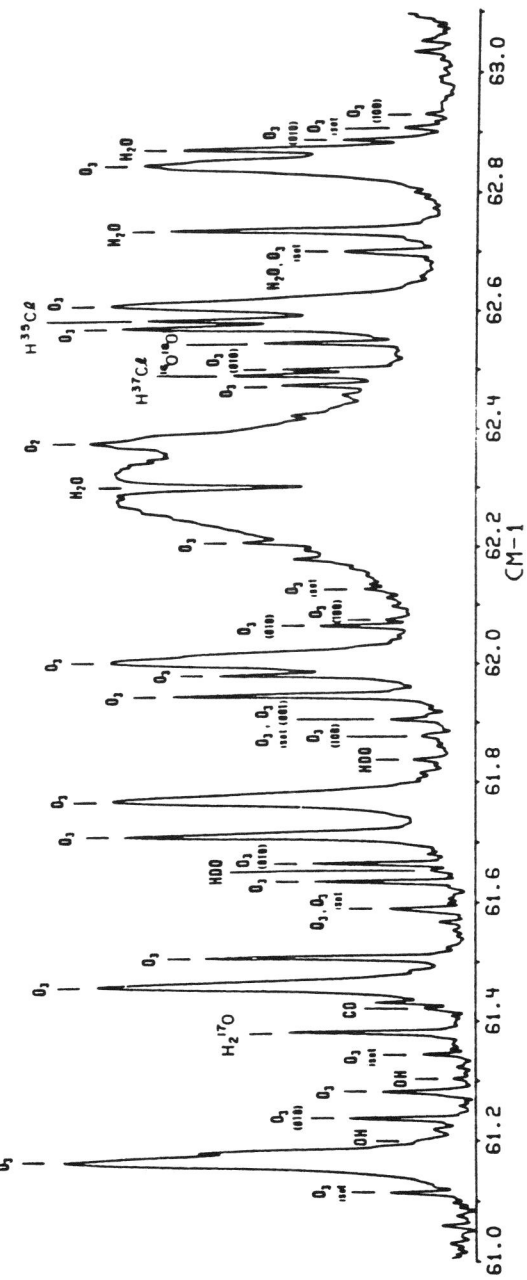

Figure 4. An expanded section of the spectrum of Figure3. The observed features are marked with the name of the molecular species responsible for the transition. The label *isot* indicates ozone with one atom of ^{18}O. For transitions due to vibrationally excited molecules the vibrational quantum numbers are indicated in parentheses.

4. Advantages of submillimetre observations

The observation the Earth's atmosphere in the submillimetre spectral region has several advantages with respect to investigations of atmospheric constituents performed in other spectral regions. The main advantages are:

- The atmosphere can be observed in emission with no constraint on either the direction, or the time of the observation;
- The atmospheric emission is in the Rayleigh-Jeans region of the black-body distribution and depends in a simple way on temperature;
- Pure rotational spectra of the ground state are the most common transitions in the submillimetre region. These transitions are well understood and less affected by systematic errors;
- The scattering of aerosol particles generated by either volcanic eruptions or polar stratospheric clouds is negligible;
- At low frequencies (below 300 GHz) the atmosphere is transparent also at low altitudes and it is possible to monitor the troposphere;
- At high frequencies (at 1 THz and above) numerous molecular constituents have their rotational spectrum and can be measured simultaneously.

5. Geometry of observation

The Earth's atmosphere can be measured either looking at the limb (limb sounding), or in the vertical direction (vertical sounding). In turn, vertical sounding can be made looking either at the nadir or at the zenith direction.

Vertical sounding exploits the fact that the atmospheric features have a altitude- dependent line shape. The signal in the wings of the observed feature originates mainly from low altitudes, and the signal in the centre of the feature originates mainly from high altitudes. In this way the observed line shape can be related to the concentration of the emitting constituent as a function of the altitude.

In the case of nadir vertical-soundings, the atmosphere is observed from above and the measurement is limited by the small contrast that exists between atmospheric and Earth's surface temperatures. Nadir soundings have been made in the millimetre region from satellites for meteorological applications.

In the case of zenith vertical-soundings, the atmosphere is observed from below, aiming at the same type of observation as in nadir soundings. Significant differences are, however, the fact that the space background provides a better contrast, and the fact that the narrow feature which originates from the farthest atmospheric layers can be covered by the broad feature originated at the near field. Therefore, with zenith soundings only relatively weak features that are not opaque in the "near field" can be observed. Often high mountains, dry sites (e.g. Antarctica), and aircraft-borne platforms are used in order to reduce water vapour attenuation.

Limb sounding is made from a high-altitude (either balloon- or space-borne) platform. The line of sight is oriented below the horizontal direction, in such a way that the Earth's surface is not reached and only the atmospheric limb is observed. Because of the Earth's curvature, the optical path reaches a minimum distance from the Earth's surface and escapes into space: this minimum distance is called the "tangent altitude". The long optical path at tangent altitude and the larger pressure of the lower layers make the emission of the atmospheric layer at tangent altitude the largest part of the observed signal. Observations at different tangent altitudes (limb scanning) can be used the for the measurement of the atmospheric composition as a function of altitude. The spectral resolution must be sufficient to resolve the selected feature from possible interfering species, but does not need to resolve the line shape as in the case of vertical sounding. The resolution of the atmospheric line shape provides, however, additional information that can be used together with that of limb scanning for a better retrieval of the vertical distribution.

6. Planned missions

The millimetre region, for its capability of penetrating the lower atmosphere unaffected by thin clouds, has been exploited by several sensors operating from space-borne platforms for the measurement of parameters of meteorological interest. The measured parameters include the temperature profile of the atmosphere, surface temperature of the sea, water-vapour and liquid-water content of the atmosphere. A review of the principles of the spacecraft radiometer system and of the early microwave sensors used for remote sensing of the Earth from space is given by Njoku [6].

The application of millimetre/submillimetre microwave techniques to the study of the composition and chemistry of the Earth's atmosphere from a satellite has started with the shuttle flights of the ATLAS (Atmospheric Laboratory for Applications and Science, which included the Millimetre-wave Atmospheric Sounder (MAS) instrument [7]), and the UARS (Upper Atmosphere Research Satellite, which had on board the Microwave Limb Sounder (MLS) instrument [8]).

MAS and MLS are similar instruments which observe the atmosphere in limb scanning at millimetre wavelengths, and have as a primary objective the measurement of ClO.

MLS consists of three radiometers at 63, 183 and 205 GHz, which observe the atmospheric limb by means of the same antenna. A filter and a polariser perform the optical splitting of the beam into the three radiometers. Each radiometer is made of two heterodyne stages. The wide spectral ranges of the three intermediate frequencies are observed in a total of six bands centred around selected lines of molecular oxygen, ozone, chlorine monoxide, hydrogen peroxide, and water vapour. Six filter banks, each with a total bandwidth of 500 MHz covered with 15 filters, are used for the simultaneous spectral analysis of the six bands.

MAS covers the same spectral regions and observe the same species with small differences in the technical solutions.

Plans exist in several countries for the development of space heterodyne-instruments that operate in the submillimetre region. The extension of this technology to shorter wavelengths would make it possible to observe spectral intervals in which stronger lines are present and more numerous species can be observed. Species that would become detectable include HCl, HOCl, CH_3Cl, HO_2, NO, NO_2, N_2O, HNO_3, CO, and HF. With a further extension to the terahertz region also OH can be observed. Furthermore, the use of short wavelengths with reduced diffraction requires a smaller antenna for an equal field-of-view.

ODIN is a Swedish small satellite project for Astronomical and Atmospheric research, which is expected to be launched in a few years from now. A small optical spectrometer and a microwave radiometer are located on the same platform. The radiometer has five receivers: one in the millimetre region (119 GHz) and four in the submillimetre region (422 GHz, 488 GHz, 553 GHz, and 575 GHz). In each receiver the spectral distribution is measured with a 1000 channel hybrid autocorrelator which can provide either a spectral resolution of 0.1 MHz in a bandwidth of 100 MHz or a spectral resolution of 1 MHz in a bandwidth of 1 Ghz.

AMLS (Advanced Microwave Limb Sounder) is an improved version of the MLS instrument and is scheduled to fly on the NASA EOS-Chem Mission. Also in this case submillimetre receivers will be used to observe a greater number of atmospheric species and exploit stronger atmospheric features. Furthermore, consideration is given to the development of a receiver in the terahertz region (either at 2.5 THz or at 3.5 THz) for the observation of OH.

In ESA three instruments are being considered. The first, named MASTER (Millimetre-wave Acquisition for Stratosphere-Troposphere Exchange Radiometer) is a millimetre wave instrument for the measurement of water vapour, ozone and CO in the lower atmosphere and for the study of the Tropospheric-Stratospheric exchange.
The second, named SOPRANO (Sub-mm Observation of Processes in the Atmosphere Noteworthy for Ozone) is a submillimetre instrument for the measurement of a comprehensive set of species of the nitrogen and chlorine family.

The third, named PIRAMHYD (Passive Infra-Red Atmospheric Measurements of Hydroxil) is an instrument that operates in the terahertz region for the measurement of OH and the hydrogen chemistry. In this case three different techniques are being considered and compared: a Fourier transform spectrometer, a Fabry-Perot spectrometer and an heterodyne receiver.

References

1. World Meteorological Organization, Atmospheric Ozone, Global Ozone Research and Monitoring Project Report No. 16 (1985), World Meteorol. Org., Geneva.
2. Carli, B., Mencaraglia, F., and Bonetti, A., (1984). *Appl. Opt.* **23**, 2594.
3. Chance, K.V., and Traub, W.A. (1987) *J. Geophys. Res.* **92**, 3061.
4. Baldecchi, M.G., Carli, B., Mencaraglia, F., Bonetti, A., and Carlotti, M., (1984) *J. Geophys. Res.* **89**, 11689.
5. Baldecchi, M.G., Carli, B., Mencaraglia, F., Barbis, A., Bonetti, A., and Carlotti, M., (1988) *J. Geophys. Res.* **93**, 5303.
6. Njoku, E.G. (1982) *Proc. I.E.E.E.* **70**, 728.
7. Croskey, C.L., Kaempfer, N., Bevilacqua, R.M., Hartmann, G.K, Kunzi, K.F., Schwartz, P.R., Olivero, J.J., Puliafito, E., Aellig, C., Umlauft, G., Waltman, W.B., and Degenhardt, W.D. (1992) The

Millimetre Wave Atmospheric Sounder (MAS): A Shuttle-Based Remote Sensing Experiment, *IEEE Trans. Microwave Theory and Techniques*, **40**, 1090-1100.

8.   Waters, J.W. *Atmos. Res.* 23, 391 (1989).

SPACECRAFT APPLICATIONS OF TERAHERTZ TECHNOLOGY

W.J. HALL
Matra Marconi Space,
FPC 320, PO Box 16, Filton, Bristol, BS12 7YB, England.

1. Introduction

The terahertz band offers exciting opportunities [1,2,3] for the remote sensing from space of atmospheric constituents with great importance for the planning of industrial and commercial activities. To exploit these opportunities requires the development of new sensor devices combining robustness with decreased dimensions and many developments of integrated receiver and antenna components are now in progress.

There is a wide range of studies and technology developments aimed at remote sensing applications in the millimetre, submillimetre and terahertz frequency bands. This paper discusses critical technology needs for remote sensing applications with the emphasis on meeting both the performance and spacecraft interface requirements.

The range of front end critical technologies includes components from antenna to intermediate frequency output. From the instrument engineering viewpoint data analysis equipments in the instrument back end are also critical items. The development of terahertz technology is based on a heritage in millimetric instruments. This is described to explain the common and novel features of terahertz applications.

Progression into the terahertz bands gives new emphasis to integrated technologies and the developments now in progress and foreseen are important for their application to spacecraft instruments being planned for uses such as the sensing of atmospheric processes relevant to the ozone cycle. Typical recent studies of submillimetric and terahertz instruments are described in the references and form the background for the technology needs described in this paper. The discussion is illustrated by examples from the literature of technology developments from Matra Marconi and other organisations.

2. Terahertz Instruments

The spacecraft applications for terahertz technology can be considered in two classes. Instruments for the study of atmospheric chemistry are now being studied in the millimetre wave and submillimetre wave bands up to 950GHz [4,5,6,7] and other instruments are foreseen in bands up to 3THz for which technology developments are already started. The submillimetre and terahertz regions are also of interest for spaceborne astronomy with instruments such as ODIN and SWAS in development to

J.M. Chamberlain and R.E. Miles (eds.), New Directions in Terahertz Technology, 257–267.
© 1997 *Kluwer Academic Publishers. Printed in the Netherlands.*

observe interstellar species such as oxygen and water which are naturally blocked by the atmosphere. A major programme of the European Space Agency comprises the FIRST mission [2] whch accesses the band up to 1.5THz.

Whilst covering widely separate fields these instruments have many common technology needs since the basic requirement is for a robust, sensitive, front end technology fed by a high precision antenna and the subsequent processing of the data to yield spectral information. To illustrate the current technology developments we will use as an example the technologies needed for submillimetric limbsounding instruments. These technologies are still in development for the submillimetric bands but the lines of future extension into the terahertz region can be discerned and the constraints posed by the spacecraft environment and performance requirement are clear.

3. Millimetric Instrument: AMSU-B

The AMSU-B instrument [8] is a nadir pointing humidity sounder supplied by Matra Marconi Space to the UK Meteorological Office for flight on the NOAA satellites. Deliveries of three flight models are complete. The instrument is a five channel radiometer using the bands 89,150, and 183GHz with spatial beamwidth of 1.1degrees. The key technologies used are a Quasi Optical demultiplexer [9] using drilled plate filters (DCP) as the diplexing elements and corrugated horn feeds. The two highest channel mixers are subharmonically pumped from Gunn diode local oscillators. They are constructed using planar double diode chips soldered to quartz substrates with signal and local oscillator coupling in rectangular waveguide. The back end of the instrument is relatively simple comprising broadband integration of the five channels. (The 183GHz channel is split into three by an intermediate frequency diplexer.) Scanning of the beam is performed by rotation about the feed axis of the primary mirror, which has a diameter of ~200mm and is a Beryllium "egg crate" structure. The calibration of the instrument is performed using an absorbing termination made of pyramidal tines which is recognisably a microwave load. The instrument and its quasi optical diplexer are shown in Figures 1 and 2. The next generation of humidity sounder, now in manufacture at Matra Marconi, is represented by the MHS instrument and also uses quasi optical demultiplexing techniques.

The instrument shows that at these frequencies construction of critical front end items is possible using relatively conventional techniques such as machining (DCP's,mirrors, mixer housing) , soldered assembly (Mixers) and electroforming (Corrugated horns). For configuration reasons some of these horns incorporate electroformed twists in the throat. Corrugated horns are recognisable in form from their microwave equivalents, although it is interesting [9] that is this apparently most conventional component which begins to show the use of chemical processing techniques such as electroforming. Other components (Quasi Optics and DCP's) are more specific to the millimetric bands but still manufactured using conventional machining. The selection of components is driven by the need to meet a strict performance specification including 95% main beam efficiency.

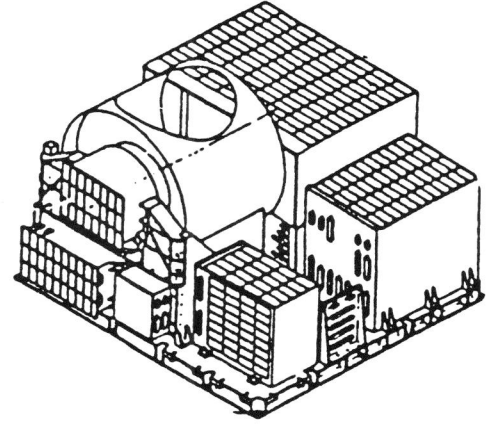

Figure 1. AMSU-B Radiometer

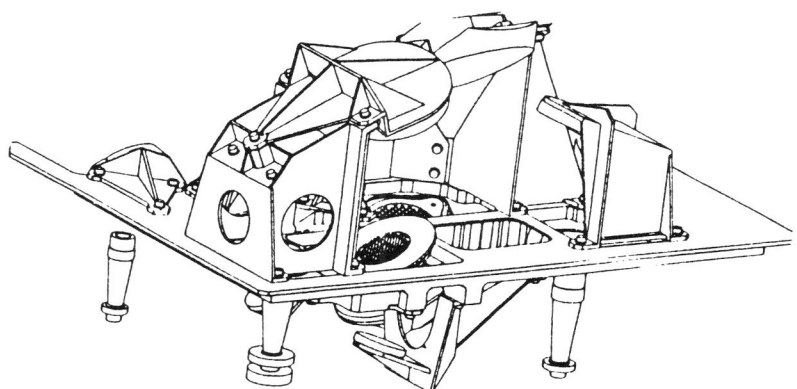

Three channel demultiplexer, 89GHz, 150GHz,183GHz Bands.
Typically Loss = 1dB, Beam efficiency >95%
1 EM 3FM delivered.

Figure 2. AMSU Quasi Optical Diplexer

A sketch of a typical subharmonic mixer configuration using a planar double diode chip is shown as Figure 3 and the measured performance of the 183GHz profiled corrugated horn from AMSU-B is shown in Figure 4.

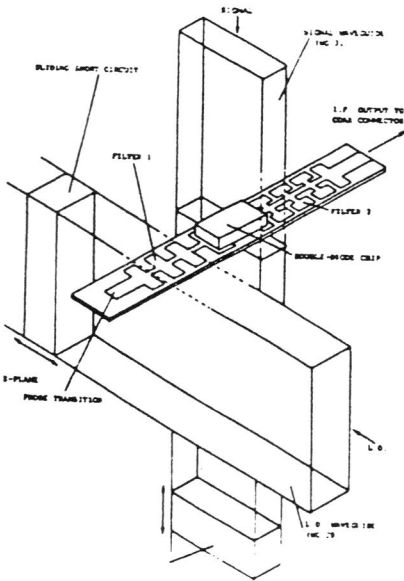

Figure 3. Typical Subharmonic Mixer

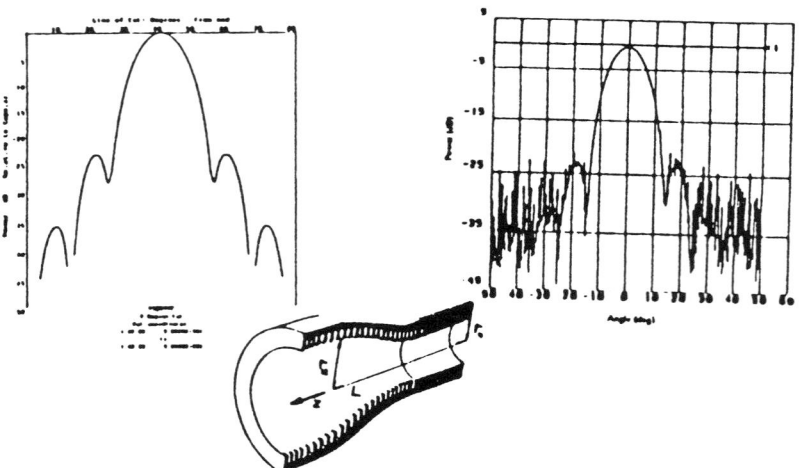

Figure 4. 183GHz Corrugated Horn.

4. Submillimetric Limb Sounder Instrument: SOPRANO

As frequency increases through the submillimetre region both new devices and new construction techniques become apparent. Two recently published studies [6,7] for the European Space Agency of the MASTER and SOPRANO instruments give good examples of the needs for technology developments in instruments based upon Schottky diode front ends. It should be noted first that whisker contacted mixer front ends with good performance are already available to about 3 THz but the prevailing trend of opinion is that integrated and planar constructions could offer important advantages of robustness and precision of manufacture [10,11,12,13,14] MASTER (~200-~400GHz) can probably be considered as definitely within the range where the front end can be constructed by separate assembly of planar diode components onto printed circuit filters so resembling existing technology as used in the MHS and AMSU-B instruments and recently demonstrated (RAL) for the AMAS 300GHz development. The 600GHz bands of SOPRANO are also accessible using this type of construction but the highest band (954GHz) shows the possible benefits of a greater integration where at least the filter circuit is integrated with the diode structure. At yet higher frequencies the advantages of further integration to include the mixer housing will become attractive.

The manufacture of mechanically drilled DCP's becomes difficult in the submillimetre band. A test piece with 0.3mm diameter holes has been made and tested at 600GHz but was considered to represent the limit for this conventional construction. Techniques for making waveguide array plates lithographically on Silicon mandrels have been described by workers at JPL so that the use of these devices, which have useful high pass characteristics can be foreseen in the THz bands. From the mm bands upwards the photolithographic etching of screens of resonant elements (Frequency Selective Surfaces - FSS) on dielectric substrates becomes a useful method of construction of quasi optical filters. The use of lithographic techniques allows precision of about 1μm and multiple screens can be used to provide sharp frequency responses. The devices are employed as band separation and image rejection filters.

Corrugated horns continue to be the most attractive feed for high performance single beam applications such as these. The methods of manufacture become more sophisticated as the size decreases but examples have been shown to 2.5THz. [15]. Other horns can give similar performance from simpler structures either for narrow bands (Potter type) or by using dielectric cores for very wide bands, although this adds a dielectric loss element to the resistive losses. Wall resistive losses in horns are concentrated in the throat region where the wave is still closely coupled to the walls.

These horns couple to waveguide mixers where there is now increasing interest in the integration first of the diode into the filter structure and then of other parts of the housing [16]. Other types of structure using integrated antenna-mixer concepts [17] have also been studied for use in the submillimetre but in general it is likely that the performance achieved will be inferior to the waveguide mixer horn combination. This is because the Open Structure type has less constrained interactions between mixer and radiator and the surrounding medium. Applications for these devices should be sought where factors such as configuration and possibly array feeds outweigh straightforward performance. When assembled as a receiver open structures will be found to need many of the elements of LO coupling, tuning and assembly which are found in the waveguide device.

In the two studies referred to subharmonic mixer receivers using Schottky multiplier chains have been proposed except for the 950GHz bands where a fundamentally pumped architecture is proposed. Eventual development using integrated construction techniques of subharmonic devices for this band would be advantageous, since the subharmonic receiver is simpler to arrange because the external local oscillator coupling elements are not needed. The presence of these elements also has the effect of reducing the receiver noise advantage of the fundamentally pumped mixer. The provision of local oscillators up to 1THz by a Schottky multiplier chain has now been demonstrated [18] and is the preferred option because power demands are low and frequency accuracy can be maintined by locking to a stable low frequency reference. The provision of Local Oscillator power for higher frequencies is a key development task, since currently available sources such as lasers and carcinotrons are bulky, expensive and demanding whilst newer device concepts (RTD's) are producing only small amounts of power [16].

Important system factors which affect the technology requirements include the sensitivity, antenna beam quality and beam configuration and bandwidth. Sensitivity can be improved by cooling of the detectors but this bears penalties in resources required. About 150K can be reached by simple passive coolers provided appropriate view factors can be obtained. 50K can be achieved by standard mechanical coolers with less restraint on configuration but at the modest penalty of about 30kg and 70W for each cooler pair.

The number of front end chains in a spectrometric instrument is not in itself a major resource requirement but the total bandwidth is. Consequently the instruments now proposed use single beams for each band and the use of multiple beam instruments is not likely without significant reduction of spectrometer resource requirements or reduction in required bandwidths. Figure 5 illustrates the proposed configuration of a typical limbsounder.

Calibration of this instrument will be performed within the quasi optic chain and the use of specular scatterers as the hot load has possible advantages, in manufacturability and characterisation compared to the absorbing tine type of load which becomes more difficult to make and whose characterisation is less easy because it scatters any reflected radiation. The submillimetre band will also see application of hot load types derived from the infrared band.

Limbsounders provide good vertical resolution and so despite the higher frequency the antenna is larger (1m-2m) than in the humidity sounders and a more significant thermal and mechanical design problem is created requiring the use of carbon fibre composites and careful compromise between stiffness under gravity, which is needed for ground testing, and thermal distortion in orbit. The testing of such electrically large antennas is currently the subject of study since new techniques combining large size with precision may be useful. [7,16,19]. The use of well known techniques such as compact ranges, phase retrieval or near field measurement may be possible but the range will need careful evaluation and design.

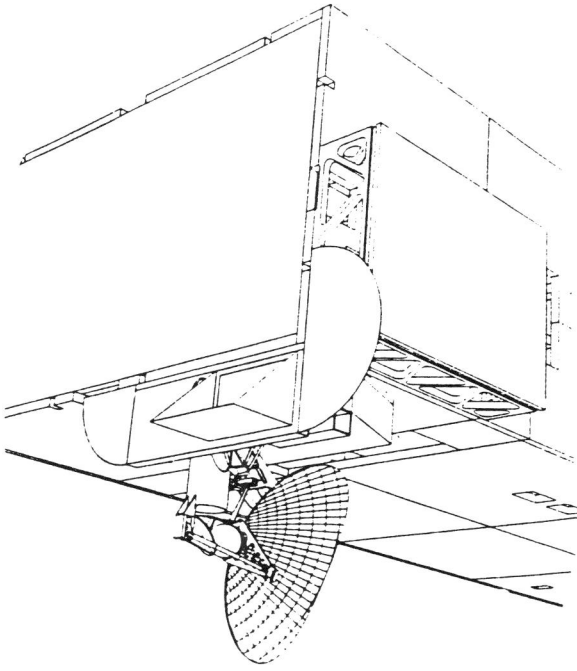

Figure 5. Limbsounding Instrument Configuration

5. Critical Technologies

Examples of the current status of some interesting technologies can be seen in the figures.

Figure 6 shows an example of the performance of a dielectric core horn [20] measured at 186GHz, which shows the ability to replicate the performance of corrugated horns. Other examples have been measured at 320GHz and made for use at 650GHz. The dielectric loaded horn is useful for applications requiring very wide bands and where simplicity of construction is of prime importance.

The corrugated horn has been demonstrated for use up to 2.5THz [15] and for applications requiring good beam quality and moderate bandwidth will be the preferred feed. For narrow band applications the Potter horn offers the prospect of simpler construction at the expense of greater sensitivity to manufacturing tolerances.

Figure 7 shows a broadband measurement of the performance of a Frequency Selective Surface. The surface is designed for use as an image rejection filter at about 300GHz where a sharp roll off will be observed and the measured characteristic is an example of the comprehensive data obtainable from a Dispersive Fourier Transform Spectrometer measurement.

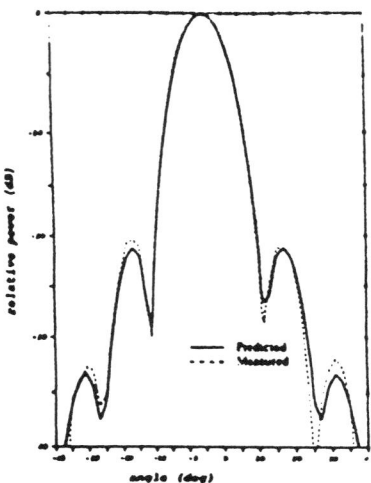

Figure 6. Performance of Dielectric Core Horn at 186GHz

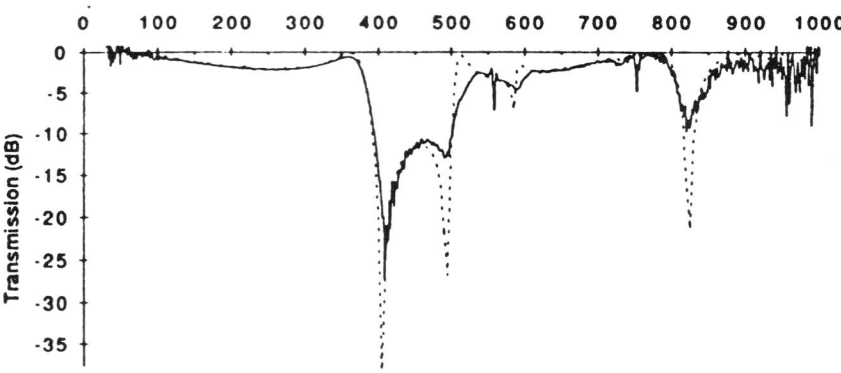

Figure 7. Broadband Measurement of Frequency Selective Performance

An adequate range of feed radiators can be seen to exist for high performance single beam applications but a key area for technology development remains in the mixer and local oscillator chain. The realisation of these components with an increased degree of integration is an important area of development. This will start from the existing planar diode technology by inreasing the scale of integration around the planar diode and is then expected to develop towards more radical concepts such as those in Figure 8 which are taken from a recent study [16].

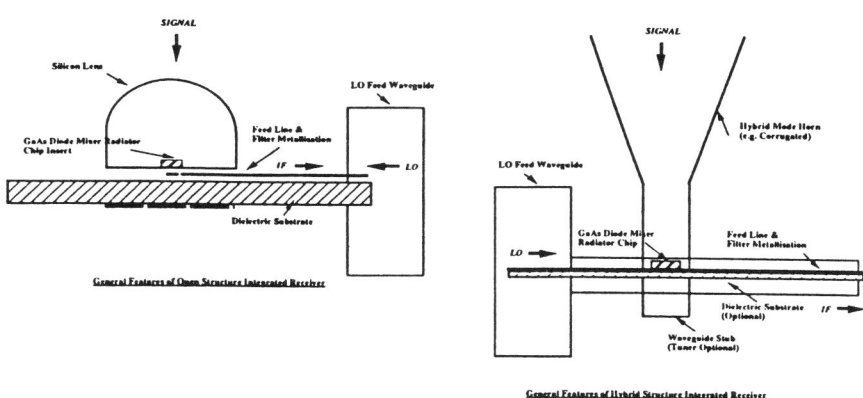

Figure 8. Examples of Integrated Receiver Construction

On a larger scale the development of the main reflector submillmetric instruments is a key task requiring a balancing of the different in orbit and ground test environments with factors such as mass and inertia. Figure 9 comprises a thermal distortion plot for a main reflector concept showing the low values which should be achievable with a single skin carbon fibre concept. The more conventional honeycomb construction will offer better gravity distortion and the compromise between these factors is an important design process.

6. Conclusions

For remote sensing applications in the terahertz bands the performance specifications are stringent so that the components required include many devices which are miniaturised version of well known microwave devices such as corrugated horns and waveguide mixers. The emphasis in development is therefore often on fabrication techniques for well understood structures, rather than on new types of structures.

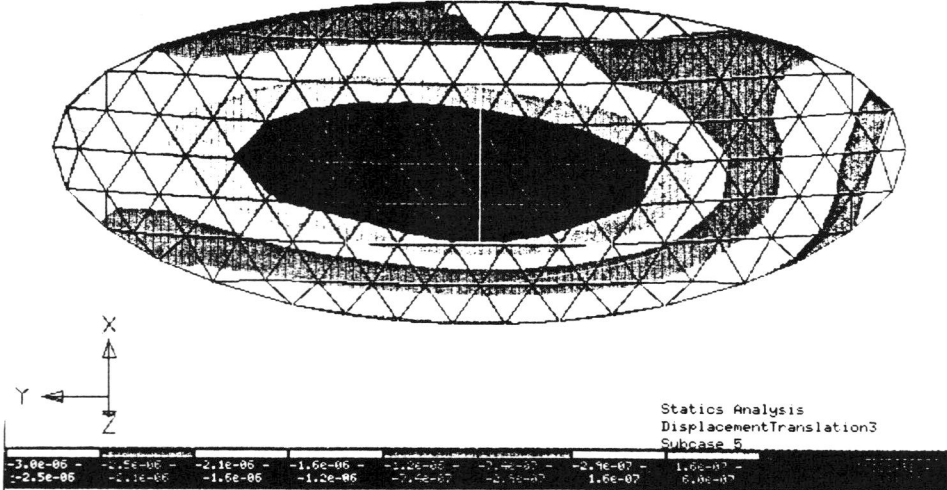

Figure 9. Predicted Thermal Distortion of 2m single skin reflector

In contrast loss requirements mean that some functions such as demultiplexing and image rejection filtering do use techniques such as quasioptics which would be cumbersome and unnecessary at lower frequencies, and technologies such as the frequency selective surface are in active development, as an extension of devices which first become commonly used in the millimetric bands.

New structures such as the dipole and slot fed lenses will tend to find applications only where configuration, scanning and adaptability outweigh straightforward performance requirements.

Up to about 1THz the use of Schottky diode multiplier chains has now been demonstrated but the provision of local oscillators with low demand for higher frequencies remains a major development task.

References

1. Meynart, R., Lamarre, D., and Lin, C.C. (1995) Millimeter-Wave Instruments for Far Future Earth Explorer Missions, *Proceedings of ESA Workshop on Millimeter Wave Technology and Applications*, ESA-WPP-098.
2. ESA Report (Anonymous) (1996) Atmospheric Chemistry Mission (Earth Explorer Assessment), ESA-SP-1196(6).
3. Carli, B. (1997) Submillimetre Measurements form Satellite, *this volume*.
4. Charlton, J.E., Lamarre, D., *et al.* (1995) MASTER: A Feasibility Assessment of a Millimetre Wave Limbsounder, *Proceedings of ESA Workshop on Millimetre Wave Technology and Applications*, ESA-WPP-098.
5. Hall, W.J., Lamarre, D., *et al.* (1995) SOPRANO Limbsounder Instruments and Technology, *ibid*.
6. ESTEC Final Report (Anonymous) (1995) TR0123 SOPRANO Limbsounder Pre-Phase A Study Extension, ESTEC Contact 10611/93/NL/SF.

7. ESTEC Final Report (Anonymous) (1996) TR0135 MASTER Limbsounder Pre-Phase A Study, ESTEC Contact 11159/94/NL/CN.
8. Charlton, J.E. and Jarrett, M.L. (1995) Radiometric Verification of AMSU-B, *Proceedings of ESA Workshop on Millimetre Wave Technology and Applications*, ESA-WPP-098.
9. Martin, R.J. (1995) The AMSU-B Quasi Optics, *ibid*.
10. Martin, R.J., Matheson, D. *et al.* (1995) A 320 Hz Tripling Multiplier, *ibid*.
11. Hall, W.J., *et al.* (1995) Submillimeter Mixer and Multiplier Diode Developments, *ibid*.
12. Krozer, V., *et al.* (1995) Fabrication and Characterisation of Planar Schottky Diodes, *ibid*.
13. Wells, J.A. and Cronin, N.J. (1993) Theoretical Analysis of Air Bridging and Back Etching Techniques on the Shunt Capacitance of Planar Subharmonic Mixer Diodes, *Proc. IEE. H* **140**, 474-480.
14. Wood, P.A.D., *et al.* (1994) GaAs Schottky Diodes for Atmospheric Measurements at 2.5 Hz, *Proc. 5th Int. Symposium on Space Terahertz Technology*, University of Michigan.
15. Ellison, B.N. (1995) Corrugated Feedhorns: Preliminary Results, *Proceedings of ESA Workshop on Millimetre Wave Technology and Applications*, ESA-WPP-098.
16. ESTEC Report (Anonymous) (1996) TR0136 Preparation of Millimetre Wave Technology Activities, ESTEC contact 11410/95/NL/CN.
17. Brown, D.A., Treen, A., and Cronin, N.J. (1994) Micromachining of Terahertz Waveguide Components with Integrated Active Devices, *Proceedings 9th Int. Conf. on IR and MM Waves*, Sendai, Japan.
18. Zimmerman, P., *et al.* (1995) Frequency Multipliers and LO Sources, *Proceedings ESA Workshop on Millimetre Wave Technology and Applications*, ESA-WPP-098.
19. Junkin, G., *et al.* (1995) Near-Field/Far-Filed Phase Retrieval Measurements of a Prototype of the Microwave Sounding Unit Antenna AMSU-B at 94 Hz, *ibid*.
20. Cahill, R. and Prior, C.J. (1993) G-Band Dielectric Core Horn, *Electronics Letters* **29**, 130-131.

MILLIMETER MEASUREMENTS

D.K.RYTTING
Santa Rosa Systems Division
Hewlett Packard
1400 Fountain Grove Parkway
Santa Rosa, California 95403

1. Introduction

This paper discusses component and device characterization for the microwave and millimeter frequency range. There will be an overview of the basic measurement techniques followed by an update on the present millimeter measurement methods and systems. Please see reference [1] for a good overview of general microwave measurement history.

There are many challenges facing the designer of modern electronic products. The market is very dynamic and new and better technology is arriving rapidly. There is a wide variety of design skills needed for these products. The primary challenges facing the designer are reducing the cost and size of the product while maintaining the performance required. As the characterization accuracy and development time is improved the product cost will certainly reduce. This requires accurate models and data for the components and devices used in the design.

A study was recently done at HP to determine the primary reason for design turnarounds. From this study of 36 different designs by different companies, the key bottleneck was inadequate models and design data. Computer-aided design tools require mathematical models for each component in a circuit. Models for non-linear active devices are often the most critical and most difficult to obtain and cause the most design trouble. The models used are developed from DC and RF measurements combined with device physics. The device model must be realistic and fit the actual behavior. Excellent measurement data is required to obtain the correct fit for the elements of the models.

Devices and components are placed in two broad categories. Linear components are passive components and active devices which are operated in the linear region. These active components must operate "small signal" so that the linear assumption is valid. The non-linear category includes power amplifiers and devices that operate mainly linear but because of efficiency requirements often start to distort at high power. Other key non-linear components are mixers used in down converters and in modulators and demodulators.

A linear device, like a filter, that has been stimulated by a sine wave will produce an output wave that has changed only in magnitude and phase. No new signals are added. Other examples of linear devices are cables, inductors, capacitors, and

269

J.M. Chamberlain and R.E. Miles (eds.), New Directions in Terahertz Technology, 269–288.
© 1997 *Kluwer Academic Publishers. Printed in the Netherlands.*

resistors. Amplifiers are designed for linear operation, but also exhibit non-linear characteristics.

A linear device is characterized by making a variety of magnitude and phase measurements. First, we have transmission measurements that compare the transmitted signal to the incident. Such measurements quantify the device gain, loss, noise figure, output power or phase distortion, for example. Reflection measurements characterize the match and impedance

A non-linear device produces output signals that include new frequency components not present at the input. Non-linear devices are designed to exhibit different behavior from a linear device. For example, a mixer must be non-linear in order to translate an input signal to the desired intermediate frequency. The output of a non-linear device is usually composed of multiple signals, due to mixing products, harmonic or intermodulation distortion, spurious responses or undesired sidebands. Some examples are mixers, diodes, and compressed amplifiers (amplifiers not in a linear region of operation).

2. Network Analyzer

Magnitude and phase (vector measurements) for linear devices can be measured using a Vector Network Analyzer (Figure 1). This tuned receiver, combined with a synthesized sweeper, is most often configured with an S-Parameter test set enabling full 2-port transmission/reflection device measurements. Vector error correction is used to provide accurate characterization of both coaxial and non-coaxial devices. The receiver's tuned IF filter provides immunity from harmonics and spurious responses to enhance measurement accuracy.

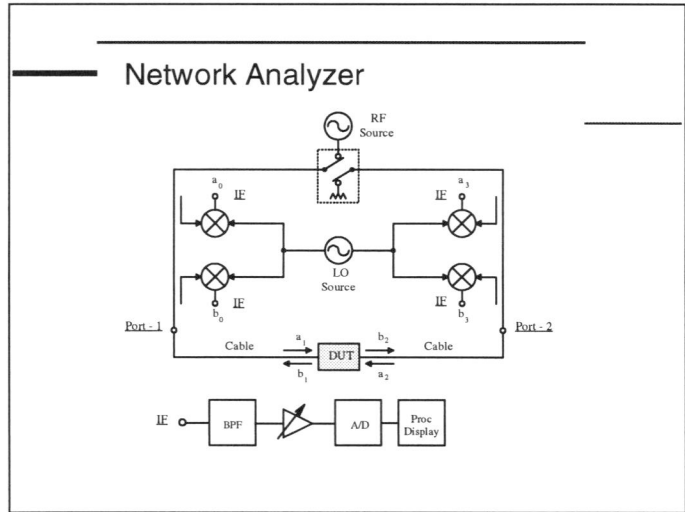

Figure 1

Modern Vector Network Analyzers can convey measurement information in a variety of display formats. Magnitude (linear, dB or SWR), phase or group delay characteristics may be displayed in log and linear frequency display formats. Magnitude and phase may also be displayed in a polar format or, for impedance, in the Smith Chart format.

3. S-Parameters

The most complete method for describing high frequency devices uses S-Parameters. We apply a stimulus to a 2-port device in both the forward and reverse directions, and measure both reflected and transmitted signals (Figure 2). A set of equations combine the resulting input and output signals to calculate the S-Parameters. S-Parameters completely describe a linear 2-port device and are easy to understand [2].

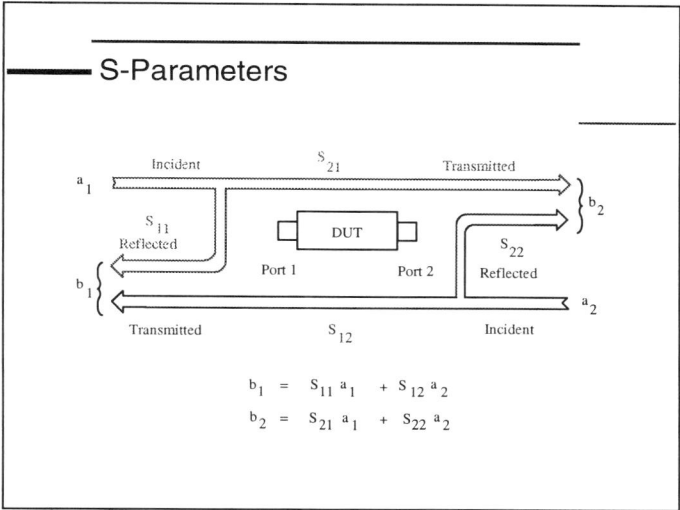

Figure 2

4. Distortionless Transmission

Even linear devices can produce distortion. One objective in device testing is to verify distortionless transmission (Figure 3). The criteria for a distortionless linear device is that the amplitude (magnitude) response must be flat, and the phase response must be linear, over the bandwidth of interest.

The phase slope technique (Figure 3) is one simple and accurate method of measuring group delay. It is a static or CW technique that approximates differentiation by measuring phase at two closely spaced frequencies (aperture), then computing the

272

slope between the points. This technique is widely used in modern vector network analyzers because it yields high resolution.

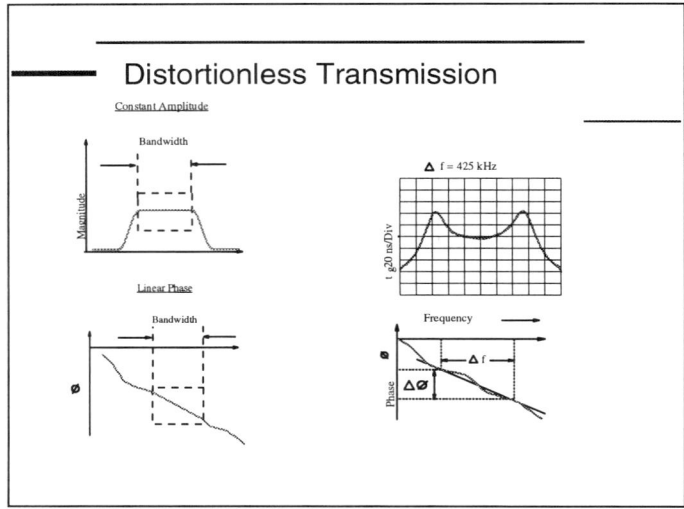

Figure 3

Narrowing the measurement aperture increases frequency resolution but will eventually result in not being able to resolve small phase variations from system noise, leading to a noisy or poor resolution group delay trace. Increasing the aperture reduces the resolution demands on the phase detectors, increasing immunity to fine grain signal and noise variations. This enables better group delay resolution. The trade-off between frequency and group delay resolution are fundamental to any group delay measurement.

5. Time Domain

With the data measured in the frequency domain using the network analyzer, the inverse Fourier transform can then be used to view the data in the time domain. The data must be properly prepared by "windowing" the frequency domain data so that the ringing (Gibbs phenomenon) is reduced when viewed in the time domain. The time domain is very useful in looking at reflections that are spaced out in time so that their unique location can be observed. The time resolution, or ability to resolve between closely spaced reflections, is improved as the frequency bandwidth is increased. The range, or how far out in time you can view before the data starts to repeat, is determined by the frequency step size. The smaller the step size the further out the alias free range will be.

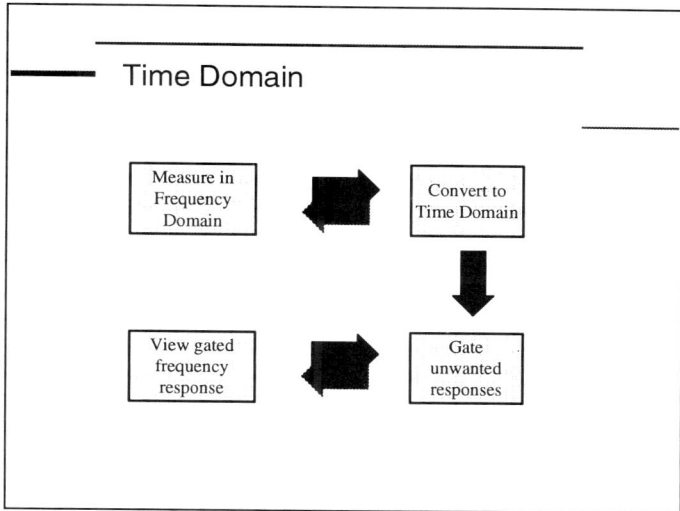

Figure 4

After the data is converted to the time domain, any unwanted reflections can be gated out (filtering in the time domain). This leaves just the responses that are desired. The data can then be transformed back to the frequency domain to observe the frequency response with the unwanted reflections removed (Figure 4). Please see references [3] and [4] for more details.

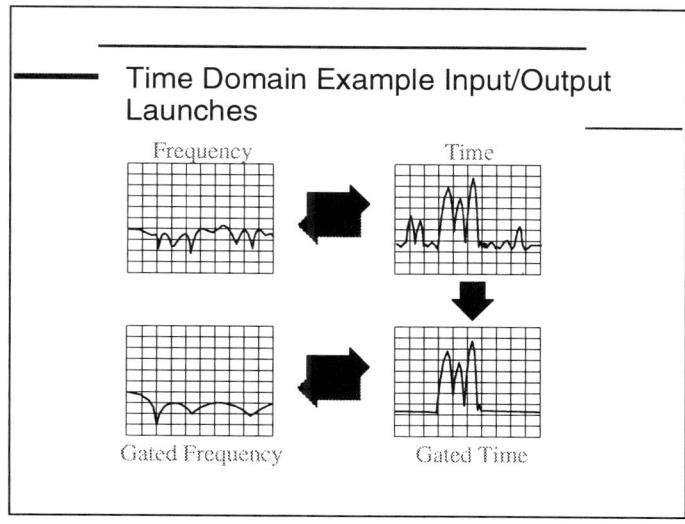

Figure 5

This example (Figure 5) is a microstrip passive component in a fixture with coax to microstrip launches at both ports. The data in the frequency domain shows a ripple response but not much insight into what is happening. When transformed to the time domain the effects of the package launches can be easily observed at the start and end of the data. The transitions are then gated out so that only the microstrip device is observed in the time domain. Then transforming back to the frequency domain gives the real data of the microstrip device without the effects of the transitions.

6. Pulse Testing

Pulsed measurements can help us to investigate and model the effects of thermal heating in high power Si and GaAs devices such as BJTs, HBTs and MESFETs. They can also help us to investigate trapping effects in GaAs material. Finally, they will allow us to characterize devices and amplifiers that are designed for pulsed applications (Figure 6).

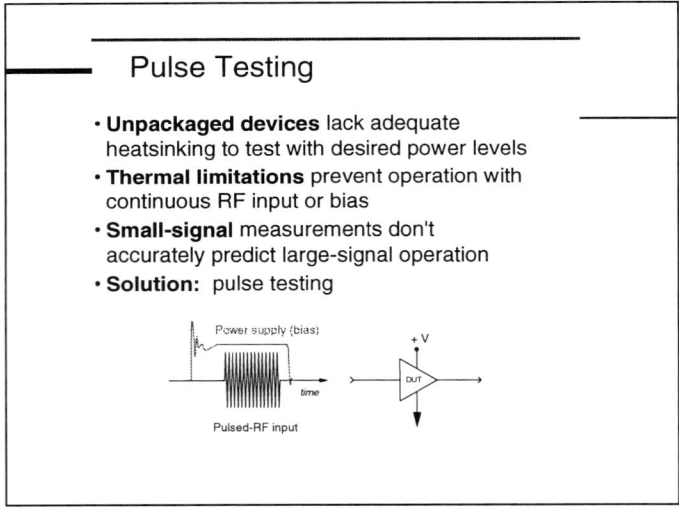

Figure 6

Pulse width determines the temperature rise within each pulse. Duty cycle affects the overall temperature stability of the device. The duty cycle must be set to allow sufficient cool down time for the device, otherwise thermal runaway may occur. The off-time required to bring the device back to ambient temperature can be as much as three to five times longer than the thermal time constant. This is independent of the pulse width (or on-time). In order to reduce the pulse width to minimize the thermal effects, the required off-time stays constant, so the duty cycle must be decreased. Typically, duty cycles on the order of 0.1% must be used for these types of measurements. By making measurements with pulse widths much narrower than the thermal and trapping time constants, we can begin to understand and ultimately model their effects.

Pulse operation requires both pulsed bias and pulsed RF and synchronization between these applied signals. Some applications may require the drain/collector to be pulsed while the gate/base is not pulsed. In applications where both terminals require pulsed bias, they may require synchronization to prevent breakdown. We must be able to precisely control the measurement trigger in order to ensure that data acquisition occurs beyond the settling times of the applied pulses.

7. Load Pull Measurements for Non-Linear Devices

The problem when dealing with a large signal amplifier is that the parameters vary as a function of input power and load impedance. These changes render small signal measurements poor for large signal operation. A load pull system measures the output impedance required for optimum output power from the amplifier. The other amplifier parameters can also be measured using load pull techniques.

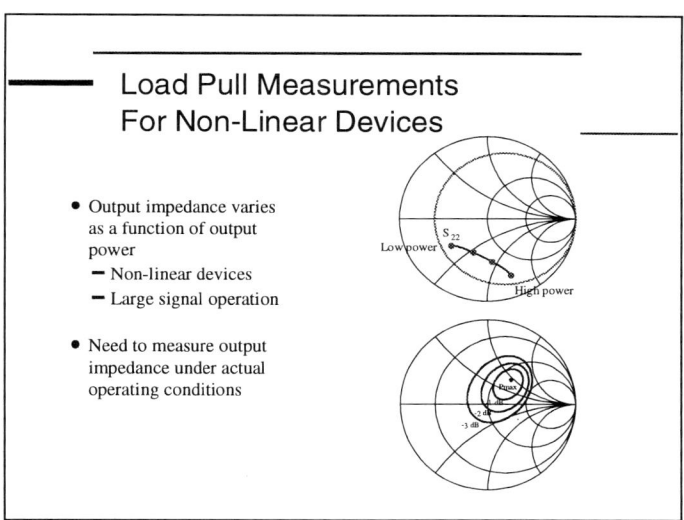

Figure 7

The most common result of a load pull measurement is a series of constant output power contours (not always circular) plotted on a Smith chart which represent all possible output impedances (Figure 7). A load pull measurement is required at each power level and frequency of interest since the output contours are sensitive to these variables for most high power amplifiers. The output power contours shown are with a constant input power level. With the many measurements to be made, it is important that a load pull measurement can be easily configured and is fast.

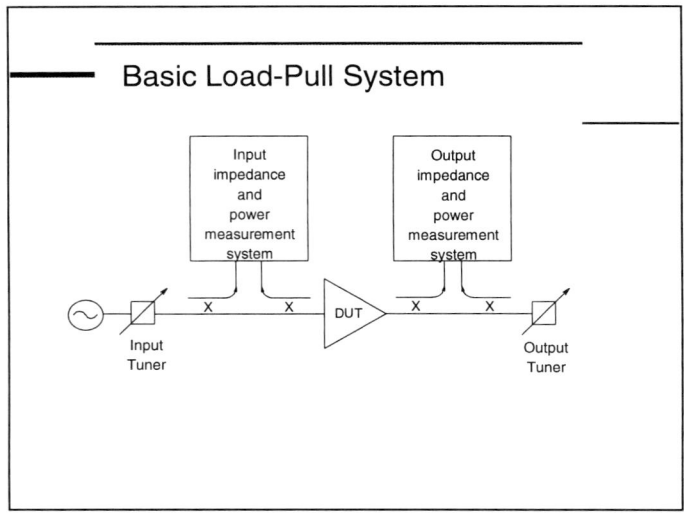

Figure 8

A basic load pull measurement system (Figure 8) typically consists of input and output load tuners, signal separation devices and impedance and power measurement capabilities. A vector network analyzer is commonly used with tunable loads as the core instruments of a load pull measurement system. Reference [5] gives a good overview of a harmonic load pull system used to measure the effects of load terminations at the harmonic frequencies.

Input and output measurements allow direct measurement of the input and output load values. As an example, the output tuner can be adjusted for maximum power transfer and measured with a network analyzer. The conjugate match is the output match of the amplifier under large signal operating conditions. The same type of measurement can be applied to the input of the amplifier to determine the input match.

Load pull data can also be obtained by driving both the input and output at the same time. The input is driven at the desired level and the output is driven to simulate the reflection from a general load termination. This technique is known as "active" load pull.

8. Measurements Errors

Systematic (repeatable) measurement errors are caused by imperfections in the measurement hardware (Figure 9). There are 6 types:

Error	Primary cause
Tracking (x2)	Frequency response of the system
Match (x2)	Port mismatch at the test ports
Directivity (x1)	Reflection leakage before device
Crosstalk (x1)	Leakage around device

In a 1-port reflection measurement, there are three systematic errors: reflection tracking, port-1 match, and directivity. For a 2-port device, measuring both reflection and transmission, there are 12 total systematic errors (6 in the forward and 6 in the reverse direction).

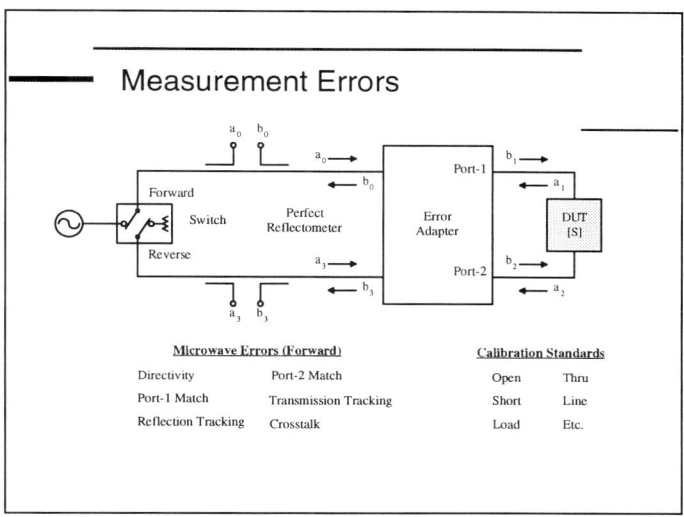

Figure 9

Calibration standards include well-defined opens, shorts, loads, "thru" connections, and transmission lines (Figure 10). Generally, correcting more errors requires more standards. Reference [6] describes the details of the calibration standards. Random errors (such as noise and connector repeatability) and drift errors (due to temperature, etc.) can never be removed with calibration. Their effects can only be minimized with careful measurement practice. A user can select from several levels of error correction depending on the device under test itself (1-port or 2-port), the degree of accuracy required, and the convenience desired. For example, a response calibration can compensate for frequency response errors, but not for port match, directivity or crosstalk errors. An S_{11}, 1-port calibration provides the most complete calibration for 1-port devices. And a full 2-port calibration compensates for both reflection and transmission errors and provides the maximum level of accuracy enhancement for 2-port devices.

There are numerous error correction methods used in the industry today. The TOSL method (Thru, Open, Short, Load) developed back in the 1960s was based on the 12 term model as described in reference [7]. There were 6 errors in the forward direction and 6 in the reverse direction. In the forward direction these errors are the directivity, port-1 match, reflection tracking, transmission tracking, port-2 match, and crosstalk. The first 3 errors can be determined by measuring 3 know calibration standards on port-1 and solving the 3 resultant simultaneous equations. Typically these standards are an open, short, and load (fixed or sliding). Then with the test ports connected, the port-2 match

278

and transmission tracking can be determined. The crosstalk is measured by placing loads on the test ports and measuring the isolation. This whole process is then repeated for the reverse direction.

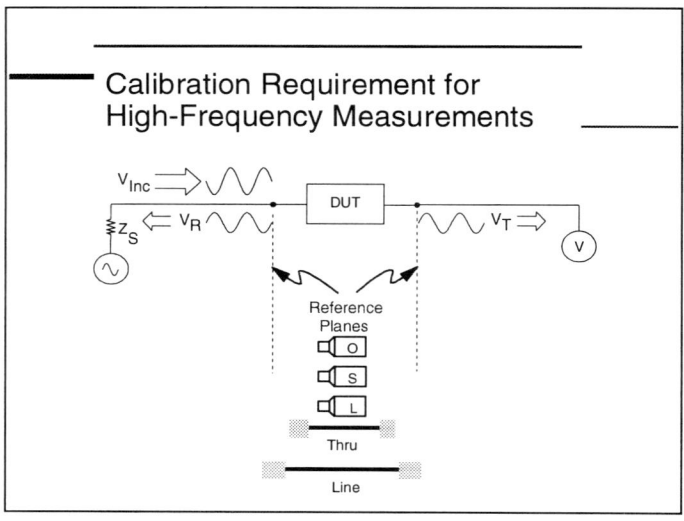

Figure 10

In the 1970s the calibration problem was looked at much closer and it was determined that the 12 term model could really be simplified to 10 terms with a model that did not change as the measurement system switched from measuring in the forward direction to the reverse direction. The two crosstalk terms could be easily measured with terminations on the ports which left only 8 terms to solve for. Since S-parameters are ratio measurements the error terms could all be referenced to one error term reducing the number of unknowns to 7. So the real calibration problem is to come up with 7 known conditions to solve for the 7 error terms of the measurement system.

The first technique to take advantage of the 7 term error model was the TRL technique (Thru, Reflect, Line) developed by Glen Engen and Cletus Hoer at NBS for the six-port network analyzer [8]. This technique used a thru connection, a delay line (ideally 1/4 wavelength long at the center of the frequency band) with known impedance (S_{11} and S_{22} known), and an equal but unknown reflection connected to each test port. These provided enough conditions to solve for the 7 error terms, and the process used also determined the remaining characteristics of the line (S_{21} and S_{12}) and the reflection coefficient of the unknown load. Multiple lines are typically used to cover a broad frequency range. The real advantage of the method is that the requirement of knowing the parameters of the open and short are removed and the errors caused by the imperfect knowledge of these standards is eliminated. The only critical standards are the impedance of the line and the requirement that the reflect be equal at each port. Measurement accuracy was greatly improved in coax and waveguide measurements. Later this approach was augmented to use another line as the thru connection whose characteristics are known or match those of the line (LRL method). The difference in the length of the two lines provided the necessary delay. The TRL and LRL methods are the most traceable techniques for measuring on wafer and are presently being formalized by NIST as to support the wafer probing industry.

It was also determined that the line could be replaced by a known match and this led to the TRM and LRM techniques. This method is inherently broad band and is very easy to use.

All these techniques require a known reference impedance as one of the standards. Also one of the calibration requirements is that the two test ports be connected for one of the measurements

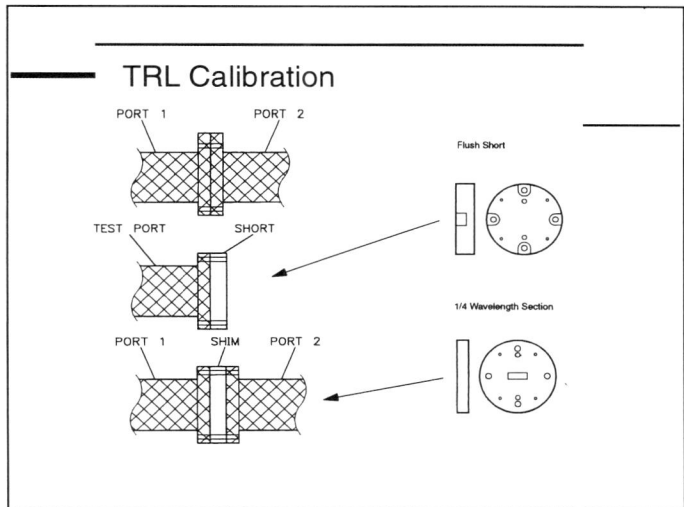

Figure 11

The TRL method is very applicable to waveguide calibration (Figure 11). First connect the thru, then the ports are separated and shorts are used on port-1 and port-2 as the high reflect standards. Finally a transmission line is connected to complete the calibration.

9. Noise Figure and Noise Parameters

Noise figure and noise parameter measurements are used to characterize the noise performance of low-noise amplifiers, transistors, mixers and receivers used in modern communications circuits. Noise figure is defined as the signal to noise ratio at the input divided by the signal to noise ratio at the output (Figure 12). An excellent primer on noise figure measurements is reference [9].

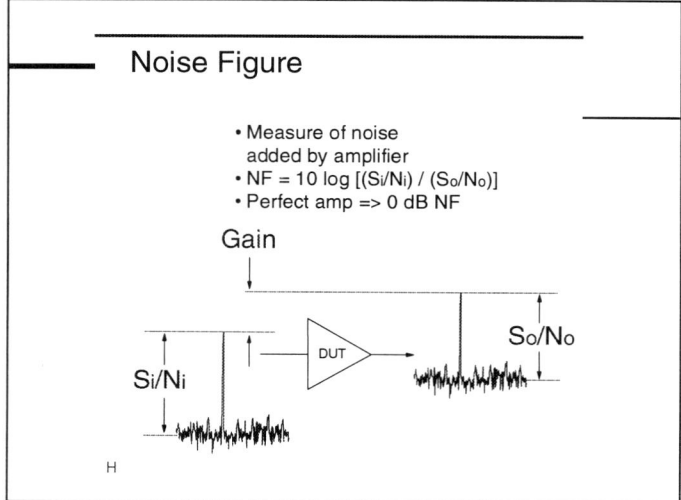

Figure 12

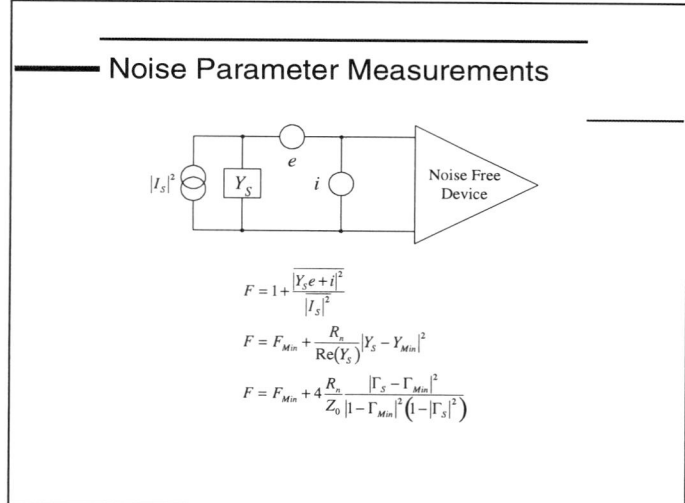

Figure 13

Amplifiers require trade-offs between the gain and its corresponding noise figure. Knowledge of how the amplifier's gain and noise figure vary as a function of the source impedance (admittance) is required.

The noise of a device or amplifier can be modeled as a noise free device preceded by the equivalent noise voltage and current added by the actual noisy device

(Figure 13). The dependence of noise figure on source impedance (admittance), for a linear two-port amplifier, is given by the following equation:

$$F = 1 + F_{Min} + \frac{R_n}{Re\ (Y_S)}|Y_S - Y_{Min}|^2 \qquad (1)$$

where the source impedance (admittance) Y_S results in the noise figure, F, expressed as a power ratio. In the equation, F_{Min} is the minimum noise figure for the amplifier, R_n is the noise resistance (the sensitivity of noise figure to impedance changes) and Y_{Min} is the optimum impedance which yields F_{Min}. These four scalar parameters are frequently referred to as the noise parameters. To measure the noise parameters, the source impedance (admittance) must be varied by using a tuner on the amplifier input.

The interactions between the noise parameters of the receiver with the source causes missmatch errors. To remove these effects it is necessary to measure the characteristics of the source, device and measurement receiver using a network analyzer and error correction techniques. Also the noise parameters of the receiver must be know to remove its interactions. With modern noise measurement systems all of these capabilities are used to yield the best possible measurement results.

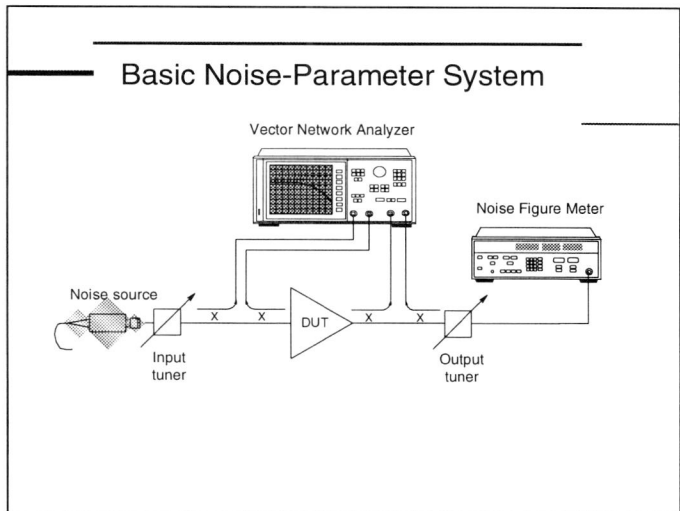

Figure 14

Noise parameter characterization relies on the determination of the four scalar noise parameters. Equation (1) shows that the source impedance Y_{Min} results in the lowest possible noise figure, F_{Min}. This suggests that if a tuner is connected to the input of the amplifier under test and adjusted until minimum noise figure is found, F_{Min} will be known (Figure 14). A second tuner at the output of the amplifier can be adjusted to determine maximum gain. After adjusting for maximum gain, the input tuner must be readjusted to again find the minimum noise figure since F_{Min} is dependent on the output match and input match. A vector network analyzer measurement of the input tuner will then determine Y_{Min} for the amplifier. Once the maximum gain and minimum noise figure are determined, a second measurement of noise figure using another source impedance value, say 50 ohms, provides enough data to determine the fourth parameter, R_n.

The most common method used today to determine the noise parameters is a computer controlled system. Four or more independent states of a programmable input tuner are set and the noise is measured at each state. This yields four equations to solve for the four noise parameters of the device.

Once the noise parameter measurements have been made, the data can be put into many forms. The output data from a noise parameter measurement, system provides both plots and graphical results. Constant noise circles (Figure 15) provide impedance information for differing values of noise figure. Also the swept frequency versus source impedance for F_{Min} is shown.

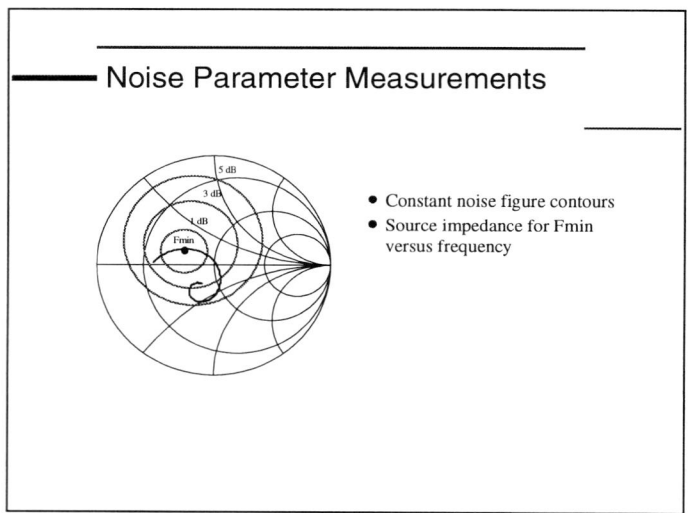

Figure 15

10. 120 GHz Measurement System

Recently there has been an increased interest in the millimeter frequency bands. Much of this has been driven by broad band device characterization and vehicular electronics. There is also a desire to have a single continuous sweep over the entire frequency range with single connection capability. The block diagram (Figure 16) has a low band portion to 50 GHz using the traditional broad band network analyzer. To this is added two waveguide bands that are combined to provide a source and receiver covering the complete frequency range. The key components in the system are the 3 way combiners/splitters, the broad band directional coupler, source multipliers, and harmonic mixers. The development of the 1.0 mm connector provides a coaxial test port for dc to

120 GHz coverage. There have also been narrow-band systems developed using this same source multiplier and harmonic mixer method to over 300 GHz.

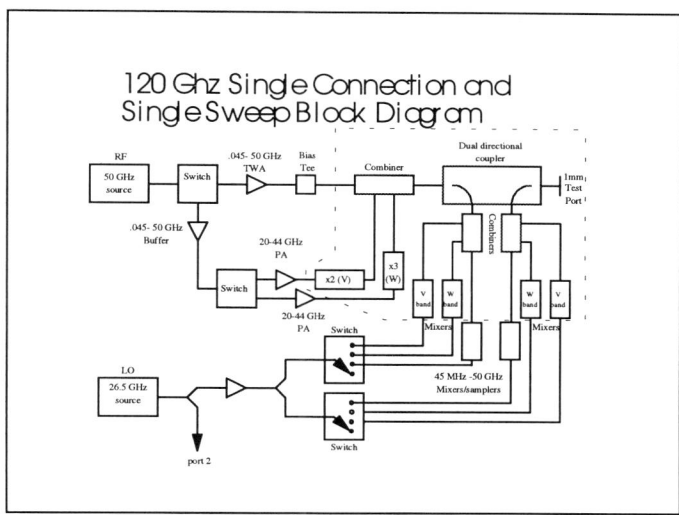

Figure 16

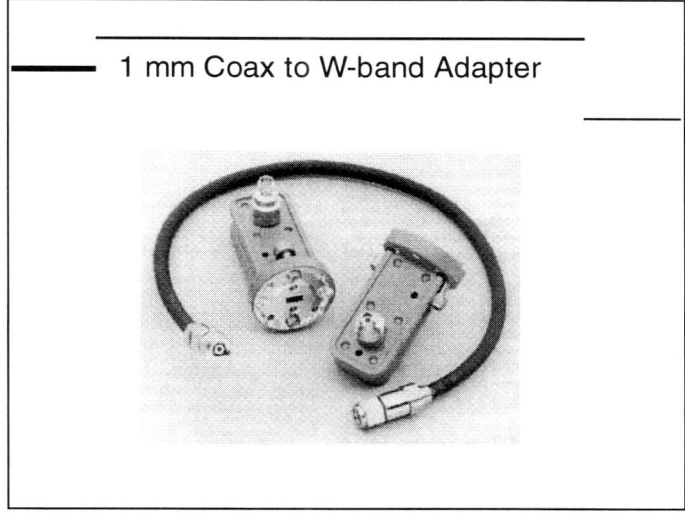

Figure 17

Figure 17 shows the interface for the 1 mm connector. At 120 GHz the losses in the cable are very high and is not practical as a low loss interconnect. However it does provide the necessary interface to waveguide and on wafer transmission lines to the broad band measurement system. Also shown in the picture of the coax to waveguide adapter is the precision waveguide interface. This interface provides precision alignment pins and an outer ring to eliminate the non-parallel interface problem caused by the flange screws.

Most of the recent millimeter designs are either on wafer or attached to a substrate. The development of a dc to 120 GHz wafer probe, by the Cascade Microtech Corporation, with less than 1 dB of loss makes accurate wafer probe measurements possible (Figure 18).

Figure 18

11. 6-Port Network Analyzer

The 6-port network analyzer (Figure 19) is a technique for measuring magnitude and phase using only power measurement at the 4 sidearms of a 6-port network. This measurement technique was developed at NBS as a high quality reflection measurement system [10] [11]. It was soon extended to two-port capability and TRL calibration was developed to enhance the accuracy. The 6-port network analyzer has only found moderate success as a commercial measurement method, but, at high millimeter frequencies it has been an excellent measurement tool, where standard heterodyne receivers have been impractical.

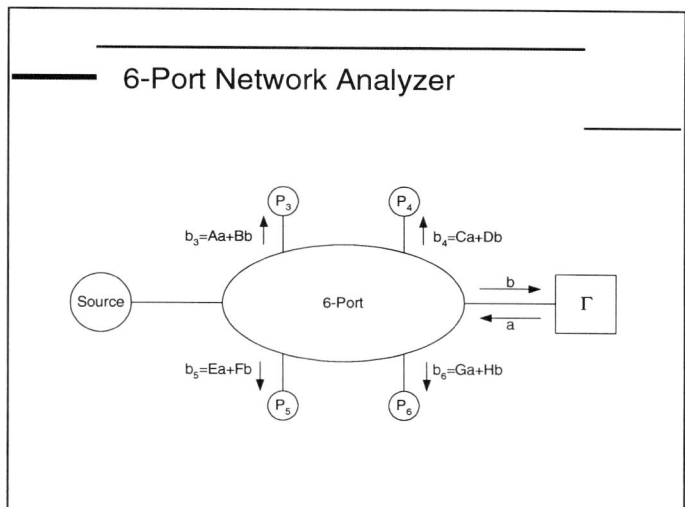

Figure 19

Figure 20 outlines the math for the 6-port measurements. The four power meters P_3, P_4, P_5, and P_6 measure different observations of the incident and reflected powers (a and b) at the device under test. There are 11 terms that describe this mapping between the test port and the power meters. These 11 terms are determined by a two stage calibration process that first calibrates the 6-port for measuring phase (5 terms) and then determines the test port characteristics (6 terms). Once these 11 terms are characterized the test port reflection coefficient Γ can be determined by measuring only 3 power ratios.

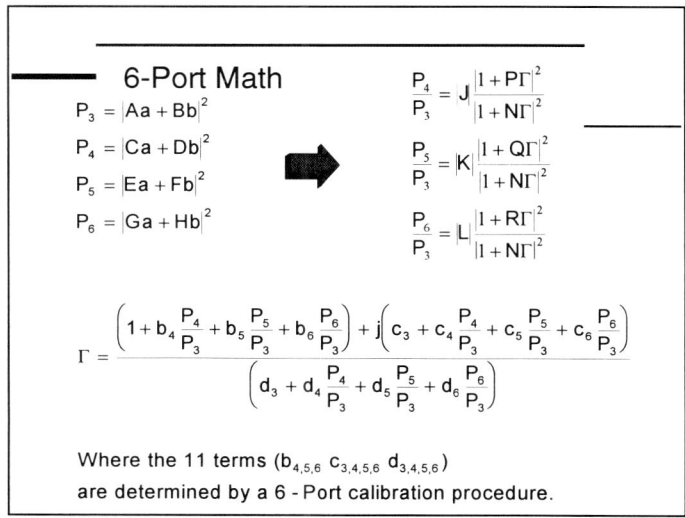

Figure 20

12. Ultra Wide Band Shockline Sampler System.

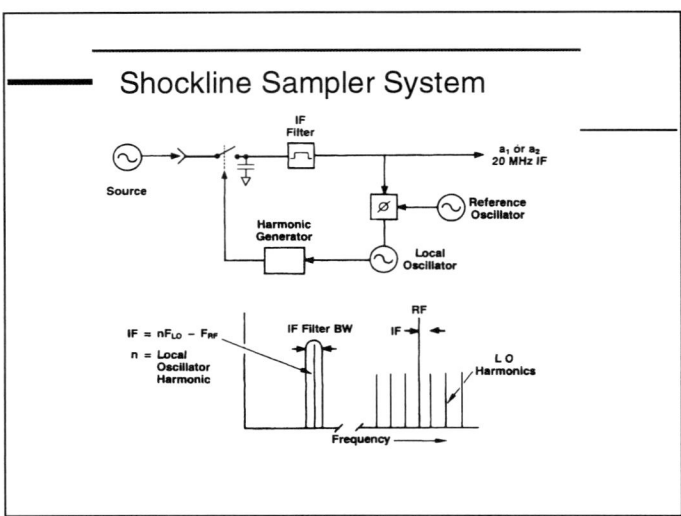

Figure 21

The development of the shockline sampler has led to very broad band coverage to over 300 GHz. The basic block diagram of the sampler system is shown in Figure 21. The sampler is a very fast switching gate that is driven at the LO frequency. The harmonic generator produces a very sharp pulse which creates a harmonic frequency comb that mixes with the RF input to produce an IF frequency output. Typically the harmonic generator consists of a step recovery diode followed by a shockline pulse sharpener. The frequency range of the sampler is determined by the effective sampler on time. A few picoseconds gate time has been achieved to allow the 300 GHz performance. The LO oscillator typically runs in the 100 MHz to a few GHz rate. The LO can be either phase locked to the IF frequency or if it is synthesized and stable enough phase locking is not required.

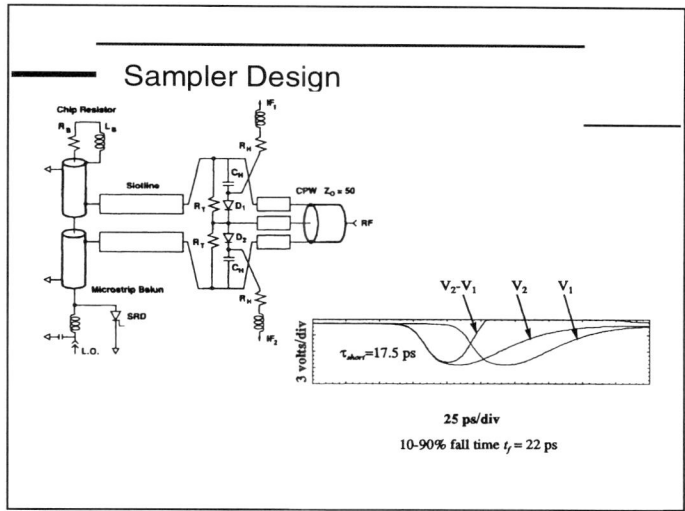

Figure 22

The Sampler is a balanced diode switch which is driven single ended by the RF from the right side of the block diagram shown in Figure 22. The LO is a pulse from the SRD (Step Recovery Diode) cascaded with a shockline fed through a single ended to balanced transformer and is launched into the slotline. The pulse propagates over the diodes turning them on and allowing the capacitors to be charged with the RF signal. The pulse then hits an effective short circuit at the RF port and reflected back out of phase and passes over the diodes the second time turning the gate off. This forms a differentiating action that further sharpens the pulse.

Experimental circuits have been developed that are mounted directly on the wafer probe to allow the lowest parasitics and the best frequency response.

The key to creating the very short gate time is the shockline (Figure 23). The shockline is a transmission line with varactor diodes, which have a inherent voltage variable capacitance, placed along the line [12]. The varactor's capacitance provides the required shunt elements for the pseudo lumped transmission line. The pulse launched into the line propagates at different velocity depending on the voltage across the varactor diodes capacitance. This causes the output to be compressed in time and provides a very short output pulse.

13. Conclusion

There is an awakening occurring in the millimeter wave arena and increased business opportunities. The question is how fast will this market develop. There are activities now in the university and research labs of various companies, but the timeline to production is still unclear and the size of the market is yet to be determined. I know that as these opportunities are clarified the necessary production level test equipment will be

288

developed to meet the needs. For now, above 120 GHz, most of the test methods will be specialized and mainly crafted for the research laboratory.

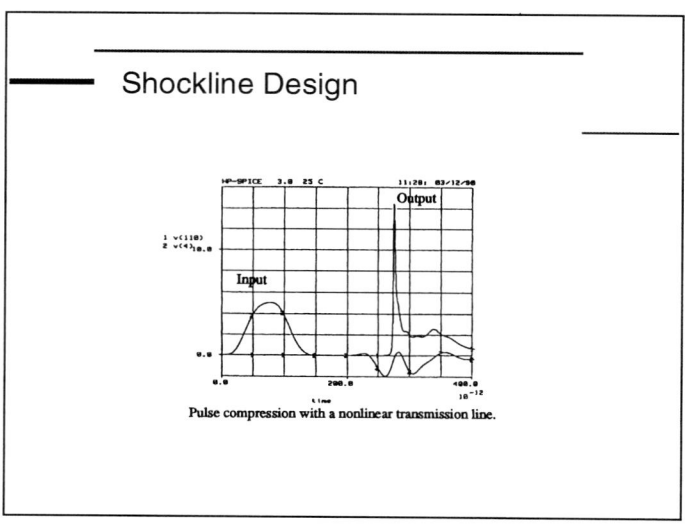

Figure 23

14. References

1. Adam, S. F., (1984) Microwave Instrumentation: An Historical Perspective, *IEEE Trans. on Microwave Theory and Techniques*, **MTT-329**, 1157 - 1160, Sept..

2. Weinberg, L. (1967) Fundamentals of Scattering Matrices, *Electro-Technology*, 55 - 72.

3. Oliver, B. M. (1964) Time Domain Reflectometry, *Hewlett Packard Journal*, Vol 15.

4. Rytting, D. K. (1984) Let Time Domain Response Provide Additional Insight Into Network Behavior, *Hewlett Packard RF & Microwave Symposium.*

5. Hughes, B., Ferrero, A. and Cognata, A. (1992) Accurate On Wafer Power and Harmonic Measurements of MM Wave Amplifiers and Devices, *IEEE MTTS Digest*, 1019.

6. Wong, K. H. (1988) Using Precision Coaxial Air Dielectric Transmission Lines as Calibration and Verification Standards, *Microwave Journal*, 83 - 92.

7. Fitzpatrick, J. (1978) Error Models for Systems Measurement, *Microwave Journal*, 63 - 66.

8. Engen, G. F. and Hoer, C. A. (1979) Thru-Reflect-Line: An Improved Technique for Calibrating the Dual 6-Port Automatic Network Analyzer, *IEEE Trans. on Microwave Theory and Techniques*, **MTT-27** (12), 987 - 993.

9. *Noise Figure Primer*, Hewlett Packard Application Note 57.

10. Engen, G. F. and Hoer, C. A. (1972) Application of an Arbitrary 6-Port Junction to Power Measurement Problems, *IEEE Trans. on Instrumentation and Measurement*, IM **21** (4), 470 - 474.

11. Hoer, C. A. (1977) A Network Analyzer Incorporating Two 6-Port Reflectometers, *IEEE Trans. on Microwave Theory and Techniques*, **MTT 25** (12),1070 - 1074.

12. Tan, M., Su, C. and Anklam, W. (1988) 7x Electrical Pulse Compression on an Inhomogeneous Nonlinear Transmission Line, *Electronic Letters*, **24**, 213-215.

WHAT FUTURE FOR WIRELESS TELECOMMUNICATIONS SYSTEMS BEYOND 60 GHz ?

D. WAKE
BT Laboratories,
Martlesham Heath, Ipswich. UK.

1. Introduction

'Mobility' and 'multimedia' are two of the most important aspects of modern telecommunications. Cellular mobile telephone networks have been around since the early 1980s, and still show enormous growth rates in terms of numbers of subscribers. For example, between early 1994 and early 1995, the number of people around the world connected to cellular mobile telephone networks rose from approximately 34 million to around 55 million. The demand for mobile telephony continues unabated as networks, services, tariffs and technology evolve. Projections of growth by Ericsson indicate that the number of mobile telephone users will increase to 350 million by the end of this century.

The vast majority of telecommunications today involve voice-only services - mainly telephony. In the future, multimedia applications and services, involving a mixture of voice, text, video, data or still image are expected to make a significant contribution to telecommunications traffic. The popularity of the World Wide Web (WWW) on the Internet and the multimedia capabilities of recent generations of WWW browsers is evidence of a demand for multimedia services. Although the Internet is really a series of interlinked computer data networks, the distinction between telecommunications and computer networks is not as clear as it once was. The boundaries between the two are beginning to merge with recent developments in switching technologies such as asynchronous transfer mode (ATM). Also, as the WWW moves towards a mass market, the trend is away from traditional computers to simple-to-use dedicated multimedia terminals, i.e. more like a telecommunication product.

If the demand for mobility and multimedia services is strong and increasing, then it is logical to suppose that there is also a latent demand for mobile multimedia services. It is this possibility that has led to research activity around the world into enabling technology for wireless networks that can deliver these high bandwidth multimedia services. High frequency radio is ideally suited as a delivery mechanism for mobile multimedia services - free space propagation allows wireless connectivity, and the large spectrum availability provides the necessary bandwidth. Frequencies up to the low millimetre-wave bands (30, 40, 60 GHz) have been considered for the longer-term future, where even more spectrum is available. For example, the RACE research project

J.M. Chamberlain and R.E. Miles (eds.), New Directions in Terahertz Technology, 289–298.

MBS (Mobile Broadband System) advocated bands around 60 GHz for a mobile system capable of providing data rates up to 155 Mb/s [1].

To some, the idea of using millimetre-wave radio for a mobile broadband network is fanciful and far-fetched. To others, it is a natural consequence of the trend towards the movement of more and more information around the world. There are many problems to be overcome, and the timescales are probably at least 15 - 20 years into the future. Infra-red systems are touted by some as a better solution, primarily because the optical transmitters and receivers are relatively cheap. Against this backdrop, the question that is addressed in this paper is: can telecommunications systems operating beyond 60 GHz eventually play a useful role in the wireless delivery of multimedia services? Before addressing this question directly, it is worth briefly setting the scene in today's telecommunications systems and then looking at the way these systems are evolving.

2. Telecommunications transmission systems - a few general points

Modern telecommunications transmission systems are based on a mix of optical fibre, copper, satellite and terrestrial radio. In the UK as well as much of the world, the core network, which links the major switching points, consists almost entirely of optical fibre. The access network, which links customers to the switching points, consists mainly of twisted-pair copper. Satellite systems are used mainly for international links, and terrestrial radio is used for a variety of mobile networks and also performs a niche role in both core and fixed access networks. Briefly, the main features which characterise these transmission media are:

- optical fibre - low loss, low dispersion, interference-free, well controlled characteristics, high capacity. The transmission medium of choice for a high capacity core network.
- copper - high loss, high dispersion, interference prone, low capacity. Used extensively in access networks of established network operators. Low cost technology is under development to use existing infrastructure for broadband services.
- satellite - limited capacity, large time delay. International links used as alternative to trans-oceanic cables.
- terrestrial radio - variable loss, variable dispersion, interference prone, multipath propagation, fast and slow fading, low capacity. Difficult to use, and quality is inferior to fixed links, but it is necessary for mobile networks. Also used in core and access networks to a limited extent.

Looking specifically at wireless systems, a few more general points are worth making:

- Although a brief overview of current wireless systems is given in the next section, table 1 shows how spectrum is used today for various system types. Most widely used systems occupy spectrum up to low microwave frequencies. Mobile radio telephony for example mainly uses spectrum around 900 MHz. In this region, electronic systems are relatively cheap to implement, but spectrum is very difficult to obtain. Systems requiring more spectrum must move up in frequency where more is available, but electronics tend to be much more expensive. These systems tend to be used for niche applications such as point-to-point high capacity links. Infra-red systems are also used currently for niche applications (commercial products are available for example for wireless local area networks). The gap between 60 GHz systems and IR systems has not been addressed to date, apart from military applications in the range 90 - 100 GHz.

TABLE 1. Frequency ranges for wireless systems

Frequency range	System type
VHF, low microwave (0.1 - 5 GHz)	mass market (e.g. mobile telephony, paging, radio access)
high microwave, low mm-wave (5 - 60 GHz)	niche (e.g. broadband radio access, inter-building links) and experimental
between low mm-wave and IR (60 GHz - 200 THz)	some specialist military applications (below 100 GHz)
infra-red (IR) (200 - 300 THz)	niche (e.g. outside broadcast, office LANs) and experimental

- Modern radio systems are based on digital transmission. In general, digital transmission achieves better bandwidth efficiency than analogue transmission mainly as a result of data compression. High tolerance to interference is a consequence of error correction in digital systems. Using the example of cellular radio network evolution in the US, analogue channels are being replaced by digital channels on a three-to-one basis.
- Wireless system architectures can be point-to-point, point-to-multipoint or cellular. Point-to-multipoint is used to reduce costs, where equipment can be shared between a number of users. Cellular architectures are used for mobile networks in order to use spectrum efficiently, and are discussed briefly in the next section.

3. Current wireless telecommunications systems

To set the scene for analysing the role of very high frequency radio systems in telecommunications networks, two important types of wireless telecommunications systems (mobile and wireless local area networks) are overviewed briefly in this section.

Mobile telecommunications systems underwent a revolution in the 1980s to become a mass market technology. These systems are still experiencing massive growth worldwide and are characterised by:

- Large scale coverage area (national / international)
- Cellular architecture; capacity is increased by dividing coverage area into cells. Frequencies can be re-used many times as long as co-channel cells are not too close to cause interference.
- Mobility management; a high amount of network intelligence is required to deal with user mobility. Mobiles must be tracked and handover of calls between cells must be managed.
- Many standards are used throughout the world, e.g. 1st generation analogue TACS (Total Access Communication System) and 2nd generation digital GSM (Global System for Mobile Communications) in UK
- Mainly voice-based services but growing use of data (9.6 kb/s on GSM)

Wireless LAN systems are gaining acceptance because they avoid the need for a cabled infrastructure in buildings. They are also used for niche applications in places such as hospitals and warehouses. These systems are characterised by:

- short range - 100m
- mainly data - Ethernet compatibility - up to 1-2 Mb/s
- ISM (Industrial, Scientific and Medical) band operation - 900 MHz (in US) and 2.4 GHz
- spread spectrum to minimise interference. This technique relies on multiplication of the data signal by a long digital code to spread the bandwidth. The receiver knows the code and can de-spread the wanted signal. Interfering signals are spread at the receiver and become low-level noise.
- Many existing products use proprietary systems, but an emerging European (ETSI) standard known as Hiperlan, operating at 5.2 GHz will allow higher data rates (20 Mb/s) [2].

4. Trends in wireless telecommunications networks

This section assesses the trends that are shaping the way current systems are evolving. Mobile and wireless LAN systems are used again as examples, and these trends are summarised in the tables below. Table 2 shows the trends for mobile systems, and table 3 shows the trends for radio LAN systems.

TABLE 2. Trends in mobile systems

	Data rate	Cell size	Frequency
present (GSM)	9.6 kb/s	macro	1 - 2 GHz
future (UMTS)	2 Mb/s	micro	2 GHz
longer term (MBS)	34 - 155 Mb/s	micro / pico	60 GHz

TABLE 3. Trends in radio LAN systems

	Data rate	Cell size	Frequency
present (proprietary)	1 Mb/s	micro	2.4 GHz
future (Hiperlan)	20 Mb/s	micro / pico	5.2 & 17 GHz
longer term	34 - 155 Mb/s	pico	60 GHz

Present (2nd generation) mobile systems such as GSM have relatively large area cells and use frequencies in the low microwave region. Standards are under consideration, UMTS (Universal Mobile Telecommunications System) in Europe for example, which aims to provide a multimedia capability up to 2 Mb/s across a range of applications and environments [3]. The cell size will have to be much smaller than current GSM cells to accommodate the highest data rates for any significant number of users. In the longer term, the RACE project MBS has studied the use of 60 GHz radio to provide higher capacity. Again, the cell sizes must be reduced still further, possibly to the picocell regime (office or smaller) to get the highest capacity.

The trends in wireless telecommunications systems can be summarised therefore as moving to higher carrier frequency and shorter range. Both of these trends are driven by the need for capacity - high carrier frequency provides a greater amount of spectrum, and small cells allow this spectrum to be re-used more efficiently.

5. Higher frequency systems

5.1. RADIO PROPAGATION CHARACTERISTICS

The propagation of radio waves from transmitter to receiver is a complex business in the majority of wireless telecommunication systems. The received signal is a combination of the direct path and other paths caused by reflection, refraction and diffraction. These combine to produce a signal that can fluctuate in power by several orders of magnitude as a function of position and frequency. Multipath propagation also causes time delays in signals which limit the achievable data rate due to intersymbol interference, although some systems use equalisation to compensate for this delay.

In general terms, signals below a frequency of around 3 GHz can be considered not to require a line-of-sight (LOS) path for successful transmission, mainly as a result of significant diffraction due to the relatively long wavelength. For indoor systems, a typical system would provide coverage over several rooms with a single station. Above this frequency, propagation becomes increasingly LOS as the coverage afforded by diffraction is scaled with wavelength. At 60 GHz, propagation is limited to LOS except for a small amount from reflection from highly reflectivity surfaces (such as chipboard) or transmission through high transmissivity surfaces (such as plasterboard) [4]. At infrared, propagation is exactly as we experience from visible light - strong shadowing behind obstacles, some reflection to improve coverage and no transmission through walls. For THz frequencies, the situation will be very similar: mainly LOS propagation with some reflection and transmission at surfaces.

5.2. 60 GHz SYSTEMS

Wireless telecommunications systems operating at frequencies around 60 GHz exist commercially from several manufacturers. These systems are relatively simple point-to-point systems, and are used for example for short-range network interconnection between buildings at data rates up to around 100 Mb/s. Several research initiatives around the world are attempting to develop more sophisticated systems operating around 60 GHz for very high data rate wireless LAN applications. In Japan for example, the Communications Research Laboratory (CRL) is developing a WLAN system operating between 59-64 GHz for data rates up to 155 Mb/s [5]. Similar projects are underway in Canada (Broadband Indoor Wireless Communications project) [6] and Australia (PLANS) [7]. In Europe, the intention of the RACE project 'Mobile Broadband system' (MBS) was to develop the necessary air interface, mobility management and radio resource management for a full mobile system offering data rates up to 155 Mb/s. The frequency bands of 62-63 (base-station to mobile) and 65-66 GHz (mobile to base-station) have been allocated for this system. A system such as this is not expected to be deployed for at least 10-15 years.

If this type of system is to be taken seriously, then the cost of each base-station must be very low, since there will be a vast number of them. Research effort at several establishments around the world is underway to develop optical technology for generating and distributing the 60 GHz carrier signals required. If the high frequency carrier is generated from a central location then the base-station can be a simple remote antenna, with no frequency translation or signal processing functions. These simple remote antennas should therefore be cheap to manufacture. Optical generation and distribution of 60 GHz signals is not straightforward since conventional laser sources have a modulation bandwidth of up to 20 - 30 GHz. A good technique for the optical generation of very high frequency signals is optical heterodyne, where the signal is generated in a fast photodiode by mixing the light from two lasers. The wavelength (or frequency) difference between the lasers is adjusted to provide the required beat frequency. Recent work at BT Labs has shown that a specially-developed laser that operates with two optical modes (rather than the usual case of one or many) is ideal for high frequency signal generation because the purity and stability of the resulting signal can be controlled very easily by electrical injection [8]. Figure 1 shows the optical and electrical (after photodetection) spectra obtained from such a component. The electrical spectrum shows a signal with very low phase noise (less than -70 dBc/Hz at 10 kHz offset). This is an example of the enabling technology under development that promises to bring the cost of 60 GHz base-stations to a low enough level for widespread deployment in the future.

optical spectrum

electrical spectrum

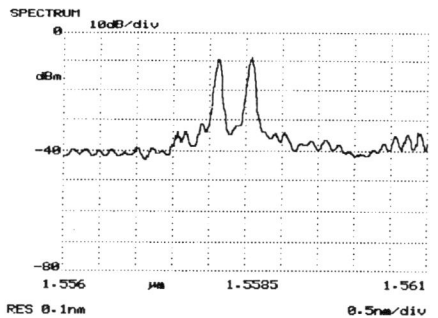
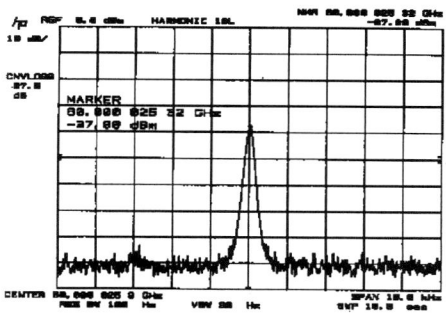

Figure 1. Optical and electrical spectra for two-moded laser with electrical injection to provide purity and stability.

5.3. INFRA-RED (IR) SYSTEMS

IR, or optical wireless, is an alternative to using radio for the wireless carrier signal. Commercial systems are available at present for use outdoors (high capacity point-to-point links) and indoors (wireless LANs). The principle characteristics of IR systems are summarised below:

- low-cost transceiver technology. Lasers, light-emitting photodiodes, and photodiodes are very cheap compared to high frequency radio upconverters and downconverters.
- unregulated spectrum. This is a major problem for radio systems, where many applications compete for a limited resource.
- high attenuation in fog (300 dB/km). This limits their use to very short-range applications unless down-time during fog can be tolerated.
- low dynamic range compared to radio systems. This is due to low transmit power (limited by safety regulations and cost of source) and high receiver noise.

IR wireless LANs can be obtained from several vendors. Generally they provide Ethernet connectivity with data rates from 1 - 10 Mb/s. Some systems use diffuse optics to illuminate a whole area, although multipath propagation and dynamic range limitations give these systems data rates at the lower end of the range. For better performance, a point-to-point architecture is used, where the optical transceivers give narrow beams which require careful alignment. The more sophisticated systems use a tracking architecture to keep alignment as the mobile is moved. For outdoors applications, again point-to-point systems can be obtained from a range of vendors. A top-of-the-range system such as Canonbeam from Canon provides 155 Mb/s data rates or four broadcast video plus 10 audio channels over a range of up to 4 km with automatic tracking. This type of system is aimed primarily at the outside broadcast market.

Research effort is underway to improve the performance and lower the cost of these systems. For example, at BT Laboratories, indoor systems have been developed with very high data rates (up to 1 Gb/s) over a range of a few metres using a tracking architecture and novel transceiver technology [9].

6. The promise of THz technology

As the frequency of operation increases, the size of components in radio systems decreases as a result of the reduction in wavelength. This can be an advantage, in terms of miniaturisation, but can also be a disadvantage because the precision engineering required to produce some of these components is expensive. If the operating frequency is pushed into the THz regime, the wavelength becomes so small that photolithographic techniques can be used to fabricate components such as rectangular metallic waveguides on substrates. Furthermore, processes can be developed in which active devices such as diode mixers and multipliers can be integrated monolithically with these waveguides. If monolithically integrated upconverters and downconverters can be realised using a mass manufacturing process, then these components should be cheap. This is the promise of THz technology.

6.1. CHOICE OF OPERATING FREQUENCY

For the short-range applications of interest, operation at a frequency corresponding to an atmospheric absorption peak is desirable to minimise interference in outdoor systems. Figure 2 shows a plot of atmospheric absorption as a function of frequency for oxygen and water vapour. Oxygen has absorption peaks at 60 and 120 GHz, while water vapour has absorption peaks at 180 and 325 GHz. For an integrated approach, the first two frequencies would require waveguides with very large dimensions, and would therefore not be practical. 180 GHz is probably the lowest frequency for which this approach is feasible.

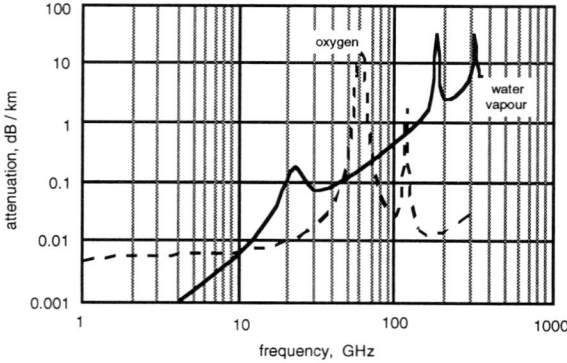

Figure 2. Atmospheric absorption as a function of frequency for oxygen and water vapour.

6.2. CURRENT STATUS OF THz TECHNOLOGY

Much of the development of terahertz technology takes a hybrid approach, where active devices are placed inside micro-machined waveguide cavities, and contacted using point-contact whiskers. A good example of this approach is the work at the Rutherford Appleton Laboratory. Components such as waveguide mixers and corrugated feedhorn antennas have been developed up to a frequency of 2.5 THz. The mixer uses a 0.5μm dot matrix diode and is contacted using novel planar whisker technology [10].

An EPSRC-funded project known as TINTIN (Terahertz Integrated Technology Initiative) was established in order to pursue the monolithic integration approach [11]. Figure 3 shows a schematic of an integrated waveguide process developed on this project. This technology is still at a very early stage of development, and considerable progress must be made before the promise of high-performance, low-cost upconverters and downconverters can be achieved.

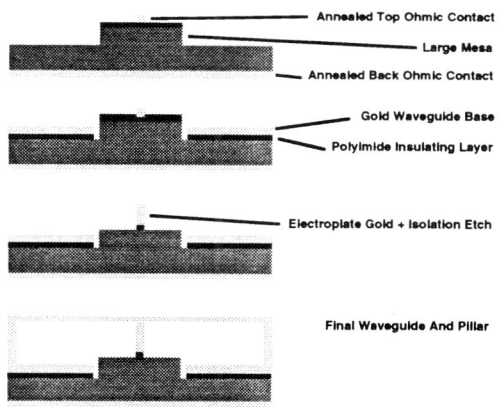

Figure 3. Schematic of an integrated waveguide developed on the TINTIN project.

7. Conclusion

We have seen that current mass-market wireless systems are relatively narrowband and use carrier frequencies in the low microwave region, where electronics are well developed and radio propagation provides large-scale coverage. We have also seen that there is a trend in wireless systems towards higher frequencies and shorter range to increase capacity. At these higher frequencies, propagation becomes line-of-sight, with little difference between 60 GHz and IR. The propagation characteristics of THz signals will also be very similar.

Research is underway to reduce the base-station cost for 60 GHz radio systems, such as the use of analogue optical fibre links to move the complexity to a central location where resources can be shared. IR transceivers are already relatively cheap and the performance is being steadily improved in the research community.

Against this backdrop, we ask the question: is there a role for THz technology in wireless telecommunications systems of the future? The answer will depend on cost. Can the cost of THz technology be low to compete with IR for example? Can it be engineered for mass market? The integrated approach to terahertz technology is the only realistic alternative for this scenario, and the TINTIN project has made a useful start in this direction. We may possibly be able to answer 'yes' to these questions ultimately, but not in the near future. This is a very long-term and speculative enterprise which will require considerable effort. It could be argued that the effort is not warranted purely for telecommunications, but if the technology is developed for other applications then telecommunications could be a grateful beneficiary.

Acknowledgements

The author would like to thank Les Westbrook and Adrian Pote for helpful comments during the preparation of this paper.

References

1. Fernandes, L. (1993) Overview of the MBS project R2067, *RACE Mobile Workshop, Metz(F)*, pp.75-79.
2. Wilkinson, T.A. (1995) HIPERLAN: An air interface designed for multimedia, *2nd International Workshop on Mobile Multi-Media Communications, Bristol(UK)* pp.S1/3/1-5
3. Swain, R.S. (1996) A vision of UMTS *Int. Conf. on Telecommunications, Istanbul(Turkey)*, pp.89-96.
4. Alexander, S.E. and Pugliese, G. (1983) Cordless communication within buildings: results of measurements at 900 MHz and 60 GHz, *Br. Telecom Technol. J.* **1**, 99-105.
5. Takimoto, Y. (1995) Recent activities on millimetre-wave indoor LAN system development in Japan, *IEEE NTC'95 The Microwave Systems Conference, Orlando(US)*, pp.7-10.
6. Prögler, M. (1994) Some notes on the broadband indoor wireless communications project of the citr. *R2067 MBS/WP2.2.2/DB035.1, MBS*
7. Skellern, D. and Percival, T.M. (1994) High-speed wireless LANs: Technologies for the missing link *IEEE Microwave and Millimetre Wave Monolithic Circuits Symposium*
8. Lima, C.R., Wake, D. and Davies, P.A. (1995) Compact optical millimetre-wave source using a dual-mode semiconductor laser, *Electron. Lett.,* **31**, 364-366.
9. Wisely, D.R. (1996) A 1 Gbit/s optical wireless tracked architecture for ATM delivery *Colloquium on Optical Free Space Communication Links (Ref. No.1996/032) London(UK)* pp.14/1-7.
10. Ellison, B.N., Maddison, B.J., Mann, C.M., Matheson, D.N., Oldfield, M.L., Marazita, S., Crowe, T.W., Maaskant, P. and Kelly, W.M. (1996) First results for a 2.5 THz Schottky diode waveguide mixer *7th Int. Symp. on Space Terahertz Technology, Virginia(US)*.
11. Parkhurst, G.M., Brown, D.A., Chamberlain, J.M., Middleton, J.R., Cronin, N.J., Collins, C., Pollard, R.D., Miles, R.E., Steenson, D.P., Bowen, J. and Henini, M. (1995) Integrated Resonant Tunnel Diode, *20th Int. Conf. on Infrared and Millimetre-waves, Orlando(US)*.

TECHNICAL ISSUES OF TERAHERTZ COMPONENT FABRICATION

or how to make the bits

R.J.WYLDE

Physics Department QMW,
Mile End Road, London, E1 4NS, UK

Thomas Keating Ltd.,
Station Mills, Billingshurst, West Sussex, RH14 9SH, UK

QMC Instruments Ltd.,
Physics Department QMW
Mile End Road, London, E1 4NS, UK

http://qmciworks.ph.qmw.ac.uk/homepage.htm

1. Outline of Talk

In remote sensing of the atmosphere - from space and from the ground - in the development of Tokamak fusion reactors for 21st Century energy production and in probing the heavens, terahertz technology can contribute to some of the 'Grand Challenges' facing mankind at the end of the 20th Century.

More consumer related uses are also planned including anti-collision radar and cruise control for cars, aircraft wind shear warning, concealed weapon detection and short range broadcasting.

This talk is about making things. And (on a terahertz reduced scale) rather large things, rather than very small things (such as Schottky diodes). They will also be linear things, so they are not involved in the actual non-linear process of generating or detecting terahertz radiation. It will not be a very exciting talk, in the context of others on active components. But it might be a useful one.

I will be talking about the manufacture, and before that, the design of a variety of terahertz components. These will include

- Beam forming and receiving components,
- Beam control components,
- Beam processing components,
- Beam absorbing components.

Inevitably, these will be drawn from systems that I have been involved in, but I will try to cover as wide a range as possible.

J.M. Chamberlain and R.E. Miles (eds.), New Directions in Terahertz Technology, 299–322.
© 1997 *Kluwer Academic Publishers. Printed in the Netherlands.*

2. Beam forming and receiving components

2.1. SINGLE MODE ANTENNAS WHERE $A\Omega = \lambda^2$

Single mode operation can be determined at a number of places within an optical system. The source can be single moded, the receive antenna can be single moded or the receiver can be a heterodyne one, which also forces single mode operation. In these cases the optical throughput, $A\Omega$ is limited to λ^2.

The range of antennas which produce single modes includes corner cubes, planar antennas and waveguide fed cones.

In the last category, corrugated horns are excellent examples of single mode defining antennas. Figures 1 and 2 show a corrugated horn integrated with an SIS mixer block working at around 250 GHz. Their operation is well understood [1] using Gaussian Beam-Mode theory [2, 3, 4]. We make corrugated horns by a lost wax, or actually, lost aluminium process.

An aluminium mandrel is formed with fins where the slots are to be. The mandrel is then plated with a thin layer of gold before copper is deposited. Cavities, which may form as the fins close over before they are completely solid, can be filled with a conducting material. The structure is then machined and the aluminium washed out with NaOH.

Waveguide structures can be spark eroded (EDM). Figures 3 and 4 show such a modern CNC spark erosion machine, with a 'C' axis which can, for example, cut spiral waveguides.

Typical antenna patterns are given in Figures 5 to 7, showing low sidelobes and high axial symmetry.

Non-linear tapers can be used to form, for example, horns with zero flare-angle at the aperture. Such horns have no phase error across their caps and have deep nulls in their antenna patterns. To see how accurate modal analysis of horns can be, look at Figures 8 and 9. The former is a prediction, showing higher sidelobes caused by the presence of HE_{1n} modes, formed by the non-linear taper. The latter is the measured result.

Although initially used in the microwave region, very high frequency horns have been made to function - workers at the Rutherford Appleton Laboratories (RAL) have experience of making these horns operate at 2.5 THz.

2.1.1. *Wideband horns*
Corrugated horns can be made to operate over quite wide frequency bands. For example, in an Electron Cyclotron Emission Experiment at the Euratom funded JET Tokamak Experiment outside Oxford, feedhorns are used which cover two bands: 120-180 GHz and 180 - 240GHz [5]. The former covers a 1:1.5 bandwidth. The internal form of the horns was designed by my colleague Prof. David Olver at QMW using computation techniques which match the field from one fin to the next. The predictions of programs based upon such techniques, as noted above, are very accurate and allow the kind of performance shown in Figures 10 to 15.

2.2. MULTIMODE ANTENNAS, $A\Omega \gg \lambda^2$

For an incoherent receiver looking at an incoherent source the number of modes ceases to be one. For example, COBE, the project which measured the Cosmic Background radiation (one of the most important experimental results of the late 20th Century) and Grating Polychromators looking at Electron Cyclotron radiation from hot Tokamak plasma come into this category.

2.2.1. *What does one mean by modes?*
In an optical fiber or a waveguide near cut on the answer is quite straightforward: There are various distributions of E and H fields whose cross-coupling integral

$$\int_S \phi_n \phi_m^\star dS = 0$$

when $n \neq m$. In mathematical terms, the functions which describe the fields are orthogonal. But what are the modes in free space which define what passes into an incoherent receiver?

In a recent paper [6], my colleagues Derek Martin and John Bowen outlined a coherent representation of partially-coherent beams. Using a Gaussian-Schell source model, they identify the number of (coherent) modes needed to describe a field as $(1 + 4/\beta^2)$, where β is the ratio of the widths of the spatial coherence function and the spectral intensity distribution. For a coherent beam, β is very large as the coherence extends to very low intensities and therefore only one mode is needed.

When more than one mode is present, the antenna patterns - the profile of reception which determines Ω, can still be determined by a a number of methods: Geometrical Optics supplemented by the Geometrical Theory of Diffraction, for example. An elegant method is outlined in a paper by Murphy and Padman [7]. This approach sums the power in all the individual higher order modes that are allowed to propagate through the system. Power is summed here - and not amplitude - as none of the modes will be coherently related to any other. This technique can be used to predict the throughput and antenna patterns of multimode systems - and therefore the number of modes involved via

$$\frac{A\Omega}{\lambda^2} = n$$

2.2.2. *Winston Cones*
Winston Cones were developed in the 1970's [8] as non-imaging light collectors for solar energy farms. They are paraboloids of revolution, with an optical axis running from one side of the entrance aperture to the other side of the smaller exit aperture. The focus of the paraboid is found at this exit aperture point. In two dimensions their optical properties are such that rays up to a specific angle pass through the exit aperture. Rays beyond that angle do not. They therefore have a nominal 'Top Hat' antenna pattern. In their practical 3D form the sharpness of the cutoff is reduced by skew rays but they produce very well defined beams and are a common feature of QMC Instruments' multi-moded bolometer systems. Figure 16 shows a dual polarisation system using Si Bolometers currently under construction for atmospheric water vapour measurements.

Our Winston Cones are formed by Electroforming onto stainless steel mandrels which are then pulled. Mandrels can be re-used.

3. Beam control components

Designers of Quasi-Optical circuits in the terahertz region have to contend with the diffractive spreading of their beams. Both lenses and mirrors can be used to control this expansion.

3.1. LENSES

Lenses have the advantage of lower cost and often produce lower cross-polar contamination. They do not disturb the direction of propagation and often lead to simpler circuits than those formed by off-axis mirrors. However lens surfaces are reflecting and reflections from these surfaces cause resonances and standing waves both within and between lenses. This can be a serious problem, particularly in coherent systems. Beams passing through lens also suffer ohmic attenuation.

We make lenses on a Hardinge CNC lathe, using profiles generated by a 'lapsed phase' method, described by John Bowen elsewhere in these proceedings. Materials include

- HDPE,
- PTFE,
- TPX.

Lenses can be impedance matched by adding layers of lower refractive index material to their surfaces, or more commonly, by cutting rings into the base lens material to a depth of $\lambda/4$ where λ is the wavelength in the blooming region. This process is known as blazing. Enough material needs to be removed to drop the average density to $\sqrt{n}$ where n is the refractive index of the material [1]. Lenses can be blazed by adding holes, annular grooves or lines.

3.2. MIRRORS

Mirrors, on the other hand, do not generate standing waves or significant ohmic losses. They do have to be operated off-axis to allow incoming and outgoing beams to be distinguished and this can lead to unwanted cross-polar and higher order mode generation. They are also more expensive to make, although CNC machine tools are reducing the absolute cost difference between lenses and mirrors.

The mirrors are often off-axis ellipsoids of revolution, whose two radii of curvature are chosen to be the local radii of curvature of the incident and reflected Gaussian beam-modes [2]. Our mirrors are machined on a NU-5 CNC Huron milling machine (see Figure 17) and

---

[1] I am assuming here that the external medium is air, or a vacuum - ie with a refractive index of essentially 1. If this is not the case, the geometrical mean needs to be taken.

[2] It is acknowledged that the choice of beam parameters when a propagating beam is decomposed into an orthonormal Gaussian-Laguerre or Gaussian-Hermite beam-mode set is arbitrary: A given field can be represented by such an orthogonal set with any choice of beamwidth and radius of curvature parameters. The choice of these parameters determines the amplitude of the, possible complex, mode coefficients needed to reconstruct the field. A good choice will minimise the number of modes which contain significant power. A poor choice will lead to significant power in higher order modes. These higher order modes suffer higher X-polar and higher-order mode cross coupling at the hands of an off-axis mirror [7]. Hence the radii of curvature for our mirror were chosen to match those from a mode set which has little power in higher order modes. (We chose beam parameters to maximise the power in the fundamental)

then hand polished. The mirrors have had RMS surface accuracy better than $10\mu m$ (less than 1/300 of a wavelength). Care is taken that the dowel hole positions which align the mirror on the baseplate are correctly placed.

In our view, QO circuits need to be made from components whose alignment aids - primarly dowels - are sufficiently accurately placed that adjustment is not required: This is especially true of complex systems whose number of degrees of freedom for adjustment precludes alignment by, say, the movement of three-point mounts.

3.3. COMPARISON OF PERFORMANCE

It is interesting to compare the performance of lenses and mirrors. Consider a simple circuit consisting of two corrugated feed horns, acting as Gaussian beam-mode launchers and two beam reforming components forming a beamwaist between them. Working with Roger Appleby at DRA Malvern such a circuit was built on a split cube circuit bench. The measurements were carried out using an HP8510C network analyser with a W band extension with a one-path-two-port test set. Results for co-polar transmission between 86 and 100 GHz are given in Figures 18 and 19.

For co-polar transmission, we see average values for bloomed and unbloomed lenses - 69% and 79% transmission. Significant resonant structure is present.

The mirror results show power losses of less than 10% (0.43 dB) through two lenses and two mirrors: This means that each horn-mirror combination is only losing about 0.2dB, quite a respectable result.

The mirror losses are significantly lower than with refracting optics, and the variations across the 12 % frequency band are also lower, indicating that standing wave effects are, as one would expect, lower.

4. Beam processing components

4.1. GRIDS

Free standing wire grids are used at the heart of many QO systems including interferometers (Figure 20), polarisation switches and network analysers. They can be made of tungsten wire, of diameters down to 10 μm with 20 μm pitching onto aluminium or stainless steel frames.

It is a non-trivial task getting the metal frames flat. It is easy to forget that thin sections of metal are full of stresses and are - in toolmaking terms - 'alive'. We use a complex and large surface grinder, shown in Figure 21 to get better than 10 μm flatness on our frames.

Glue is placed on the wires and a keeper frame used to ensure that the wires are close up to the frames. We can wind up to 550mm diameter grids.

To operate at cryogenic temperatures, we are forced to solder the wires onto the frame as glues fail after repeated thermal stressing. Here we use Invar frames and gold plated tungsten wire. The Au plating allows the soldering to work. Invar is used so that the wire tension increases as the grids are cooled (the tungsten contracts faster than the Invar).

4.2. SINGLE SIDEBAND FILTERING

As an example of the use of grids, I will now describe how they can be used to form Single Sideband Filters.

One can sense the presence of trace elements in the atmosphere from their vibrational and rotational emission spectra. For this task it is important that the passive radiometer, whose signal frequency is carefully adjusted to that of the wanted emission line, does not see contamination in the receivers image band.

For example MASTER, a planned ESA THz sounder, needs single sideband filtering with imageband rejection > 30dB while keeping inband losses below 0.5dB. Dichroic plates, formed by drill plates or mesh layers, are often proposed to provide such filtering. An alternative is the use of single-pass interferometric filters of the polarising Martin-Puplett type [9, 10].

The Martin-Puplett filter is a version of a Martin-Puplett interferometer which itself is a polarising version of the Michelson interferometer. The incoming signal is split into two beams which are recombined after traveling along paths of differing lengths. The combined signal is therefore modified in a way that depends upon the *reduced* path difference. By '*reduced* path difference' one means the path difference in wavelength terms. Such modification can be used to filter desired signals from undesired ones.

A drawing of two such Martin-Puplett filters (MPF's) in a chain being fed by a QO beam is given in Figure 22. As coherent receivers are to be used, polarisation coding of frequency bands is acceptable because the receiving process will force single polarisation operation.

The power output of MP device follows a simple $sin^2(\theta)$ law, where θ is a function of the reduced path difference in the two arms.

4.3. THE OPERATION OF A MPF

Figure 23 shows the operation of a MPF [9]. One can view its workings in two stages: The first section, formed by a crossed grid and two roof mirrors, acts as a frequency-dependent polarisation rotator. Then the second stage - the single grid - analyses the effect of the first stage, allowing the wanted signal to pass and the unwanted signal is reflected.

Looking at Figure 25a-e:

- Figure a shows a polarised beam at +45 degrees arriving at the crossed grid.
- Figure b shows that the cross grid has resolved this into vertical and horizontal beams moving in opposite directions down two arms.
- Figure c shows the roof mirrors, with edges set at 45 degrees to the horizontal, flipping the direction of polarisation by 90 degrees and returning them to the crossed grid. The horizontally polarised beam returns as a vertical polarised beam, and vice versa. Then the beam that had been reflected by one arm of the crossed grid is transmitted by that arm and reflected by the other.
- Figure d show the recombined beam heading towards the single grid in a polarisation state determined by the 'reduced phase lag' between the two arms.
- Figures ei, with a 2π phase lag, and eii, with a π phase lag, show the grid resolving the recombined beam, with unwanted signal being passed to a dump.

A recent practical advance, made by John Payne at NRAO in the US, allows the construction of crossed grids (Figure 24). A MP filter made from such a crossed grid does not change the direction of beam propagation. This advance will allow the construction of smaller MP filter circuits. Conventional designs required the output beam to be bent by 90 degrees to the input.

Returning to Figure 22 we can see the whole circuit. In a time-reversed view a corrugated horn feeds the off-axis ellipsoidal mirror. The beam passes through a cross grid and roof mirror assembly where the frequency dependent polarisation is generated. Unwanted frequencies are picked out and dumped by the second grid. A second MPF repeats the process leading to the double filter action.

A breadboard (Figure 25) was delivered to Matra Marconi Space in Toulouse in July 1996 and experiments to see how it performs will start shortly.

4.4. ISOLATORS

Isolators, common non-reciprocal devices found in the microwave region are not often found in the THz region. But free space isolators with quite reasonable performance can be made. As an example of their use, I will describe a two colour Interferometer measuring $\int N_e dl$ in the diverter region of JET [11].

The interferometer works at two frequencies, or colours. These colours allow compensation for vibration and mechanical movement between the instrument and the Tokamak, to be made. The specification for the instrument is

- One MHz sampling rate and 7 degree resolution (0.02 fringe)
- Plasma arm attenuation 65 dB, generated by the complex path taken by the signal to the Tokamak
- Crosstalk less than 83 dB
- $4W_{optics}$, Transmit and Receive systems on different sides of plates

The QO circuit is shown in Figure 26 and a section in more detail from our Gaussian Beam-Mode Program is shown in Figure 27.

The need to keep crosstalk below 83 dB encouraged the use of isolators to stop any reference signal beam from reaching the detectors fed directly from the plasma. Isolators were formed by the use of two polarising grids and a ferrite based Faraday rotator. Looking at Figure 28 one sees an incoming beam passing through a vertical grid. The beam then enters the ferrite via a blooming layer and suffers a rotation of 45 degrees. [3]

The beam can then pass through a grid set at 45 degrees without loss. A returning beam, on the other hand, will suffer rotation in the same sense as before and find itself rotated a further 45 degrees. With its polarisation now parallel to the vertical grid's wires it is reflected into the dump.

For the JET Interferometer we used ferrite sheets 100mm by 100mm square an insertion loss of 1dB or less and isolation of more than 17dB at 130 GHz were obtained.

---

[3]This rotation angle is only slightly dependent on frequency in the THz region because the gyromagnetic frequency lies in the microwave region. In contrast, isolators in the microwave region are often narrow banded because the rotation angle changes strongly with frequency.

5. Beam absorbing components

5.1. GENERAL COMMENTS ON ABSORBERS

An ideal absorber, often called RAM (Radar Absorbing Material) would absorb all radiation from all angles, without scattering or re-radiating any power. RAM materials are, of course, not ideal, and energy which is not absorbed will either be scattered or suffer specular reflection.

The operation of an absorber can best be thought of as a two stage process. First the radiation must be allowed or coaxed into entering the material. Then it must be absorbed and turned into heat.

Unfortunately these two requirements - getting the power in and absorbing it - tend to pull the RAM designer in different directions. The reflectivity of a material in air is a function of its impedance: For normal incidence, the amplitude reflectivity R is given by

$$R = \frac{Z_{air} - Z_{material}}{Z_{air} + Z_{material}}$$

where Z is the impedance of the material - the ratio of E to H field in the sample. Z_{air} is close to that of the vacuum, 377Ω.

The impedance of the material is given by

$$Z = c\mu/\tilde{n} = (\frac{\mu}{\epsilon - j\sigma/\omega})^{1/2}$$

To minimise reflections one either needs to use materials with non-unity permeability ("magnetically loaded") or with low permeability/refractive index. But materials with low permeability will tend not to absorb very well.

Flat surface absorbers are sometimes used, and have the advantage that it is easy to predict (and therefore control) where any specular reflection will go. However the single surface performance tends not to be too good.

Complex structures are often used which maximise, from a geometric optics point-of-view, the number of reflections suffered by an incident wave before leaving the structure. The structure can significantly increase the fraction of incident power entering the material at the expense of losing control of the direction of scattered radiation. Such structures, such as wedges and pyramids, can often produce peaks in reflectivity caused by constructive interference of small reflections into Bragg peaks. Foam structures can also be employed, such as found in AN72 [12].

Loading materials are often carbon and various forms of iron and iron oxides.

5.1.1. *Desirable properties of an absorber*
It is not only electromagnetic properties which are of importance in choosing a RAM. For example if it is to be flown in space it should be clean, and not shed carbon. It is important too that it does not outgas or store water. This will be crucial in a system which has to be evacuated or filled with nitrogen to remove atmospheric absorption.

Thus the properties of interest in a RAM material, apart from its quality as a terahertz absorber include

- weight

− thickness
− surface integrity
− vacuum integrity
− cryogenic properties
− cost

There is a significant lack of information on the performance of absorbers, even in the microwave region, where there is a significant commercial market. Often a single absorption figure is quoted [See, for example[13]). A full treatment should show reflected power as a function of both incident and emergent angles - ie a four dimensional surface. Such a surface will contain both specular and scattering information. A reduced form of such a data set would be found by integrating over all reflected angles to give the total absorption.

Different applications will place different demands on the performance of absorbers. Some will need to work at normal incidence, with low normal incidence reflection, but with tolerance to high-angle scatter. Other applications will demand very low scatter, but may not need good normal incidence performance. More than one type of absorber will often be needed with any terahertz system.

5.1.2. *Currently Available Materials*

The Radar Modelling community - people who want to know the antenna pattern or Radar Cross Section (RCS) of ships, aircraft and satellites - has placed the greatest demand on FIR absorbers as it has very stringent requirements on minimising unwanted reflections in the face of small target RCS. What little development and measurement effort to produce terahertz RAM there has been has had the requirements of RCS measurements in mind.

At low frequencies, materials such as AN72 from Grace Materials [12] are popular, but are actually rather poor absorbers, showing near normal incidence specular reflectivity worse than -20 dB at 100 GHz

Commercially produced Axminster Carpet made from wool has been used above 500 GHz in RCS ranges. Other man-made carpet materials have been tried, but have not proved as effective.

Thorn-EMI have developed a pyramidal based RAM [14]. With a quoted performance at 693 GHz (a popular FIR Laser line) of -39dB at an angle of incidence of 20 degrees.

The Lowell Research Foundation has developed a range of iron-oxide loaded silicone-based absorbers, using wedge-type surface geometries [15]. The quoted performance for a material called FIRAM-500 at 500 GHz for narrow angle near-monostatic reflectivity is -46dB [16]. Given the wedge geometry, there is a significant difference in reflectivity as a function of angle, and some very low minima have been observed, along with some Bragg diffraction peaks.

TK has been developing small RAM tiles which can be clipped together to form larger surfaces. The tiles have a square pyramidal surface and are polypropylene based. They are vacuum compatitible and cryogenically stable and are unlikely to absorb water. 500 GHz narrow angle near-monostatic reflectivity is -58dB [16]. Specular plus near-in scatter at 40 degrees and 585 GHz has been measured to be better than -42dB.

5.2. CALIBRATED LOADS

TK along with AEA Technology is developing a Calibrated Hot Load (CHL) under contract to ESTEC for future atmospheric and astronomical missions

Calibrated loads are needed to define a reference temperature for calibrating passive radiometers. They need to have an accurately known temperature - although their absolute temperature is often unimportant - and they must be very black. By black, I mean have a high emissivity. By reciprocity this means that the load must be a good absorber [4].

5.3. STRUCTURE OF POSSIBLE HOT LOADS

A range of different types of calibrated load have been used in the THz region.

5.3.1. *Brewster-angle based*

We have Brewster-angle-based loads for QO fusion diagnostic projects. They must be operated at a single polarisation. For coherent systems this is not a problem as all coherent receivers will operate at a signal polarisation. But it does limit the QO configuration of the receiver and polarisation coding of channels will not be possible. Reflection from the orthogonal polarzation will be very high: See Figure 1.12 in [17]. Assuming that polarisation concerns can be accepted, the angular spread of the (in a time reversed view) incident beam will determine performance. A beam of finite width, in the 'Angular Spectrum of Plane Waves' view[2], is made up of a spread of plane waves, centred on the direction of propagation. The spread is given by the well known Gaussian beam-mode formula [3]:

$$\theta = \frac{\lambda}{\pi W_0}$$

where W_0 is the beamwaist of the beam. (Note that this is not the beam size at the hot load).

If the beamwaist is too small, the performance will suffer, because the components of the incident beam diverge too far from the Brewster angle. In addition alignment in polarisation and - even more critically - setting the incidence angle are especially problematical. The refractive index of the material must be precisely known, must not vary with frequency *and be stable over long time periods*.

5.3.2. *Structured Cavity based design*

Anechoic chambers have been used for many years, and the absorber material has also been employed for centimeter, millimeter and submillimeter-wavelength calibration loads [18, 19].

SWAS, a NASA astronomical satellite operating in the 0.5 THz region, has an ambient temperature load with four sided pyramids cast into a mould. We are informed [16] that it had a -40dB power reflectivity at 500 GHz with a moderate sized Gaussian beam.

---

[4]Reciprocity is a very useful concept to appeal to when designing loads: The principle is that a perfect absorber must be a perfect emitter, so that we can conveniently view the Hot Load as an absorber of the reference beam rather than as the source of radiation.

Goldsmith and Kot [20] also report the use of such absorbers in a calibration hot target for the frequency range 85-95 GHz yielding between -33 and -28 dB reflected power according to the polarisation angle. However the authors point out some problems in addition to reflections, such as cross-polarisation and frequency and structure dependent reflections, even across 10 % bandwidth.

5.3.3. *Scaleless Cavity based design*

The design of near infrared (NIR) hot loads with emissivities ≈ 1 have been reported by several authors [21, 22, 23, 24]. Whilst there is no complete quantitative agreement between the various models, it would appear possible to reach such emissivities in practice, by careful design and good temperture control. At submillimeter wavelengths the energy radiated from a black body emanates from a thicker layer of absorber - usually a plastic or non-metal loaded with iron or carbon - so that the behaviour is somewhat different from radiation-heated NIR black bodies:

– Temperature gradients must be avoided through the absorber.
– The quasi-optical (gaussian) beam must be able to enter the aperture without truncation ($\leq 50dB$)

Carli [25] reports a sub-millimeter-wavelength calibration load of moderate bandwidths, which achieves $\epsilon_{eff} = 0.999$ in a narrow frequency range, but which is unfortunatly unsuitable in design for space operation. We have hence adopted as our baseline method a variation of the NIR hot loads.

5.3.4. *Scaleless cone design*

This proposed solution to ESA (See Figure 29) uses a long narrow cone coated with a thin layer of submillimeter-wavelength absorbing material. This permits excellent temperature uniformity, and radiation will experience many specular reflections and re-radiations within the core before it emerges from the cone. A shroud is provided to further contain NIR radiation.

In the event of reflections from the dielectric of the submillimeter -absorber (in particular at low angles of incidence) the beam would only emerge after a large number of reflections: taking geometric optics (albeit a first approximation), an axial beam will only emerge from a cone with 6.5 degree half-angle after 14 specular reflections [5] so that a specular reflectivity as high as 0.6 would yield an effective emissivity of greater then 0.999. There would be a slight deterioration for the first and last reflections, due to the higher reflectivity of the dielectric at these angles [17]: assuming (pessimistically) full reflection in these two cases, the effective emissivity would only decrease to about 0.998. Since specular reflectivity well below 0.6 is expected$\epsilon_{eff} \geq 0.999$ should be achieved.

6. Conclusion

The design and construction of terahertz components is a challenging area to practice science and engineering - 'engineering physics' if you like. It remains an underused part

---

[5]The, formula, given in [26] is $N = 180/\theta$ where N is the number of reflections and θ is the cone full angle in degrees

of the electromagnetic spectrum. In contrast to the microwave and NIR regions, which lie either side of our region, THz components and systems are still improving at a rate at which one does need statistics to notice the advance.

References

1. Wylde, R.J. (1984), Millimetre-wave Gaussian beam-mode optics and corrugated feed horns, *Proc IEE* **Part H(131)**, pp 258-262.
2. Martin, D.H. (Sept 1995) *Lecture notes on Millimetre-wave and Terahertz Optics, Lecture 1*, & Clemmow.,P.C. *The Plane-Wave Spectrum Representation of Electromagnetic Fields*, Pergamon.
3. Martin, D.H. and Lesurf, J.C.G. (1978) Submillimetre-wave Optics, *Infrared Physics* **18**, 405-412; Kogelnik, H. and Li, T. (1966) Laser Beams and Resonators, *App.Optics* **5**, 1550-1567.
4. Foster, R.P. and Wylde, R.J. (1992) The Effect of Quasi-optics Errors on Reflector Antenna Performance, *IEEE Trans* **MTT40**, 1318-1322.
5. Private communication: Bartlett, D and Smith, R.C.
6. Martin,. D.H. and Bowen, J.W. (1993) Long Wave Optics., *IEEE Trans* **MTT41**, 1676-1690.
7. Murphy, J.A. and Padman,R. (1991) Radiation Patterns of few-moded horns and condencing lightpipes, *Infrared Physics* **31(3)**, 291-299.
8. Harper, D.A.,Hidebrand. R.H., Steining.R. and Winston. R. (Jan 1976) Heat Trap: an optimized far infrared field optics system, *Applied Optics* **15(1)**.
9. Button, K. (1992) *Infrared and Millimeter Waves* Vol **6** Chap **2**, Academic Press, ISBN 0-12-147706-1 (details on the operation of Martin-Puplett Interferometers).
10. Lesurf, J.C.G. (1990) *Millimetre-wave Optics, Devices and Systems* Chap **9**, Adam Hilger, ISBN 0-85274-129-4.
11. Prentice, R., Edlington,T., Smith, R.T.C., Trotman, D.L, Wylde,R.J. and Zimmermann, P. (Feb 1995) A two colour mm-wave interferomter for the JET divertor, *Rev. Sci. Instrum.* **66(2)**.
12. Grace,N.V., Nijverheidstraat 7, B-2260 Westerlo Belgium. Phone ++ 32 14 57 56 11.
13. GEC-Marconi Materials Technology Application note on PA3 Broadband Absorber.
14. European Patent Application No 88309752.9 filed 18 November 1988.
15. Giles,R.H. *Silicone-based Anechoics at terahertz Frequencies* Horgan,T.M. and Waldman,J. (Dec 10-14 1992) *17th International Conference on Infrared and Millimeter Waves (Conference Digest), Pasadena, CA*, SPIE.
16. Private communication: SWAS team from Lowell Research Foundation.
17. Born, M. and Wolf, E. (1980) *Priciples of Optics, 6th edition, Figure 1.12*, Pergamon Press, ISBN 0-08-026482-4.
18. Keen, N.J. (1975) The Calibration of a Large Paraboloidal Antenna Using a Cryogenic Black-body Enclosure, *Proc. European Microwave Conference, Hamburg*, 668-671.
19. Erickson, N.R, (Nov 1985) A very Low-Noise Single-Sideband Receiver for 200-260 GHz, *IEEE Trans* **MTT33**, 1179-1188.
20. Goldsmith, P.F and Kot,R.A. (Sep 1979) Microwave radiometer blackbody calibration standard for use at millimeter wavelengths, *Rev. Sci. Instum.* **50(9)**.
21. Gouffe, A. (Jan-Mar 1945) Corrections d'ouverture des corps-noirs artificiels compte tenu des diffusions multiples internes, *Revue D'Optique*.
22. Williams, C.S. (1961) Discussion of the Theories of Cavity-Type Sources of Radiant Energy, *J. Opt. Soc. Am.* **51**, 566.
23. Sparrow, E.M., and Jonsson, V.K. (Jul 1963) Radiant Emission Characteristics of Diffuse Conical Cavities, *J. Opt. Soc. Am.* Vol **53(7)**.
24. Campanaro, P. and Ricolfi, T. (Jan 1967) New determination of the Total Normal Emissivity of Cylindrical and Conic Cavities, *J. Opt. Soc. Am.* **57(1)**.
25. Carli, B. (Dec 1974) Design of a Blackbody Reference Standard for the Submillimeter region, *IEEE Trans* **MTT22(12)**.
26. Janz, S., Boyd, D.A. and Ellis, R.F. Reflectance characterstics in the submillimeter and millimeter wavelength region of a vacuum compatable absorber.

Figure 1.

Figure 2.

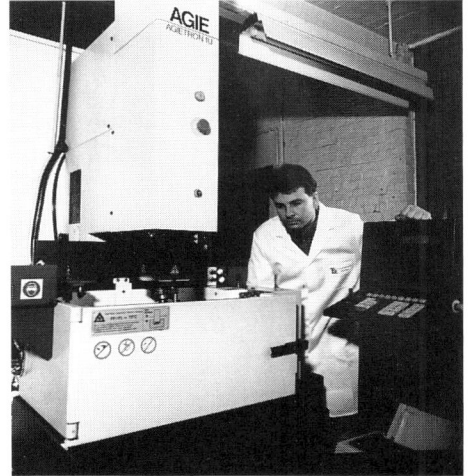

Figure 3.

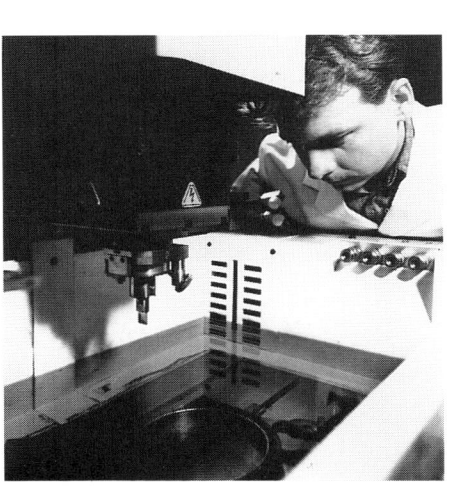

Figure 4.

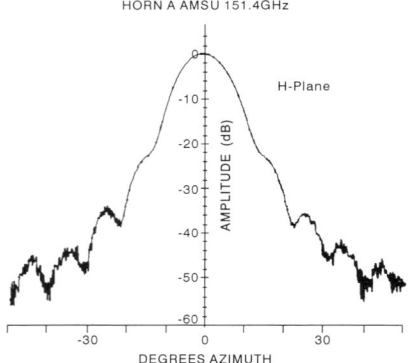

Figure 5.

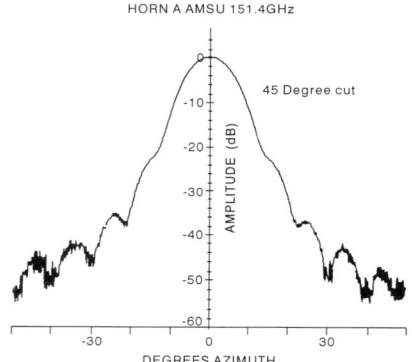

Figure 6.

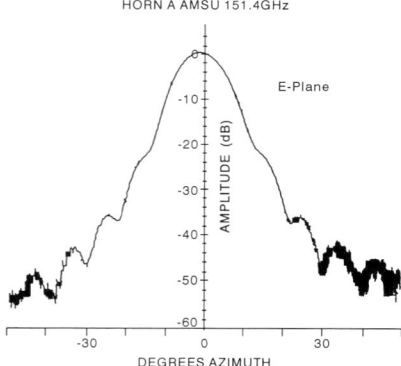

Figure 7.

H1 EM HORN 183 PROFILE H PLANE CUT

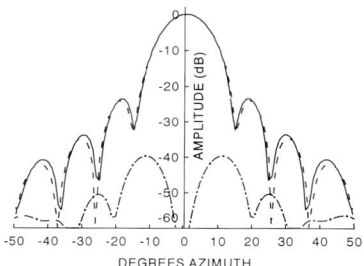

Figure 8.

H1 EM HORN 183 PROFILE H PLANE CUT

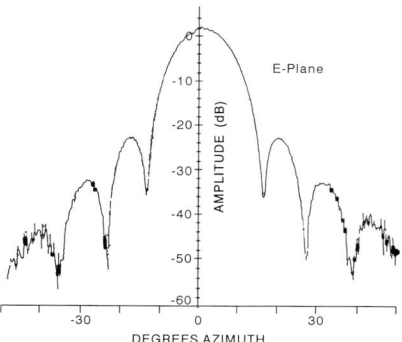

Figure 9.

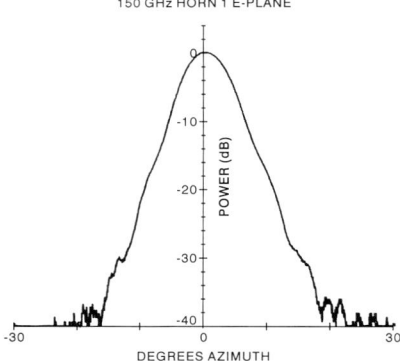

Figure 10.

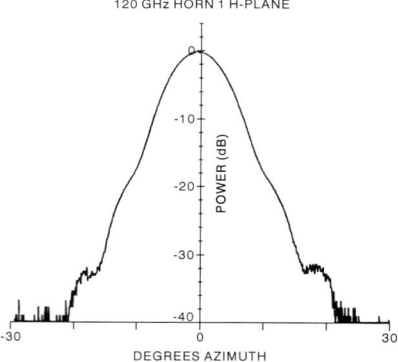

Figure 11.

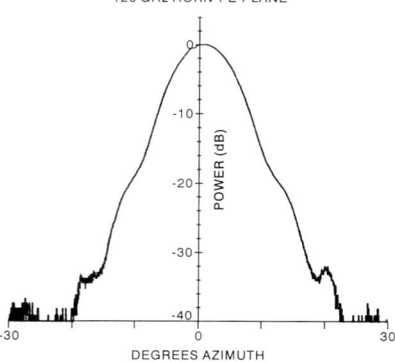

Figure 12.

315

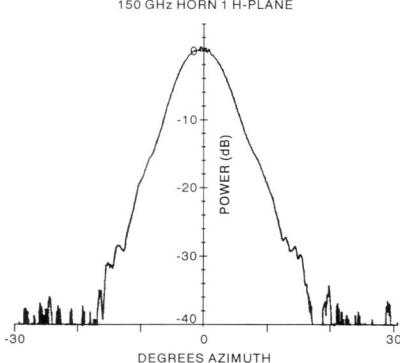

Figure 13.

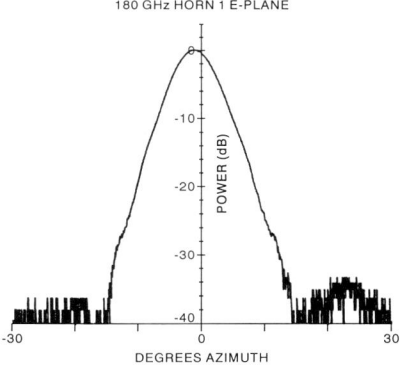

Figure 14.

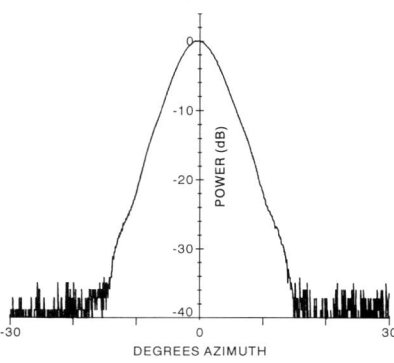

Figure 15.

316

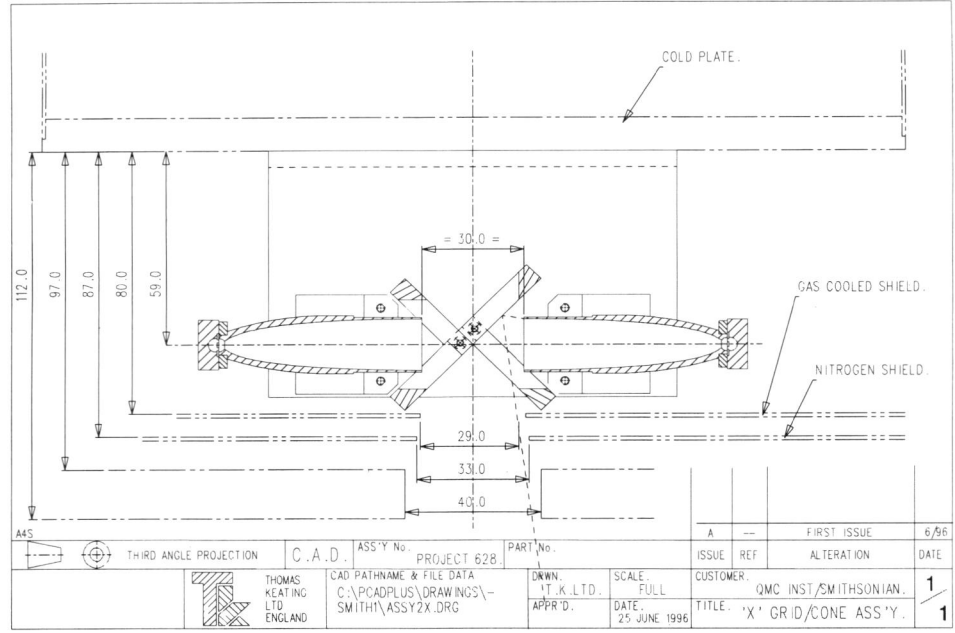

Figure 16.

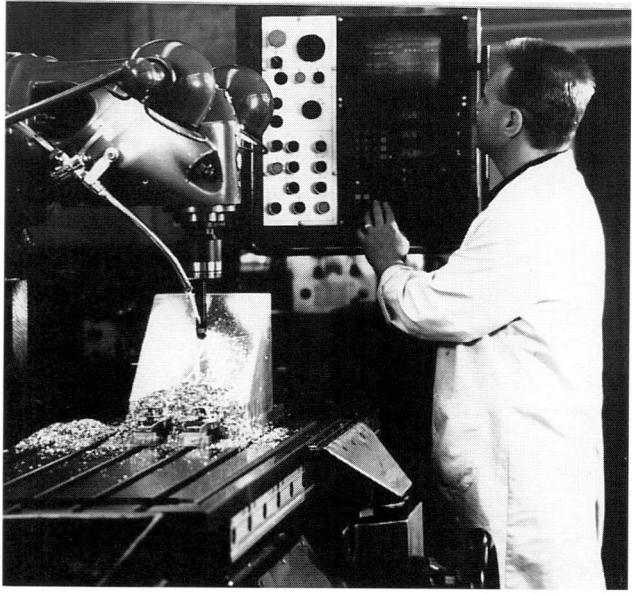

Figure 17.

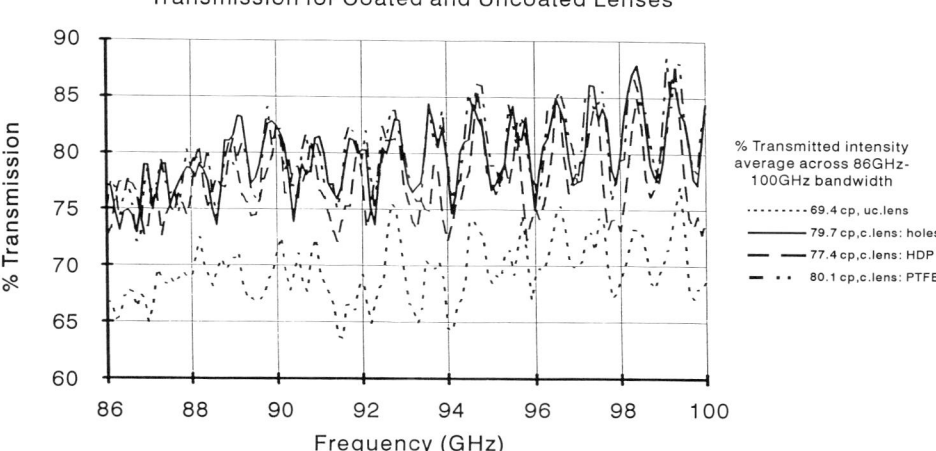

Figure 18.

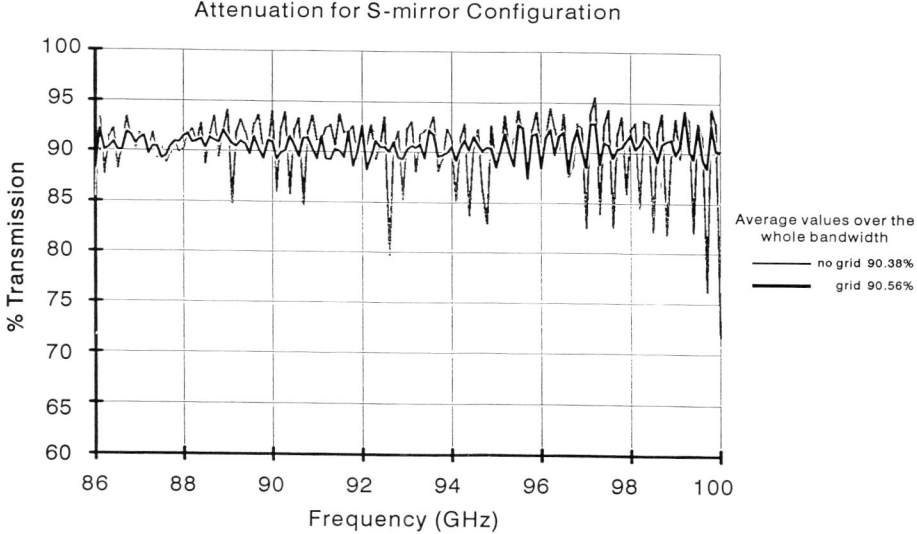

Figure 19.

318

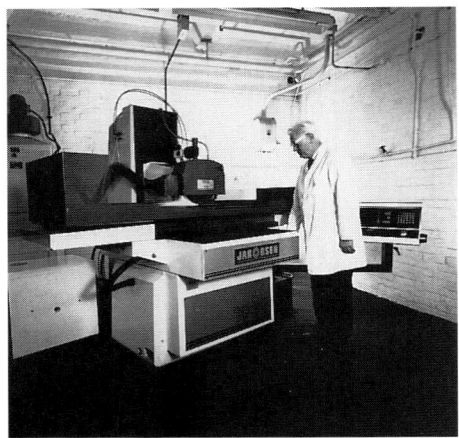

Figure 20.   *Figure 21.*

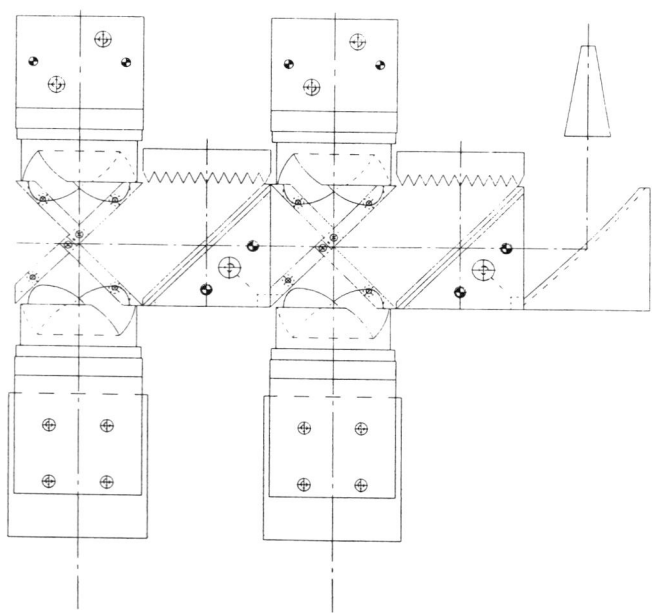

Figure 22.

319

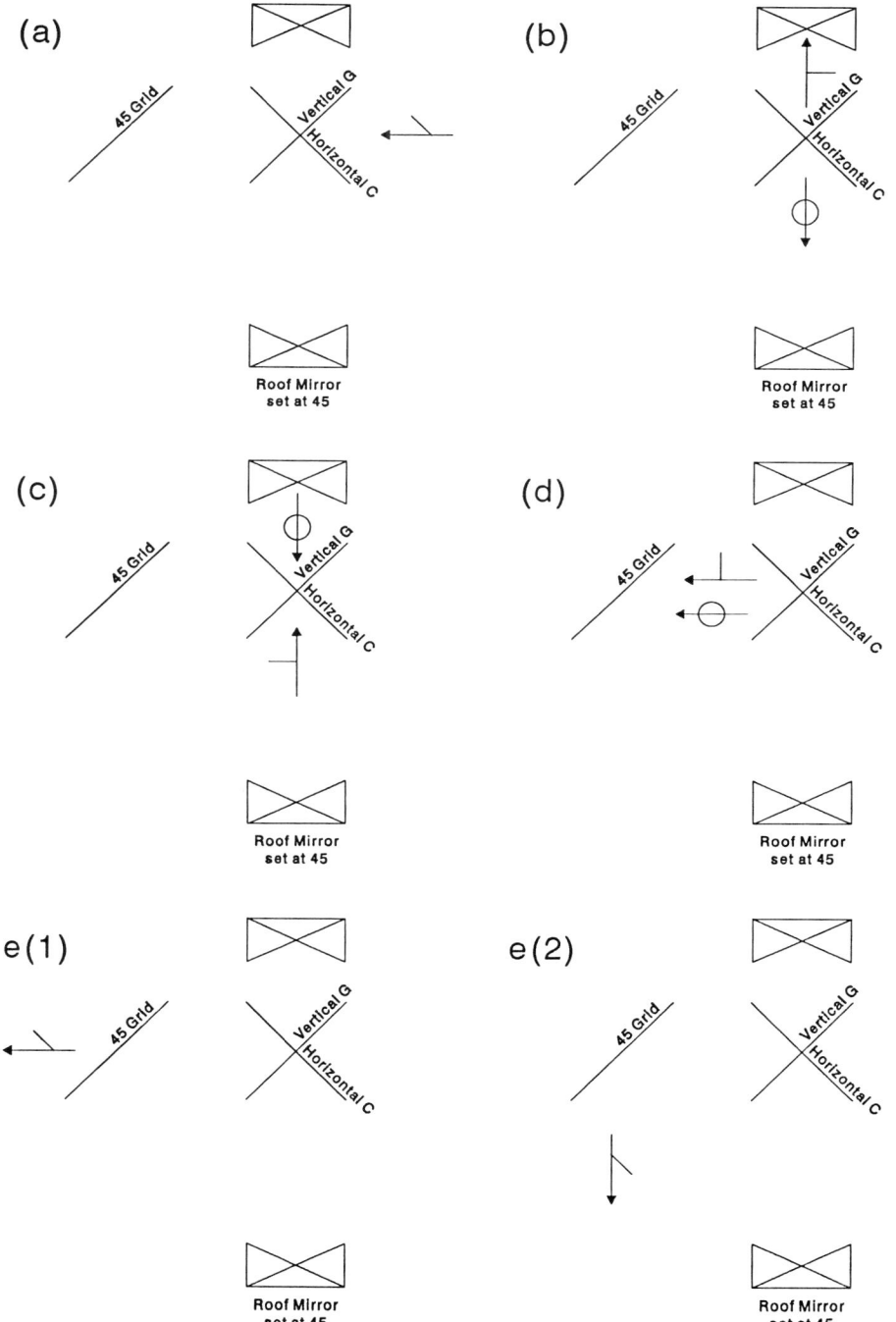

Figure 23.

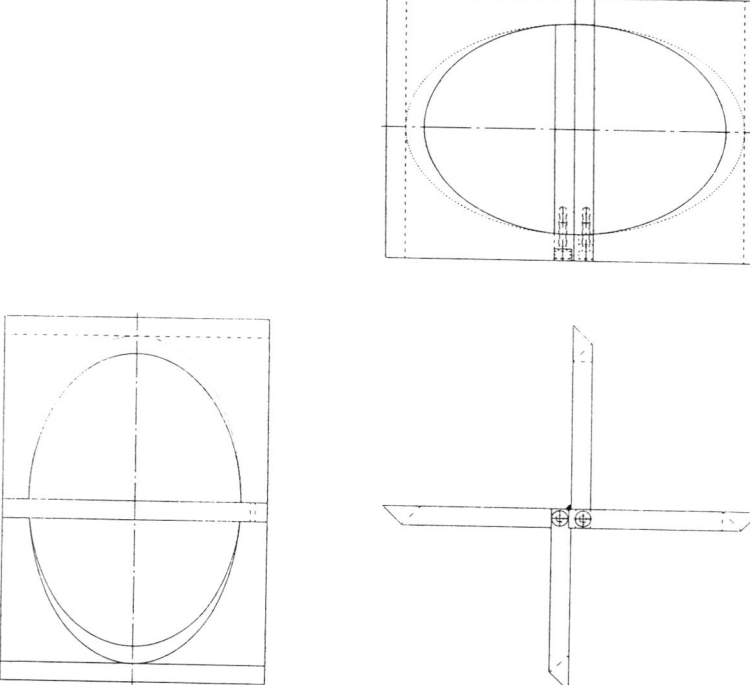

Figure 24.

Figure 25.

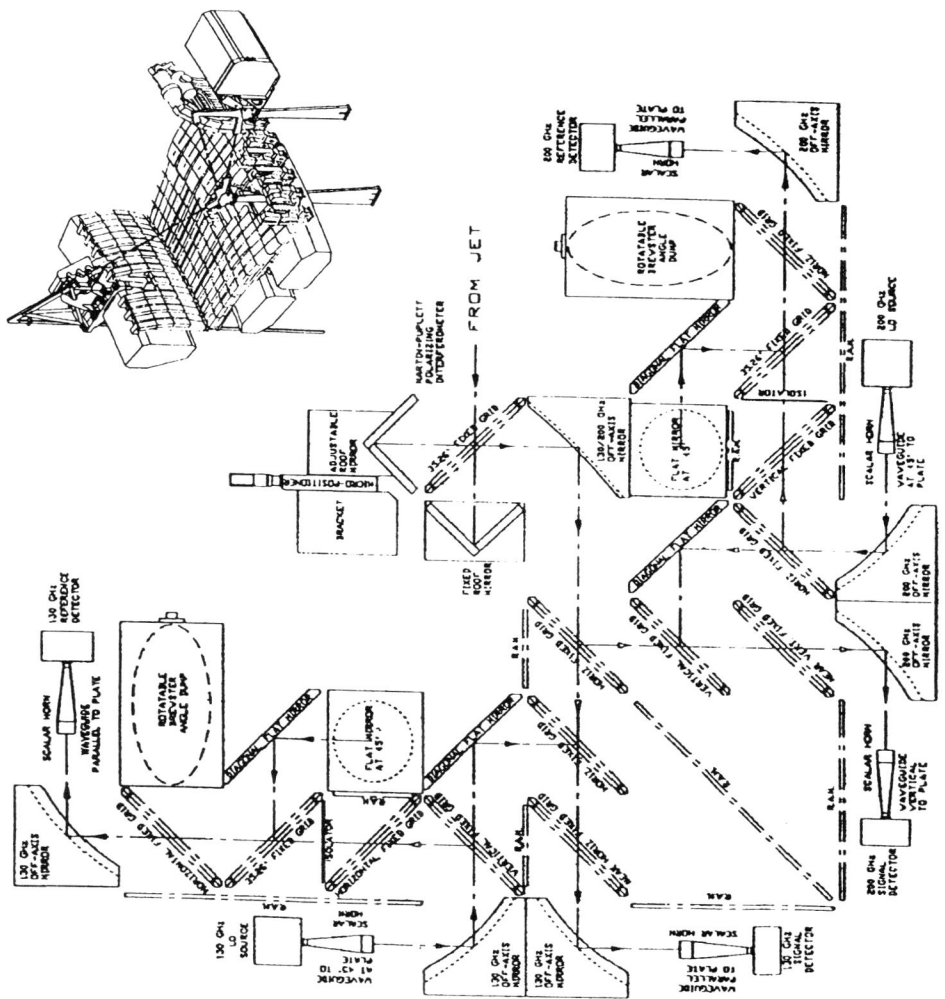

Figure 26.

SNAPS
GAUSSIAN B-M

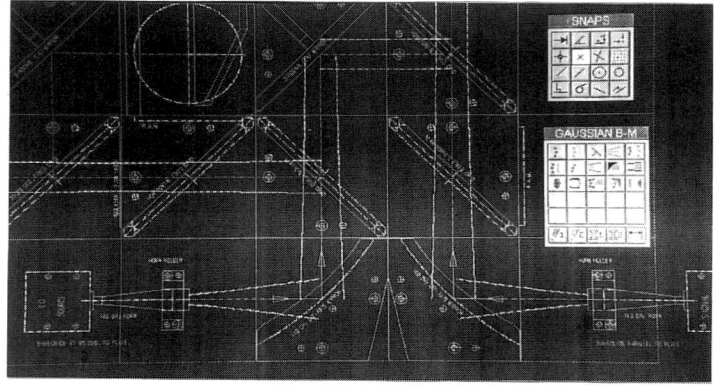

Figure 27.

Ecosorb
Vertical grid
45° grid
Fixed horn
Rotating horn
Ecosorb
ISOLATOR TEST CIRCUIT

Figure 28.

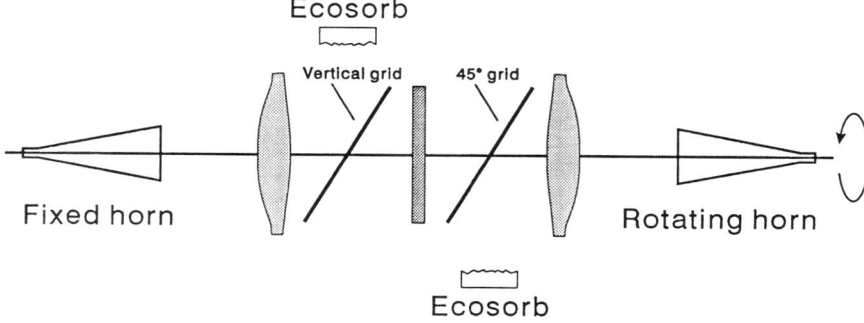

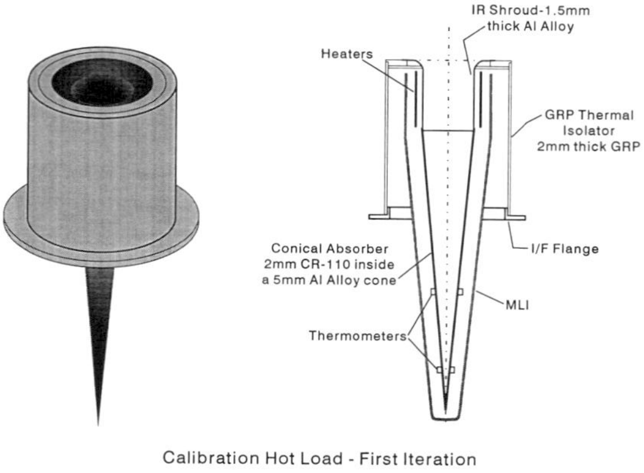

IR Shroud-1.5mm thick Al Alloy
Heaters
GRP Thermal Isolator 2mm thick GRP
I/F Flange
Conical Absorber 2mm CR-110 inside a 5mm Al Alloy cone
MLI
Thermometers
Calibration Hot Load - First Iteration

Figure 29.

VECTOR MEASUREMENTS FROM 8 GHz TO THE THz RANGE, OBTAINED IN A REAL LIFE EXPERIMENT

Experiment conducted on July 9th 1996, at the Château de Bonas, from 9H35 to 11H25 AM, during the NATO Conference "New directions in THz technology".

P. GOY,
AB MILLIMETRE,
52 rue Lhomond,
75005 Paris, France.

M. GROSS,
Laboratoire Kastler-Brossel,
Département de Physique de l'Ecole normale supérieure,
24 rue Lhomond,
75231 Paris Cedex 05, France.

Abstract

The Millimeter-Submillimeter Waves Vector Network Analyzer MVNA-8-350 can cover the frequencies from 8 GHz to the Terahertz range. It had been installed in the conference room. Transmission and transmission-reflexion experiments in waveguide devices and in free space have been performed and are described thereafter, with all results obtained in less than two hours. The frequencies of interest were successively 480, 34-36, 285, 140, 50-70, 19, 68-112 and 94 GHz. Due to the phase information, a better understanding of physical phenomena was demonstrated. Measured devices included Michelson and Fabry-Perot interferometers, cylindrical and whispering-gallery cavities, transparent or absorbing dielectric slabs.

1. Introduction

Millimeter waves can be generated, and detected, by non-linear devices such as Schottky diodes powered from centimeter sources. A very simple arrangement involving two centimeter sweepers, the second being maintained at a constant frequency difference from the first, permitted us to develop, since 1985, a scalar analyzer [1,2]. In 1989, the use of a vector receiver, taking directly its reference onto the main quartz oscillator, transformed the scalar analyzer onto a vector one [3,4,5]. Improvements in the efficiency of different parts of the analyzer pushed up its frequency limit more or

323

J.M. Chamberlain and R.E. Miles (eds.), New Directions in Terahertz Technology, 323–340.
© 1997 *Kluwer Academic Publishers. Printed in the Netherlands.*

less linearly with time, from 110 GHz in 1985 to about 900 GHz in 1995. In particular, submillimeter waves have been generated and detected by Schottky diodes powered by millimeter sources (Gunn oscillators, used as extensions). The dynamic range can be very large, ca 120 dB below 70 GHz (analyzer alone) or up to 440 GHz (analyzer with extensions). Let us define a minimum dynamic range of 60 dB. Then, one can distinguish three frequency domains of the analyzer:

> 8-180 Ghz,   where the analyzer can work without extension,
> 180-500 GHz,   with an extension using a Gunn oscillator,
> 500-800 GHz,   with an extension using two Gunn oscillators.

A common nickname of the analyzer is "the French radio". This nickname can be justified by the fact that its receiver is a heterodyne receiver, just like in a radio receiver. As a consequence, the detection is sensitive, linear and vector. These two last aspects have the following origin: for each downconversion in a mixer (or harmonic mixer) powered by the appropriate Local Oscillator (LO), the output IF signal reproduces the input RF signal, in amplitude and phase.

> RF Input: $A \cos(\omega t - \phi)$;   IF Output: $A' \cos(\omega' t - \phi')$

$$A'/A = constant \tag{1}$$
$$\phi' - \phi = constant \tag{2}$$

However, in the analyzer there is a source and a detector, so that one could speak of a "transceiver", as well. The distance range of this "transceiver" is most frequently limited to the size of a table (2 m long, 75 cm wide at the Château de Bonas).

2. Transmission at 480 GHz, in Air and through a Dielectric Material

On the source side, a 96 GHz Gunn oscillator supplies a Schottky multiplier working as a quintupler powering a small feed horn. On the detection side, a similar feed horn faces the first at a distance of about 10 cm, and sends the collected microwaves into a Schottky harmonic mixer. After downconversion at 9 MHz of the 480 GHz wave in the mixer, the signal is processed by the heterodyne vector receiver of the analyzer, in which the last downconversion produces a 500 Hz signal. This signal is sent into an oscilloscope, on which one can "see" the millimeter wave, reproduced in amplitude, phase and signal-to-noise ratio, thanks to the linearity of the heterodyne receiver (Eqs.1-2). At the same time, the 9 MHz signal is sent into a short wave radio receiver, so that one can "hear" the millimeter wave. In particular, a slow change of the source-detector distance produces a corresponding move of the sine wave on the scope, and a change in tone in the sound from the radio receiver. The analyzer makes visible and audible a Doppler effect where the speed of move, v, is below a few cm/s, compared to the light velocity c=300,000km/s, corresponding to a relative change ΔF of a few Hz compared to the frequency F=480 GHz.

$$v/c = \Delta F/F < E\text{-}10 \tag{3}$$

A 10.95 mm thick bevelled shape dielectric sample made from nylon is then introduced between the two horns. The observed drop in amplitude is -26 dB, and the observed change in phase is about 13 turns. The loss tangent, tanδ, can be deduced from the formula:

$$tan\delta = \frac{(1.1\alpha)}{(nF)} \tag{4}$$

where n is the refractive index of the dielectric, and α the loss per centimeter (dB/cm) and F is measured in Ghz. The refractive index is related to the permittivity ε':

$$\varepsilon' = n^2 \tag{5}$$

and to the number of phase turns $\Delta\phi/360°$:

$$(n\text{-}1)e/\lambda = (\Delta\phi/360°) \tag{6}$$

where e is the sample thickness, and λ the wavelength in vacuum.

From Eqs.4-6, one deduces the values n=1.74 (ε'=3.03) and tanδ=0.03 measured in nylon at 480 GHz, to be compared to the values n=1.74 and tanδ=0.013 measured up to 110 GHz. This behaviour has been observed in many polymers: the real part of the permittivity does not change much; on the contrary, the loss increases rapidly when entering into the submillimeter domain [6].

3. Free Space Reflexion in the 34-36 GHz Interval

In the previous example (section 2.), the phase appeared to be useful for refractive index measurements (Eq.6). This is a very general aspect of the phase measurement: the linear phase evolution corresponds to a change in distance (or "optical" distance) between source and detection. There is also a very straightforward means of measuring a distance from a vector measurement: after a frequency sweep, the Fourier Transform FT calculation gives access to the time domain. Knowing the speed of light, the time domain can also be viewed as a distance domain. Two small horns, attached to a source and a detector in the Q-band (33-50 GHz) were put on top of the analyzer in the direction of the back of the conference room. The reflexion from the room is obtained in a one-thousand point 34-36 GHz sweep. The FT calculation shows a few broad peaks around 5 meters, corresponding to the location of a slide projector and its table, and a sharp peak at 8 meters, corresponding to the back wall of the conference room.

4. Michelson Interfermoeter at 285 and 140 GHz

In a small Michelson interferometer (Thomson-CSF), which used to be employed as a wavemeter for submillimeter wave tubes "Carcinotrons" (also produced by Thomson-CSF), the incoming wave is split by a 45° grid. Each of the two produced waves is then reflected onto a mirror. The first mirror is standing. The second can be moved by turning a micrometer. The grid-beam splitter recombines the two waves, and their sum is extracted and detected. The maximum output is obtained when the difference in path of the two waves is an integer multiple of λ. Since the wave goes to, and comes back from, the movable mirror, the period in micrometer position is $\lambda/2$. After normalisation on their respective maxima, transmissions across the Michelson were observed at 285 and 140 GHz when moving as regularly as possible by hand, the micrometer (a metronome beat of 1 second for a move of 0.1 mm). The observed amplitude variation and periodicity correspond to what is expected (Figure 1). If the two branches were exactly balanced, the minimum output could be zero, when the difference of path of the two waves is an odd number of $\lambda/2$. This is not observed, since the signal in the moving mirror branch is larger than in the other branch, so that the "unwrapped" phase describes the average move of this mirror (360° for $\lambda/2$ mirror move, see Figure 2). The polar plot of the results (Figure 3, where one of the traces has been shifted by 180° for clarity's sake) gives a direct "view" of the phenomenon, the detected signal OB being the sum of two vectors, OA, standing, and AB, moving.

5. Transmission through a Cylindrical Cavity from 50 to 70 GHz

The source and the detector in the V-Band (where the dynamic range is larger than 90 dB) are attached to each other via a 30 dB attenuator which avoids the saturation of the detection, and reduces the standing waves effects in the propagation line. Then a cylindrical cavity, about 7 mm diameter and 7 mm long, coupled via holes pierced at the center of each flat surface, is introduced between the source and the detector. The 50-70 GHz frequency sweep, with 5 MHz steps (4,000 points) shows four resonances, A, B, C, D, see Figure 4 (rectangular plot), and Figure 5 (polar plot). Each Lorentzian resonance appears as a circle in the polar plane. The upper frequency resonance D (which is the TE121 mode) presents the highest quality factor Q, and shows a polygonal shape in the polar plane, since the frequency steps (5 MHz) are not small compared to the cavity width (13 MHz). The software of the analyzer calculates the quality factors $Q_c = 1,100$ and $Q_d = .5,100$. These fits are shown in Figure 6, where are also visible the FT of the resonances (straight lines in dB versus time). The exponential decay times τ of the energy through the cavity at resonance are the times for the amplitude to decrease by $1/e = 4.343$ dB. Here $\tau C = 2.75$ns, and $\tau D = 11.8$ns. They are proportional to the quality factors Q, and obey the law:

$$Q = 2\pi F \tau \qquad (7)$$

Mr Israel Galin, from Aerojet, Azuza CA, made the remark that the operated characterization measures the loaded cavity, not the intrinsic quality factor. That is

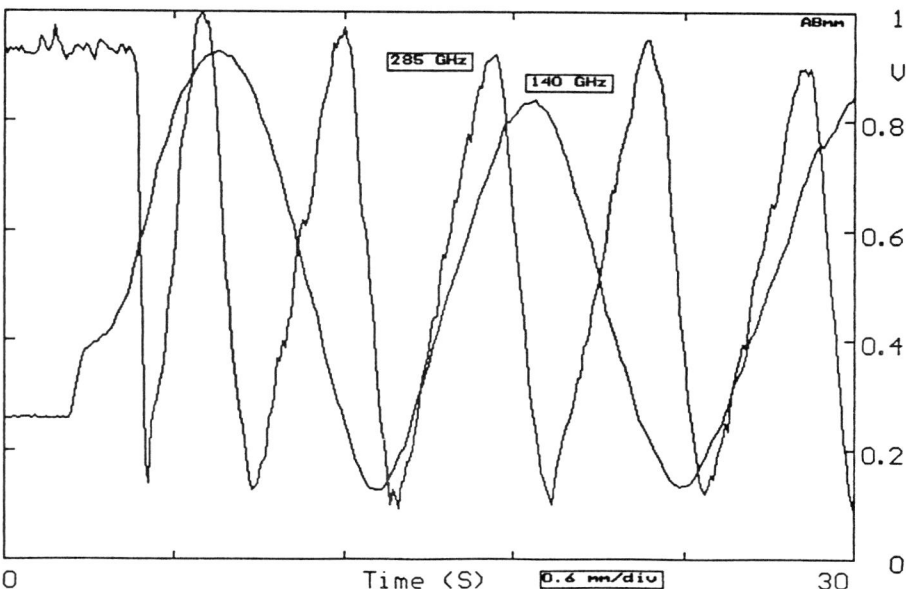

Figure 1. Transmitted amplitudes (shown in a linear scale) at 285 and 140 GHz across a Michelson interferometer, the moving mirror being moved by hand at 0.1mm/s, thus making a move of 0.6mm/div. The periodicity is, naturally, $\lambda/2$.

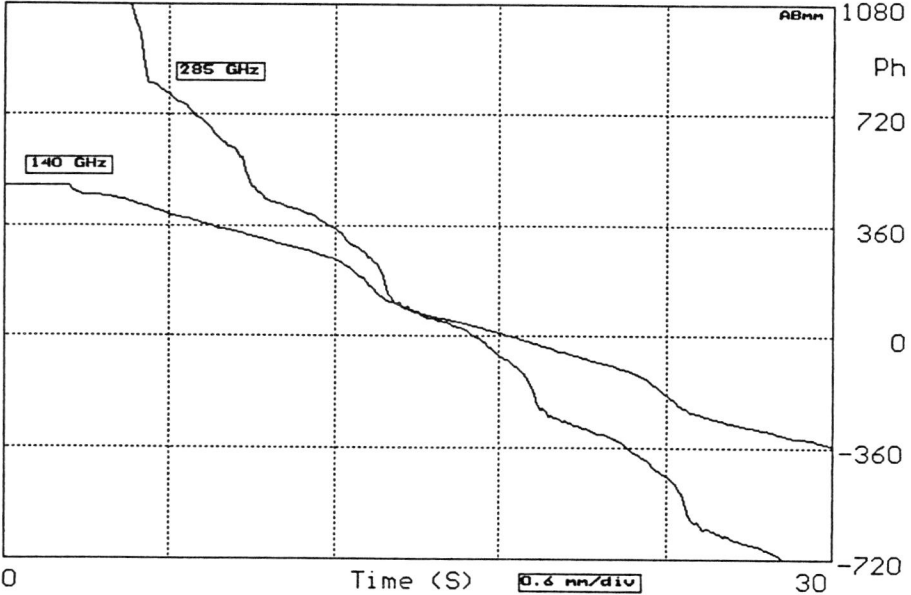

Figure 2. Same as Figure 1, showing the "unwrapped phase", since the phase variation limited to -180°/+180° has no physical significance.

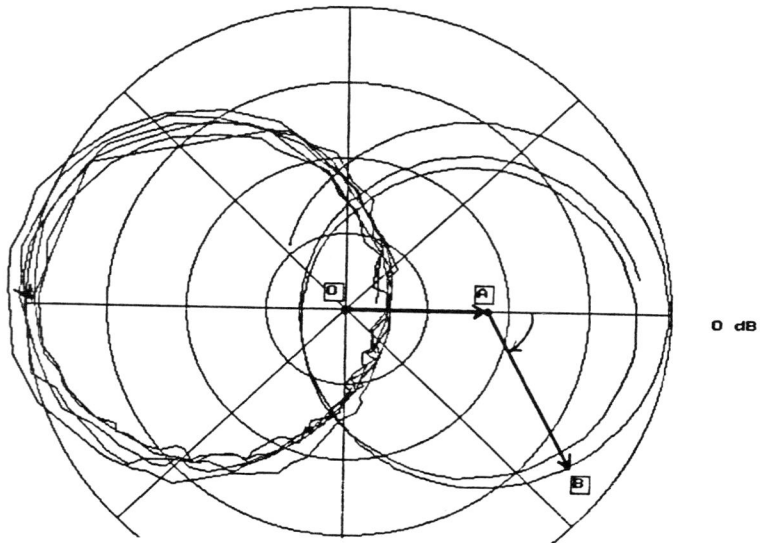

Figure 3. All the polar plots in the following Figures represent the detected microwave field amplitude as the radius (it is a linear scale, not a dB scale), and the detected phase as the angle. Here are presented the polar plots of Figs.1-2. For clarity's sake, the 285 GHz signal (left) has been shifted by 180° with respect to the 140 GHz signal (right). The circle shape comes from the sum of a standing vector, such as OA, plus a moving vector, such as AB.

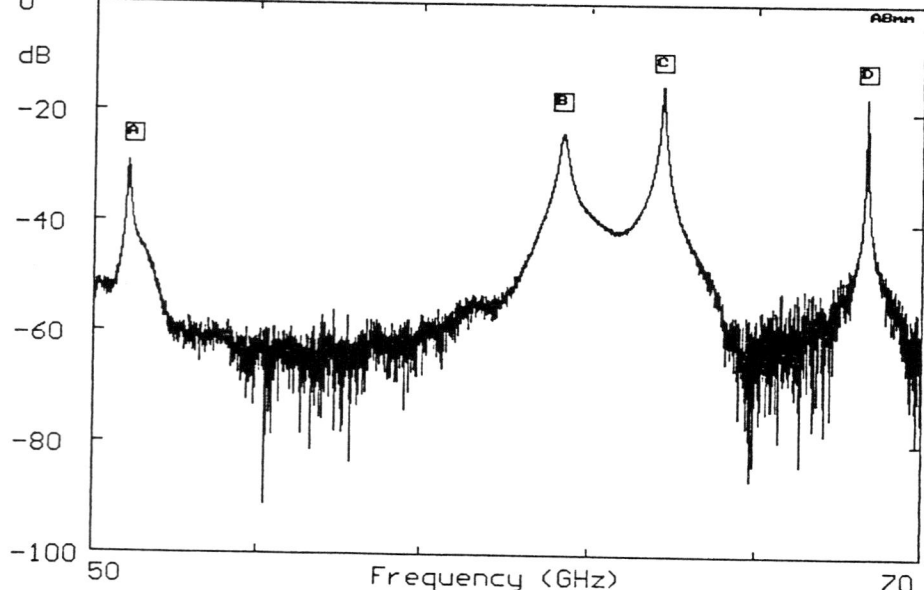

Figure 4. Transmission through a cylindrical cavity (shown in a dB scale). The observed resonant modes are labelled A, B, C and D.

correct, and the coupling holes, for instance, modify slightly the frequency resonance, and damp the quality factor. For that reason, we presented a second cavity measurement (section 6) in which we could get rid of the modifications induced by the coupling.

6. Sapphire Made Whispering-Gallery Mode Cavity at 19 GHz

A 9.97 mm thick sapphire slab of 60 mm diameter, with C axis perpendicular to the plane surfaces, is coupled along its cylindrical surface to the evanescent microwave field coming out from a total reflexion prism. The wave transmitted through the prism shows a minimum when the whispering gallery mode resonance is excited into the sapphire cylinder. Thanks to a micrometer, the distance from the prism to the side of the sapphire cylinder can be changed. In Figure 7 one sees the experimental traces and their Lorentzian fits. From right to left, the increase of the distance permits to decrease the coupling from overcoupling, critical coupling, undercoupling and large undercoupling, where the intrinsic properties of the resonance (position F_o, quality factor Q_o) can be obtained. On Figure 8 are shown the polar plots of the fits, with correponding quality factor Q=4,320 (overcoupled), 8,650 (close to critical coupling), 10,900 and 15,650 (undercoupled), and Qo=19,390 (very much undercoupled). The loss tangent of this (very low loss) dielectric material is simply the inverse of Qo:

$$\tan\delta = 1/Q_o = 0.000\ 05 \tag{8}$$

7. Transmission-Reflexion from Dielectrics in the Band 68 to 112 GHz

In the previous experiments (sections 2-6), one was looking at a single detected signal at a time. However the analyzer can operate a double detection, and observe, as an example, transmission-reflexion at the same time. For characterizing dielectric materials, a measurement bench is made of a source (a frequency sextupler) feeding a directional coupler at the direct output of which is an emitting feed horn, and at the side of which is the detector for reflected waves, connected to channel II of the receiver. The Gaussian beam from the emitting horn is refocused by a 100 mm focus lens onto the sample. The transmission detection is operated symetrically by a detector connected to channel I, fed by a horn collecting the Gaussian beam from a second lens similar to the first. The distances are 200 mm from the emitting horn to the first lens, 200 mm from this lens to the sample, 200 mm from the sample to the second lens and 200 mm from the second lens to the transmission detection horn.

There exist sophisticated calibrations of a vector analyzer, especially for the waveguide propagation. These calibrations can be adapted to the free space propagation, as in this case. However, these sophisticated calibrations take time. Therefore, the calibrations which have been used in this section are the simplest ones: transmission calibration is obtained in a sweep with nothing at the sample location; reflexion calibration is obtained in a sweep with a metal plate at the sample location. Due to these

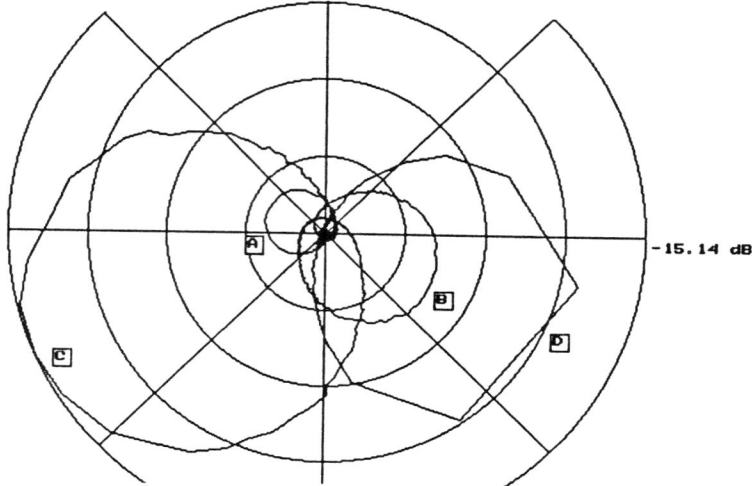

Figure 5. Polar plot of Figure 4. A Lorentzian resonance, such as the cavity resonances A,B, C, D, appears as a circle in the polar plane. The polygonal shape of resonance D comes from the fact that the 5 MHz sweep steps are not small compared to the D-resonance width, of the order of 13 MHz.

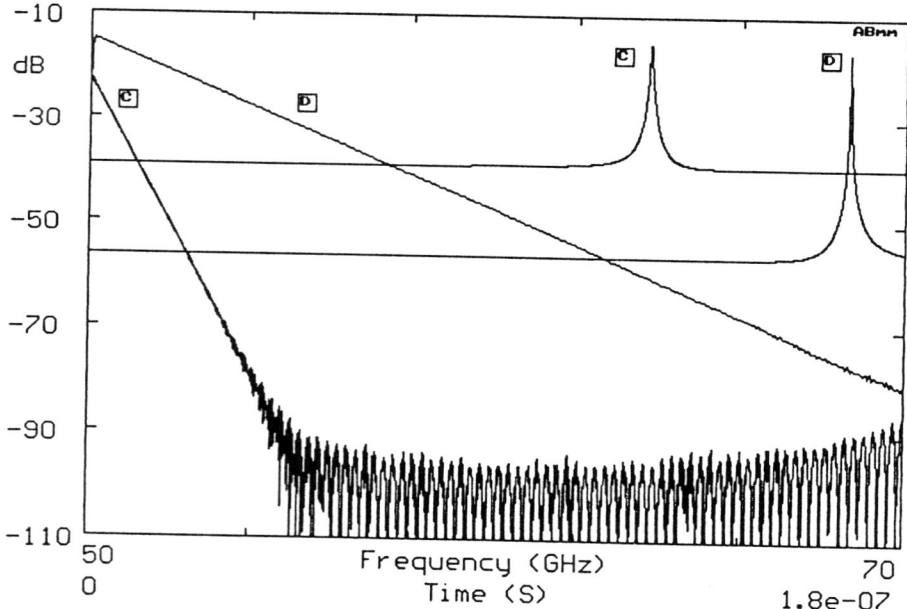

Figure 6. Top: Lorentzian fits of the resonances C and D of Figs.4-5. Bottom: FT of the resonances, showing the decay time of the modes C and D, the lower Q (C) showing a shorter time than the higher Q (D).

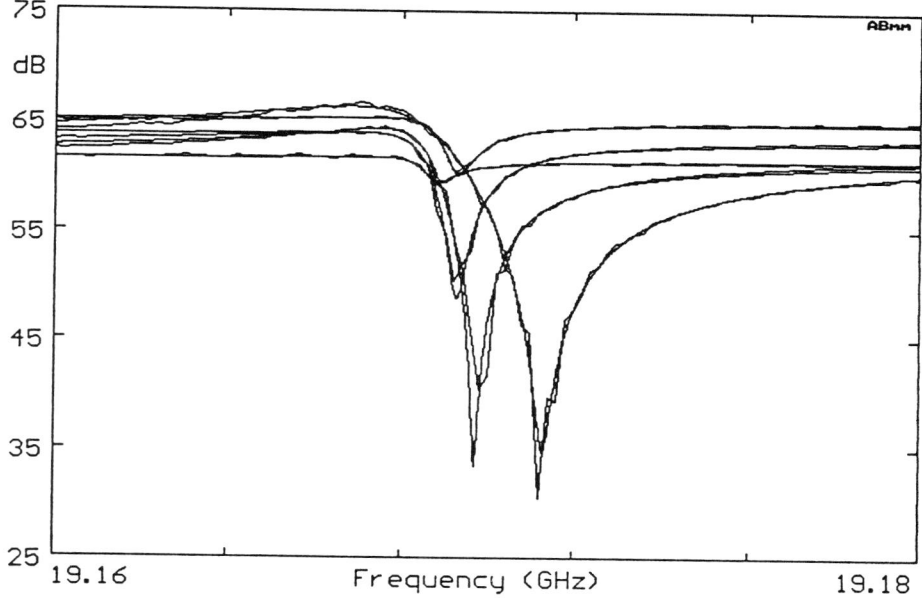

Figure 7. Whispering gallery resonances observed in a sapphire cylinder and their Lorentzian fits, with a coupling reduced from right to left.

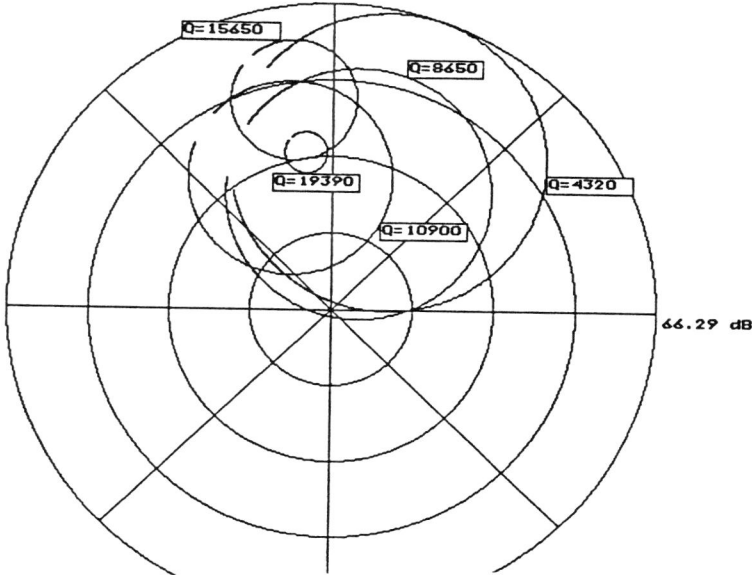

Figure 8. Polar plots of the Lorentzian fits of the resonances observed in Figure 7.

simple calibrations, standing-wave effects can pollute the results. However, very effective filtering, operated thanks to the FT calculations, permits the recovery of a measurement practically free from standing-wave effects, see section 7.1. After these simple calibrations in two sweeps, all samples have been rapidly measured successively, and the data treatments have been done together. In the next sections however, each sample will be presented separately, for clarity's sake.

7.1. SAPPHIRE CHARACTERIZATION

A 9.97mm thick sapphire slab, already used for observing the whispering gallery mode resonance (section 6) is characterized by transmission-reflexion in the band 68-112 GHz. Parasitic standing-wave effects can be almost as large as the effects to be observed, as can be seen on the experimental traces of the transmitted-reflected amplitudes in Figure 9, and on the polar plots in Figure 10. On the calculated FT, one can cut all signal contributions coming from these parasitic reflexions (which arrive later than from the sample), so that the inverse FT permits to recover the pure sample effects, as shown in Figures11 and 12. Figure 13 is a rectangular plot of the transmitted signal, in which are shown the actual phase evolution "Ph T", and the phase evolution "Ph T-L" in which is substracted (by the software) the linear variation. The circle "T-L" of Figure 12 is a polar plot of this calculation. The trefoil shape of the transmitted signal (labelled "T" in Figure 12) is generated by the vector sum of (1) a point moving on the unity circle (since the low-loss sapphire is very much transparent but causes a phase change proportional to the frequency) and (2) a point moving on the small circle (labelled "T-L") describing the interference effects in transmission. The phase evolution strictly obeys Eq.6 only for the frequencies at which the transmission is maximum, or minimum. From these measurements and Eq.6, the sapphire refractive index is n=3.064 and the permittivity (Eq.5) $\varepsilon'=9.391$. The maximum transmission being of the order of 0 dB, this sample presents too small a loss to be measured simply by transmission. The maximum reflected amplitude r obeys the law:

$$r = (\varepsilon'-1)/(\varepsilon'+1) \tag{9}$$

giving here r=0.807, corresponding to -1.86 dB, which is observed (Figures 11-13). From conservation energy considerations, the minimum transmitted amplitude t is such that:

$$r^2 + t^2 = 1 \tag{10}$$

giving in sapphire the predicted value -4.58 dB. One observes a close value of -4.4 dB (Figure 11).

7.2. ARALDITE CHARACTERIZATION

Contrary to sapphire, Araldite (which is an epoxy) is an absorbing material. A 20.95 mm thick slab of Araldite damps almost entirely the standing waves inside the sample,

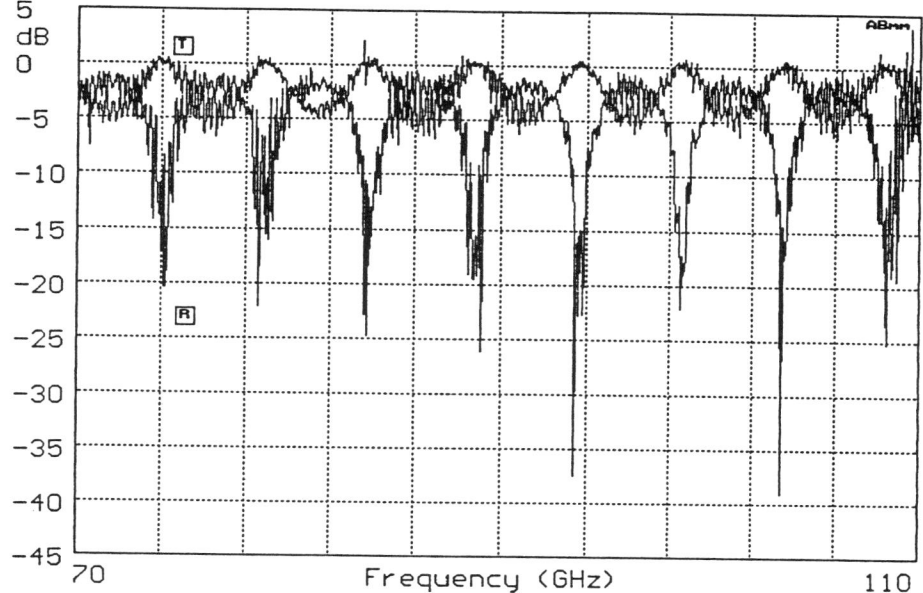

Figure 9. Transmitted (upper curve) and reflected (lower curve) signals from a 9.97 mm thick sapphire sample. The rapid oscillations are due to parasitic standing waves.

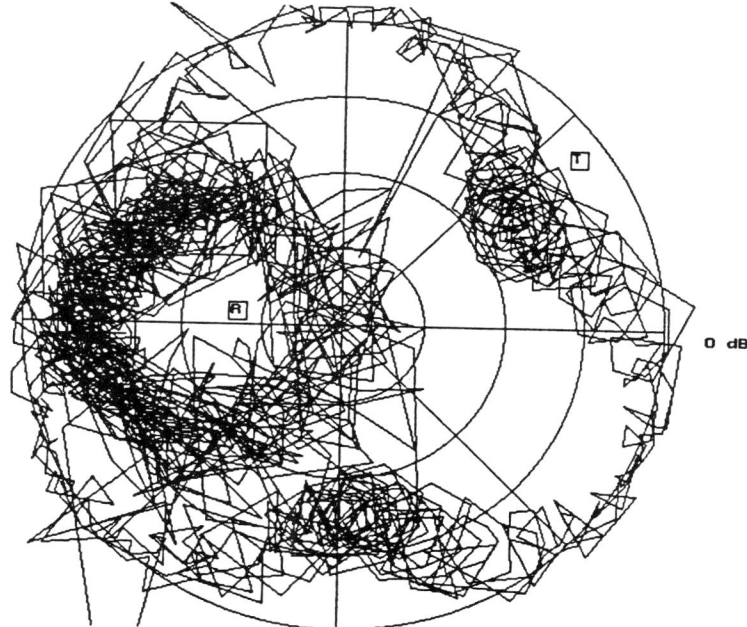

Figure 10. Polar plot of Figure 8.

334

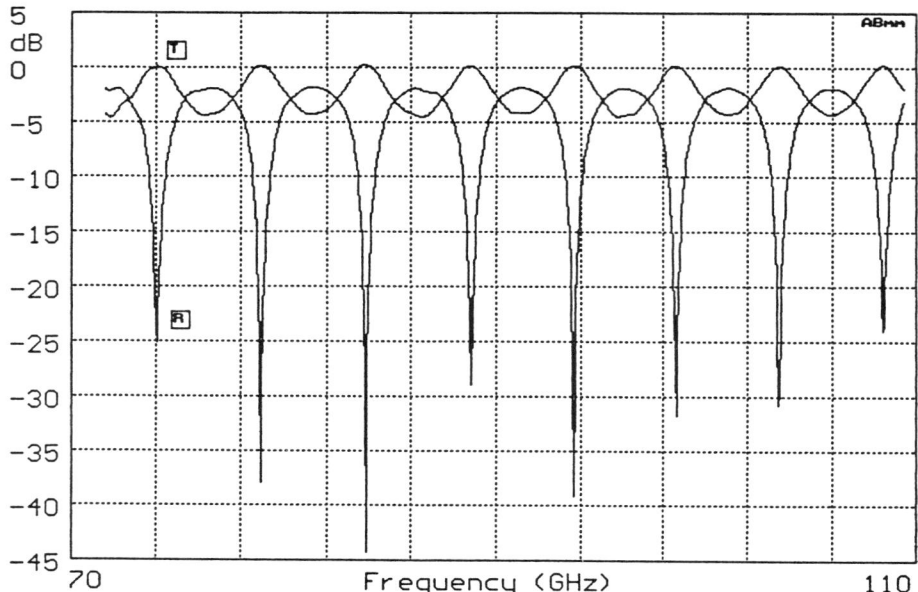

Figure 11. Same as Figure 9, after FT filtering of the parasitic standing waves. The remaining oscillations are due to the standing waves inside the sapphire sample.

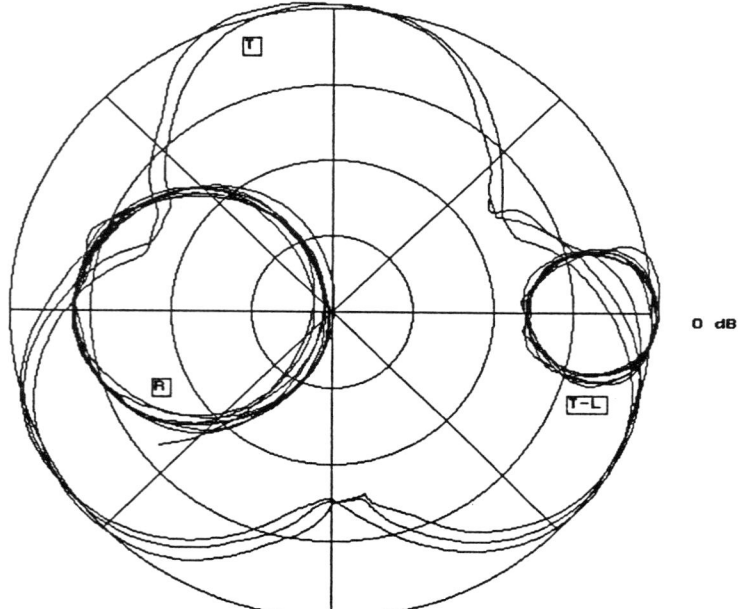

Figure 12. Polar plot of Figure 11. The circle "R" and the trefoil "T" are actual reflected and transmitted results. The circle "T-L" is obtained from the trefoil "T" after substraction of the linear phase variation.

and the transmitted dB amplitude is practically a straight line decreasing linearly with frequency ("T dB", Figure 14, top). On the reflected wave appear small standing-waves oscillations (Figure 14, center). The phase evolution ("T Ph", Figure 14) and Eq. (6) give n=1.701, and Eq.(5) gives ε'=2.893. The amplitude evolution and Eq. (4) give tanδ =0.021. On the polar plot (Figure 15), the reflected amplitude seems to converge towards a point just above the value 0.25 (the inner circle). For an infinite dielectric thickness, the reflected amplitude R is given by:

$$R = (n-1)/(n+1) \tag{11}.$$

For araldite, one obtains R=0.259, corresponding to -11.7 dB, which is practically the "asymptotic" value observed on "R dB", Figure 14. Transmission results are presented in Figure 16 with a scale going to zero degree phase (with a reversed sense for clarity's sake) and zero dB amplitude. One sees that the observed linear dB and Ph variations extrapolate to zero, which is normal, according to Eqs. (4-6), for constant ε' and tanδ.

7.3. BIREFRINGENT MATERIAL

A 13.05 mm thick slab of material called C-stock 265 (Cuming Corp.) appeared to be birefringent, in other words to have two different refractive indices, depending on the electrical field orientation, which will be aligned along the Large index axis, or along the Small index axis (perpendicular to the first). This material consists of glass spheres and aluminium flake mixed in a binder to give an artificial dielectric with expected ε' in the range 13-15 at 10 GHz. Transmission amplitudes and phases shown in Figures 17 and 18 correspond to the average values nL=3.24 (ε'=10.5), nS=2.96 (ε'=8.76), tanδ L=0.10 and tanδS=0.08. According to Eq.11, this would correspond to reflected amplitudes RL=-5.5dB and RS=-6.1dB. The observed reflected amplitudes RL' and RS' are extremely close one to the other, and around -5dB (Figure 17, top). The polar plot of the transmission shows nice spirals (Figure 19). However, the rectangular large scale plot of the transmission, Figure 20, shows that the observed straight dB and Ph dependences do not extrapolate to zero (contrary to the "normal" dielectric of Figure 16). Looking in detail at the measurements, one observes that the high-frequency loss data (around 110 GHz) are about 8% below the average in the band, and the the low-frequency ones (around 70 GHz) are 8% above. Similarly, the high-frequency refractive-index-n data are 0.4% below the average, and the low-frequency ones 0.4% above. This, and the claimed value circa ε'=14 at 10 GHz, are indications that one cannot extrapolate the 70-110 GHz measurements towards other frequencies.

8. Fabry-Perot Cavity in Air at 94 GHz

A semiconfocal Fabry-Pérot cavity is made from a spherical mirror of about 300 mm radius of curvature, facing a plane mirror placed at about 150 mm. The quality factor is good, of the order of 160,000, making the device very sensitive to any change in the

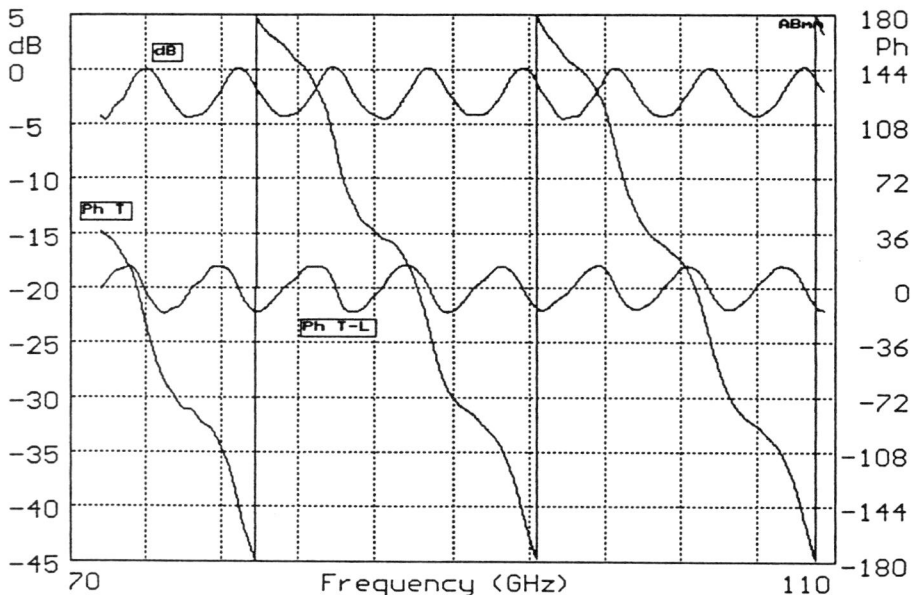

Figure 13. Signal transmitted across the sapphire sample. The actual phase is shown in "Ph T", and the phase oscillations around zero "Ph T-L", obtained after substraction of the linear phase variation from "Ph T", are shown at the center.

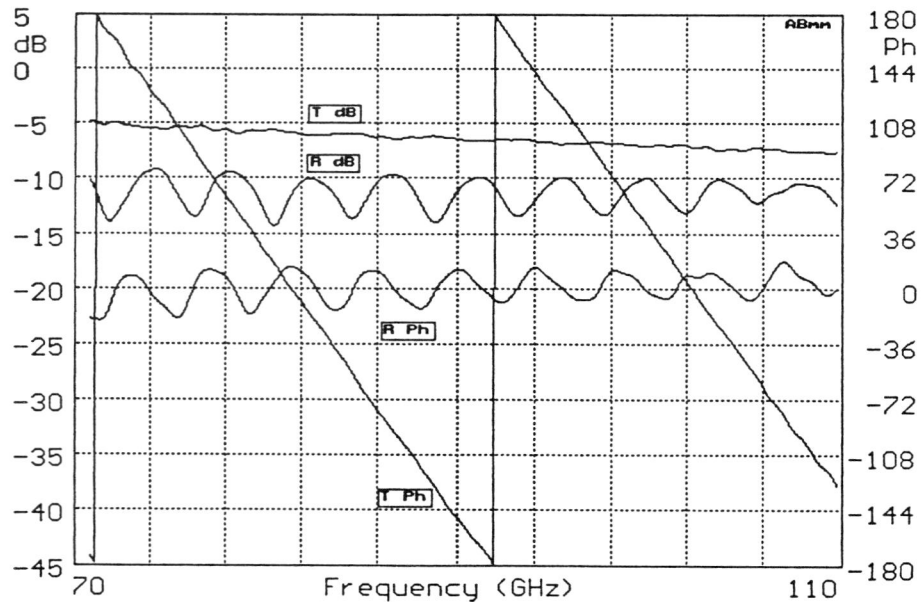

Figure 14. Transmission-reflexion observed with a 20.95 mm thick Araldite sample.

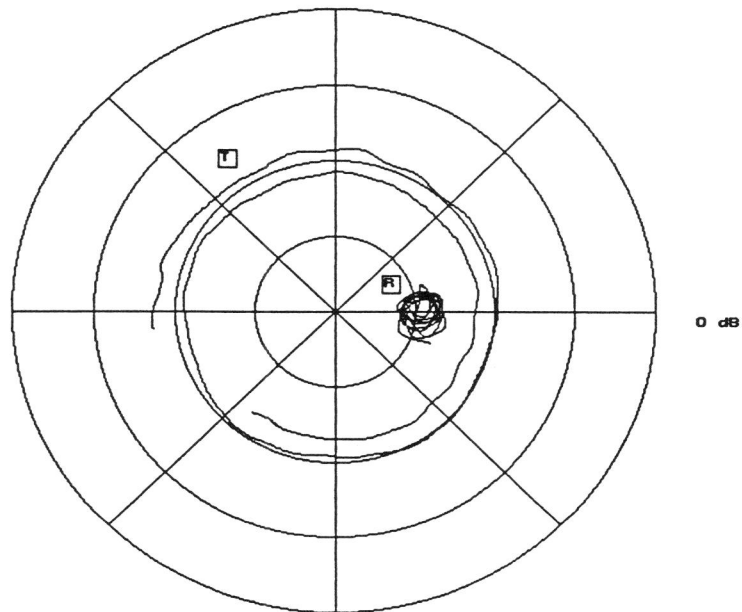

Figure 15. Polar plot of Figure 14.

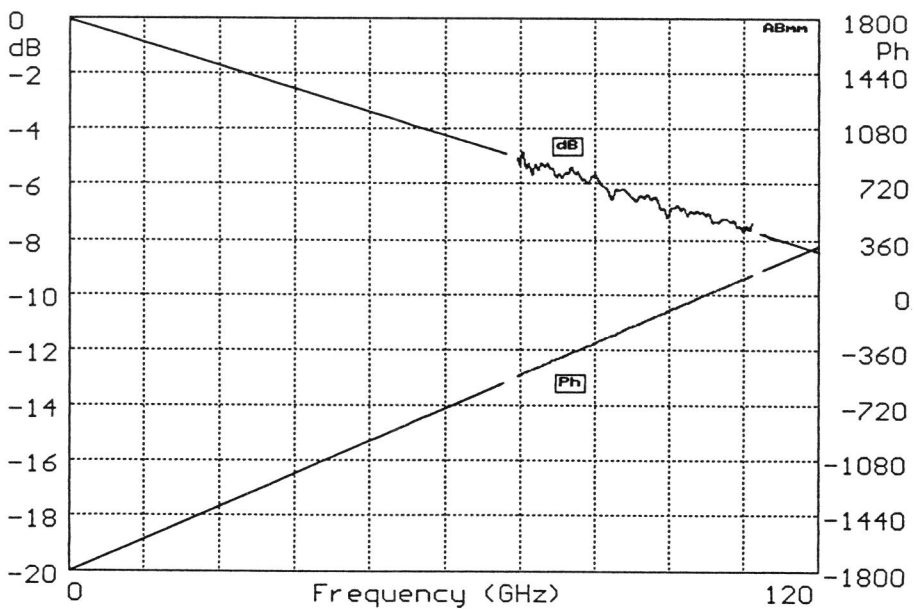

Figure 16. Transmission through Araldite. Extrapolations to zero frequency are at zero dB and zero phase (i.e. integer number of 360°).

propagation medium between the mirrors. This cavity is mostly used for the characterization of low-loss dielectrics, since there is a large number of crossings of the wave across the sample placed into the cavity. The very simple experiment, performed at the end of the presentation at the Château de Bonas, consisted just in blowing the experimentalist's breath into the cavity tuned exactly at 94.25 GHz. Then the phase made a clockwise change. To the question: "Was this rotation due to the fact that the breathing air was hot?", the answer was to be: "No!". Hot air would have made the air less dense, and would have made the cavity tuning higher in frequency than the applied 94.25 GHz, therefore inducing a counter-clockwise phase change. The observed clockwise change was due to the water-vapour content of human breathing. The proof of these assumptions was finally made with an air-dryer, the hot air from which created a counter-clockwise phase rotation, indeed. Lucky were we not blowing the delicate Château de Bonas fuses with this dangerous extra-electrical equipment!

Aknowledgements

The authors would like to thank the conference organisers, especially Professors Tatsuo Itoh, Erik Kollberg, Roger Pollard and Dr. Martyn Chamberlain for having given them the opportunity of presenting this very special communication. Thanks are due to Jean-Claude Buisson for having carefully read the manuscript.

References

1. Goy, P., Gross, M. and Mauc, C., (1986) Un banc millimétrique de 27 à 110 GHz complet, simple et économique, *Onde électrique* **66**, *121-133.*
2. Goy, P., Gross, M., Raimond, J.M. and Buisson, J.C. (1988) Analyseur scalaire de réseau millimétrique couvrant de 16 à 250 GHz, *Ann. Télécommun.* **43**, *331-340.*
    (a) Goy, P., Gross, M. and Raimond, J.M., (1990) 8-1000 GHz vector network analyzer, in R.E.Temkin (ed.), *Proc. 15th Conf. on IR and MM Waves,* SPIE **1514**, *172-173.*
    (b) same authors (1991) Vector measurements in the millimeter and submillimeter, in M.R.Siergrist, M.W.Tran and T.M.Tran (eds.), *Proc. 16th Conf. on IR and MM Waves,* SPIE **1576**, *354-355.*
3. Goy, P., Gross, M. and Beck, F., (1993) Transmission-Reflection measurements from 8 GHz to the THz, in J.R.Birch and T.J.Parker (eds.), *Proc. of the 18th Int. Conf. on IR and MM Waves,* SPIE **2104**, *487-488.*
4. Goy, P. and Gross M. French Patent CNRS-ENS (Sep 1, 1989); European Patent EP 0 420 767 (Apr 3, 1991); US Patent N° 5 119 035 (Jun 2, 1992).
5. Afsar, M.N. (1991) Precision Millimeter-Wave Measurements of complex Refractive Index of common Polymers, *IEE Transactions on Instrumentation and Measurements,* **1M-36**, *530-559.*

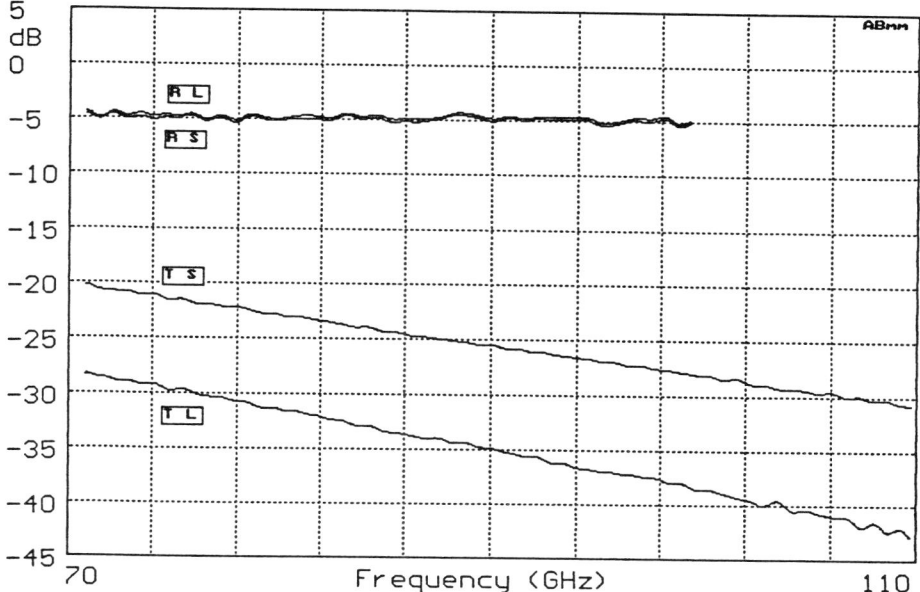

Figure 17. Amplitude transmission (labelled T)-reflexion (labelled R) with a birefringent sample (labelled S at the small index orientation, L at the large index orientation).

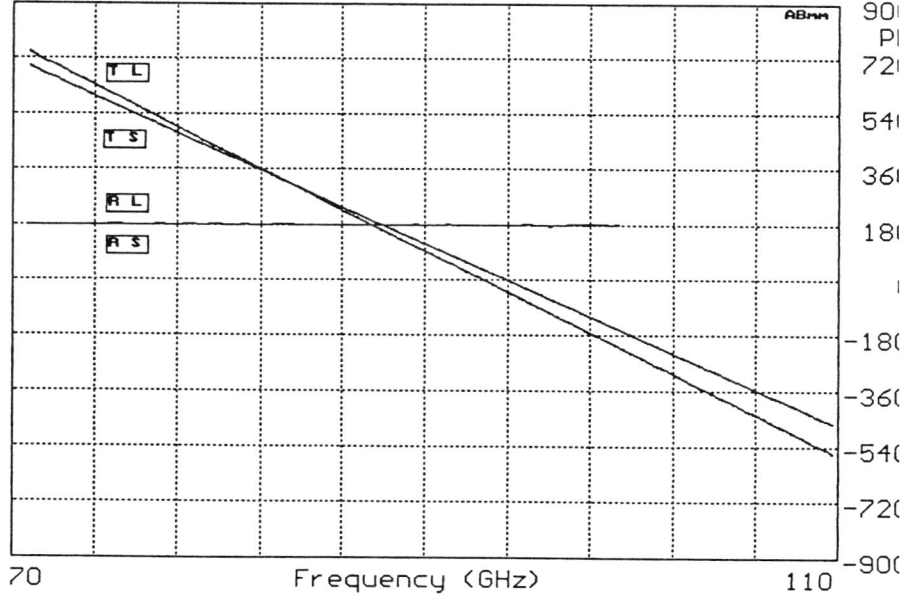

Figure 18. Same as Figure 17, for the phase.

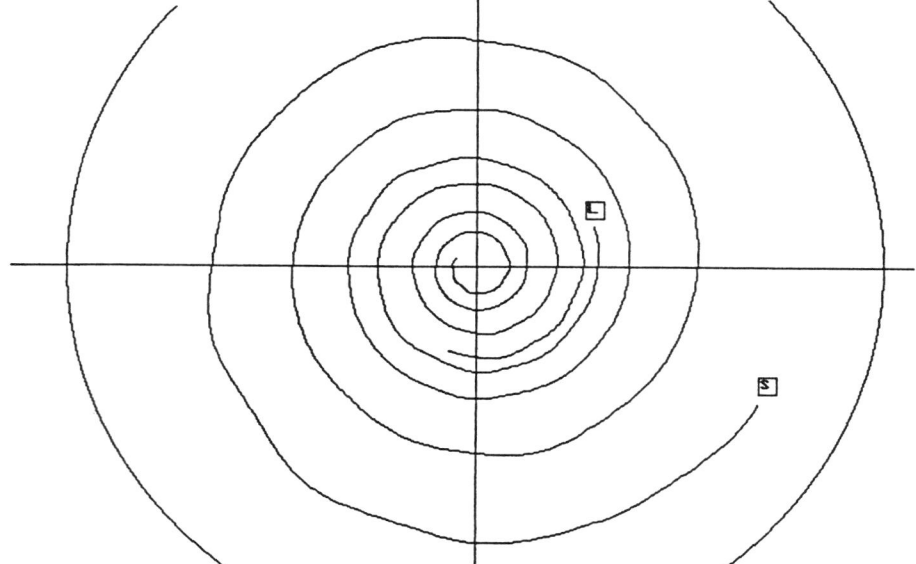

Figure 19. Polar plot of the transmitted signals across the birefringent sample. S is for the orientation along the small index axis, L is for the orientation along the large index axis. The outer circle is for an amplitude of 1/8, the center is for a zero amplitude.

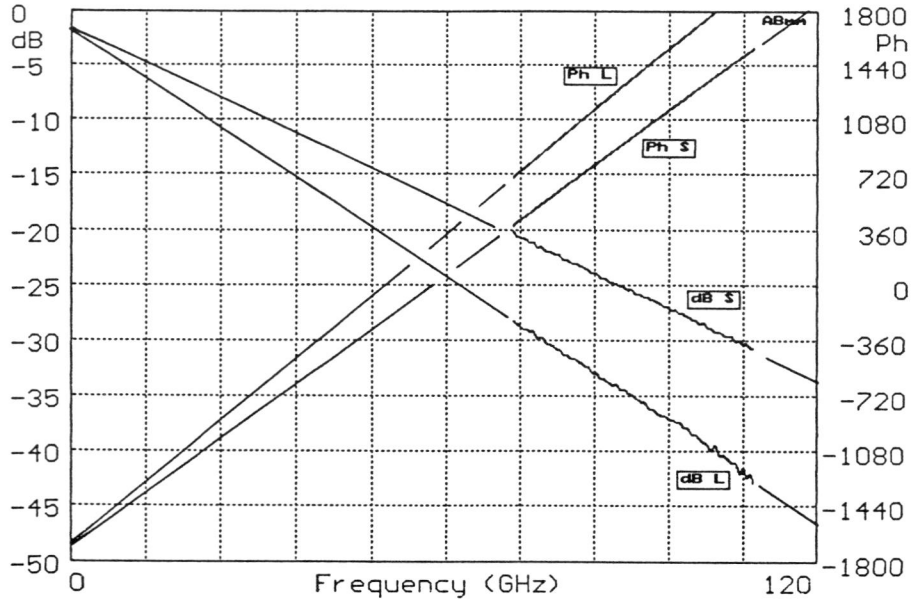

Figure 20. Observed transmitted amplitudes (dB S and dB L for the small, and large, index axis orientations) and phases (Ph S and Ph L, with an upside-down direction for clarity's sake) across the birefringent sample. Linear extrapolations do not go to zero dB nor zero phase.

Theme 5

Lightwave/Terahertz Interaction

LIGHTWAVE/TERAHERTZ INTERACTION

H.R.FETTERMAN
Department of Electrical Engineering
University of California
Los Angeles, California 90095
USA

Abstract

The use of Lightwaves to generate, transmit and process Terahertz radiation has now reached a critical stage of development. This is a result of major advances in laser technology and fiber optic devices. Soon these developments will permit realistic Terahertz systems to be fabricated, for the first time, for specific applications areas.

1. Introduction

Lightwave techniques have demonstrated their unique capabilities over the last several years in systems that use their short pulse, Femtosecond, capabilities. These include systems which generate Terahertz radiation using electronic switches and antennas [1], transient currents in bulk materials [2] and optical rectification [3]. Scattering from electron-phonon excitations [4] and four wave mixing [5] are also being explored. A number of new materials ranging from DAST [6] to poled polymers [7] have been used in these experiments. The lasers used here range from Ti:Sapphire systems to novel forms of semiconductor mode locked systems [8].

Other experiments include CW mixing experiments in MSM systems with semiconductors having high recombination rates such as implanted InP and low temperature GaAs [9]. These systems are usually linked to antennas and are capable of generating high quality, narrow band, Terahertz radiation. The lasers used here also include semiconductor lasers which are tunable and can be mixed. Another variation of the semiconductor mode locked system uses the beating between modes in a single laser [10] to generate extremely stable Terahertz radiation. In addition to generation and detection of Terahertz radiation the use of Lightwave technology offers the concepts of modulation, phase control and transmission over relatively long paths. Systems which can work at submillimeter wavelengths can thus be made free from the constraints of atmospheric absorption and lack of active components. All of the major operational functions can be implemented at Lightwave frequencies and then transformed into the Terahertz regime at the final stage.

J.M. Chamberlain and R.E. Miles (eds.), New Directions in Terahertz Technology, 343–358.

Examples of the use of mode locked lasers systems and the generation of Terahertz pulses using optical rectification are discussed in detail by other researchers in this workshop. In this section we look at several important technologies which will support this use of Lightwave technology for Terahertz applications. The second part of this discussion will deal with three Lightwave systems as examples of what can be done with the new technologies available.

An appropriate place to start in this approach is an electro-optic oscillator discussed several years ago (Plant et al., 1992) in which a 60 GHz mode locked semiconductor laser was used to optically drive a heterojunction bipolar transistor (HBT)connected to a broad band antenna. In Figure 1a the basic system is shown and the results are displayed in Figure 1b. The system generated narrow band Terahertz radiation and was tunable by either changing the laser cavity or by adding sidebands as shown. In some sense this is a transit time device using light pulses instead of avalanche pulses as in an IMPATT or domains as in a Gunn device. This work has since been extended to higher frequencies using mode locked lasers with capabilities as high as 350 GHz in fundamental operation, and higher in harmonic modes [11]. Other related work, using mixing between lasers, has generated coherent power at frequencies ranging up to 5 Thz [12].

Investigating how this work can be extended up from millimeterwave frequencies in a quasi-monolithic approach we have looked into several new areas. An

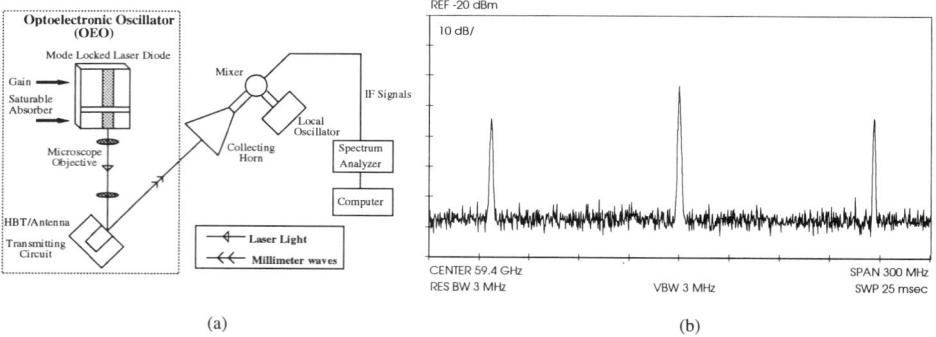

(a)   (b)

Figure 1. (a) A mode locked semiconductor laser is used to drive an HBT phototransistor. (b) Radiated 60 GHz signal is detected with information sidebands applied using the phototransistor's base.

example of new monolithic capability is the use of epitaxial lift-off to bring together several different material systems and devices on substrates suitable for Terahertz applications. In terms of new device technology we have now started looking into a new generation of traveling wave phototransistors which have the potential for operation up to hundreds of GHz. Since they have transistor action they also have gain and the possibility of using the third terminal for incorporating information. Finally, we discuss the latest advances in optical modulators using poled polymer structures. This a vital step in being able to insert and control high frequency signals on our optical carriers.

2. Examples of New Technology

2.1. EPITAXIAL LIFT-OFF HEMTS TO 94 GHZ

Epitaxial lift-off (ELO) has become an increasingly important tool to integrate devices of dissimilar material systems. High electron mobility transistors (HEMTs) have also shown great promise as high frequency photodetectors [13]. Currently we are exploring the use of optically driven ELO HEMTs to generate millimeter wave radiation up to 94 GHz. Our results demonstrate the viability of using ELO techniques for integration of high frequency HEMT photodetectors in novel new optoelectronic systems. The devices we are working on have f_T s of over 500 GHz, with gates lengths of less than 0.1 micron, and should permit operation well into the submillimeter regime.

In our previous experiments, quantitative comparisons of optical mixing signals to 22 GHz on 3 mm GaAs ELO devices were presented before and after lift-off under both backside and frontside illumination [14]. It was found that ELO devices outperformed bulk devices in quantum efficiency, responsivity and optical heterodyne strength. This improvement was attributed to improved optical coupling efficiency, optical backgating and a reduction in substrate leakage. In our latest experiments, these results have been extended to 60 and 94 GHz on 1 μm InP ELO devices.

The ELO process consists of two steps: first, detaching the epitaxial film from the growth substrate and second reattaching the film to the host substrate. While the attachment process is the same for GaAs and InP devices, the detachment process changes with the substrate. For our GaAs devices, ELO was achieved by undercutting a thin AlAs layer with a highly selective etch in HF. For the new InP devices, the entire backside of the substrate is etched in HCl. Both processes are generally referred to as ELO. Van der Waals bonding was used to attach the devices to the host.

Initially, the electrical response of our ELO devices were measured on a vector network analyzer. The device was biased at gate and drain voltages for peak transconductance. Although the initial devices showed f_Ts of only 18 GHz, the maximum frequency of oscillation for these devices were well over 100 GHz. The experimental setup of our optical measurements is shown in Figure 2. The device under test is illuminated with two collinear optical beams obtained from a single frequency temperature stabilized HeNe (632.991 nm, 0.5 mW) and a tunable continuous wave ring dye laser (600-640 nm, 100 mW). This ensures several hundreds of GHz of tunability of the heterodyne difference frequency. Coplanar probes (Picoprobe Model 67A & 120) were used to contact the device. The optically generated millimeter waves were launched into waveguides and then fed into a millimeter wave receiver system for downconversion to an intermediate frequency which could be seen on a spectrum analyzer. The receiver consisted of an external mixer and a local oscillator. The downconverted 94 GHz mixing signal of the ELO HEMTs are shown in Figure 3. Our results lend themselves to making compact, tunable, Terahertz wave sources by using two tunable semiconductor lasers or a mode-locked semiconductor laser in conjunction with our ELO HEMT photodetectors integrated with radiating antenna.

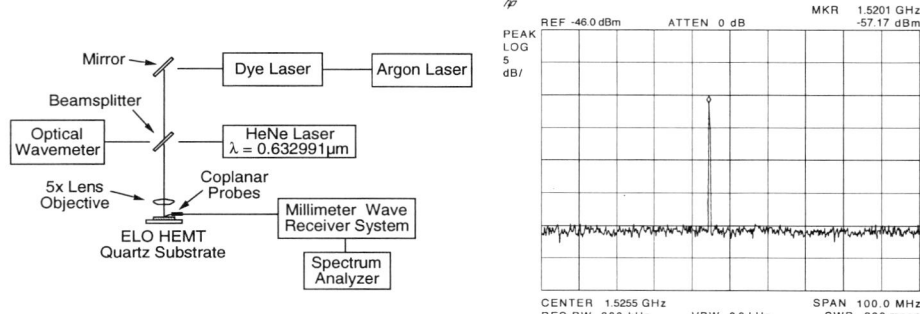

Figure 2. Experimental setup of optical heterodyne measurements.

Figure 3. Optically generated 90 GHz mixing signal (downconverted to IF).

The excellent optical frequency response and signal-to-noise ratios obtained from these devices indicate that they can be major additions to new optoelectronic integrated systems that exploit ELO technology. Current efforts in this area include extending the optical measurements to 200 GHz (0.05 mm InP HEMTs) using electro-optic sampling techniques, testing at 1.3 mm for applications in communication systems, and integration with on-wafer polyimide waveguides.

2.2. HETEROJUNCTION BIPOLAR PHOTOTRANSISTORS

In realizing efficient heterojunction photo-transistors (HPTs), classic design conflicts arise between the simultaneous optimization of high-frequency performance and optical coupling efficiency. Lumped element HPTs with RC bandwidth limitations need to be scaled down in size for high speed operation. This results in poor optical coupling efficiency and reduced optical absorption. Furthermore, these small devices saturate at low input optical power levels. Current efforts are to resolve these issues by utilizing a novel traveling wave HPT (TW-HPT) configuration in which optical power transfer occurs along the sides of the device from a polyimide channel waveguide. Unlike p-i-n or MSM traveling wave implementations, this structure has gain at the IF frequencies. While allowing for independent optimization of speed and optical power transfer, the polyimide waveguides bring added functionality making this approach very attractive for OEICs. We present proof-of-concept results demonstrating the viability of the TW-HPT. This system uses HBT technology which has been demonstrated as viable for hundreds of Ghz [15]. It also lends itself to a generation of balanced mixers and related configurations which will bring dramatic enhancements to the field.

Initially, we measured the optical performance at 1.3 µm of traditional lumped-element HPTs using optical mixing techniques. The HPT was first biased as a diode by shorting the base to the emitter. The optical response in this configuration was used to normalize the response of the HPT when biased in the common-emitter amplification mode. This effectively calibrates out all optical coupling and absorption losses revealing the actual internal optical gain. The measurements show that the optical response closely resembles the device's electrical response measured from a network analyzer. Therefore, the lumped-element HPT exhibited high speed and high optical gain, but suffers from poor optical coupling arising from mismatch in the optical mode size and the absorption cross-section making the device an ineffective optical detector for real commercial systems. To overcome this design conflict, we have gone to a novel traveling wave structure in which bandwidth and optical absorption can be simultaneously optimized.

The TW-HPTs used in this study consisted of an InP based double heterojunction layer structure with AlInAs emitter and collector and a 30 nm InGaAs base. Metal pads in the lumped-element design are now replaced by transmission lines for the emitter, base and collector. A low loss polyimide optical waveguide [16], with 10x10 mm cross-section, is defined along the side of the base and emitter mesas using standard photolithography. The intensity modulated light coupled into the polyimide waveguide leaks into the base and subcollector regions along the length of the device as shown in Figure 4. The bandwidth limitation is now based on velocity mismatch between the injected optical and the induced electrical waves and is no longer limited by the total junction area. In this effort, we have designed the electrodes for a 50 ohm characteristic impedance at 60 GHz using HP's High Frequency Structure Simulator. In our typical prototype device the base, collector and emitter electrodes are formed by three tightly coupled microstrip lines which are 20 mm wide with a separation of 15 mm. The optical waveguide is designed to mode match the input optical beam properly and the device length is adjusted to optimize absorption. Simulations using RSoft's BeamProp program show that 90% of the optical power can be transferred within a device length of 800 μm. Devices of lengths varying from 20 to 2000 μm were fabricated and are being tested. Figure 5 is a photograph of a 200 μm long fabricated device. Preliminary results show good frequency response at millimeter wave

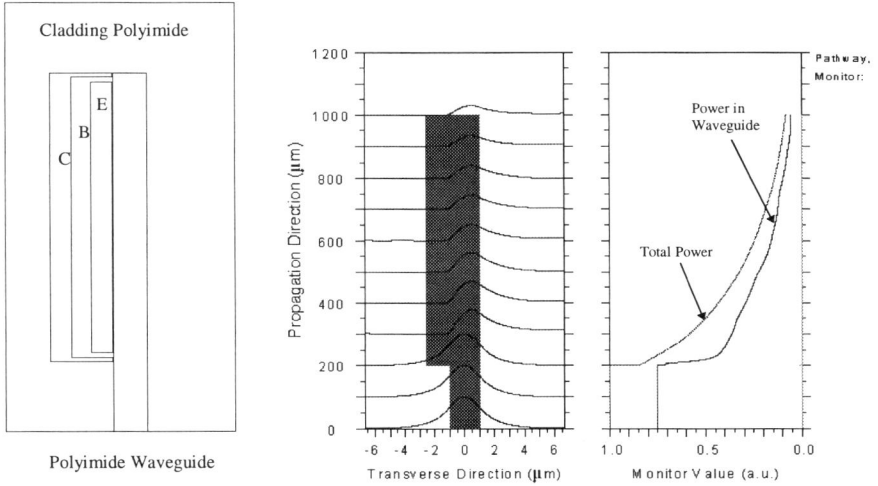

Figure 4. Schematic of the side coupled TW-HPT and BeamProp simulation showing 800 μm coupling length.

348

frequencies at 1.3 μm with a signal to noise ratio of 25 dB using a 2000 μm long device. Coupling efficiency, power saturation and optical gain are currently being characterized and extension to 60 GHz is being actively pursued. This approach lends itself to high frequency converters from optical signals to Terahertz frequencies in integrated systems which require gain and flexibility.

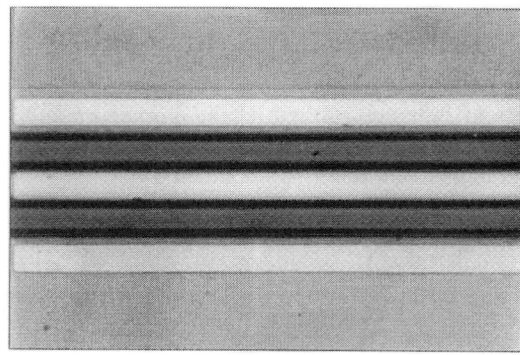

Figure 5. 200 μm long fabricated TW-HPT. From top to bottom are the base, emitter and collector electrodes. In between the electrodes are the polyimide waveguides.

2.3. HIGH BANDWIDTH POLYMER MODULATORS

Nonlinear electro-optic polymer materials have the advantages of fast response and low dispersion. Traveling wave devices made from these polymer materials have an intrinsically higher bandwidth compared with competing material systems. We reported earlier the demonstration of 60 GHz polymer electro-optic modulators [17], and have now extended this development to 94 GHz devices. Measurements using Femtosecond pulses [18] have shown that these systems will work at Terahertz frequencies because of their optoelectronic mechanism of operation. These devices will provide an important link between the submillimeter frequencies and their optical carriers.

Circuit performance is the limiting factor in extending the frequency response of these devices. The resistive loss of the electrodes becomes higher as the frequency increases. Using optimized device parameters and improved processing techniques, we have designed and fabricated two types of improved electrodes which have low losses at high frequency. Our coplanar strips configuration requires in-plane poling while our new microstrip line electrode works with our standard corona poling process. We have characterized and tested both electrodes with fin line transitions at 94 GHz to couple efficiently the millimeter wave driving power into the modulators. Figure 6 shows the overview of the polymer modulator, and Figure 7 shows S_{21} of the microstrip line electrode with anti-podal fin line transitions from and to the W-band microwave waveguides.

Effective characterization of the high speed electro-optic phase modulator is also becoming a more difficult issue as the frequency is raised into the millimeter and submillimeter wavelength ranges. We have developed a very sensitive heterodyne

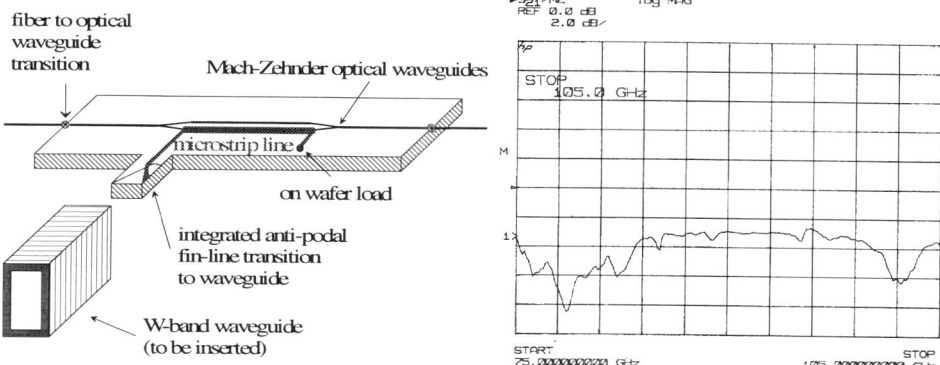

Figure 6. Overview of the polymer modulator with transition to waveguide.

Figure 7. S$_{21}$ of microstrip line with 2 anti-podal transitions to and from W-band waveguide.

detection system to characterize the device at high frequency. Figure 8 shows the optical heterodyne detection system with an external-cavity semiconductor tunable laser. In this optical heterodyne detection technique a second YAG laser is used as a local oscillator to down convert the high frequency phase modulated signal to a much lower frequency amplitude modulated signal. Complex and costly high frequency optical signal detection and millimeter wave instrumentation can be avoided in this sensitive detection approach. The external cavity, locked semiconductor tunable laser has a wide wavelength capability, making it possible to characterize the optical phase modulation up to a few THz.

The fabrication procedures for these devices have become more mature and the optical insertion losses have been significantly reduced. Mach-Zehnder optical interferometer waveguides have been fabricated which use the maximum available nonlinearity and directly give an amplitude modulation output. The modulators have been fabricated on silicon wafers for on-chip device driver integration and optical waveguide end surface cleaving techniques have been developed. Thus the polymer modulators are now entering a new phase in which we can project systems that work at extremely high frequencies, in arrays, and with a high level of systems integration with logic and driver elements.

3. Applications of Lightwave/Terahertz Interactions

In the first section we have discussed three new device concepts that will be able to enhance the use of Lightwave technology in the generation, detection and transmission of Terahertz signals. This section discusses the actual implementation of systems which are dependent on this type of technology. The first considers a form of optoelectronic oscillator developed at lower frequencies at JPL [19]. Next we consider the application of this form of technology to phase array radars and imaging systems. Finally we consider how this approach can be used to implement an artificial nonlinear material that effectively performs phase conjugation using four wave degenerate mixing.

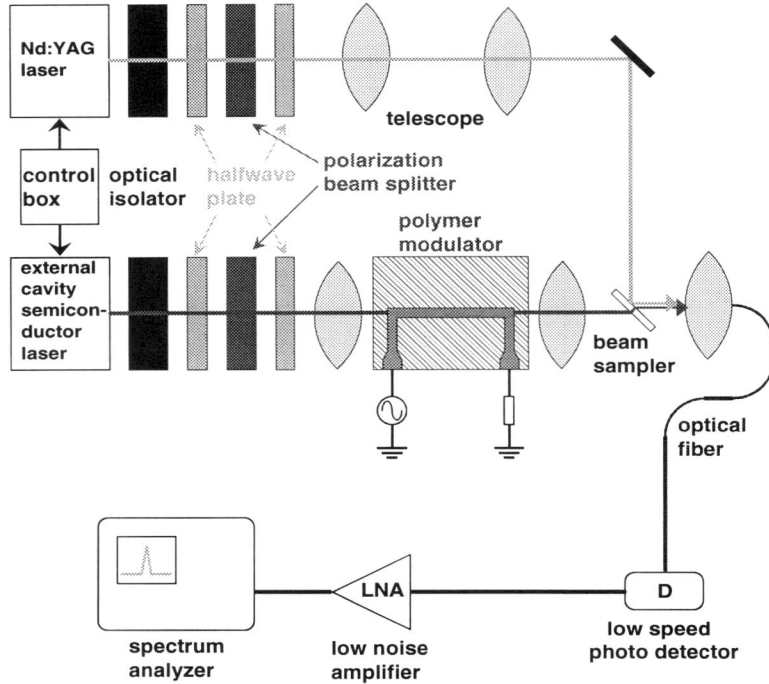

Figure 8. Heterodyne setup with a semiconductor external cavity tunable lasers.

3.1. OPTOELECTRONIC OSCILLATOR

This novel optoelectronic oscillator converts continuous light energy into highly stable and spectrally pure millimeter/submillimeter wave signals. It consists of a pump laser, a feedback circuit including an intensity modulator, an optical delay line, a photodetector and a filter as shown in Figure 9. In previous work it was shown that the oscillation frequency is limited only by the speed of the modulator, and the fundamental noise in the oscillator is limited by elements which do not have dispersive losses in the RF. Thus, the OEO holds the potential for providing signals with unprecedented spectral purity at millimeter and submillimeter frequencies. The system we have indicated is monolithic using the polymer modulator, on wafer polyimide guide, and HBT phototransistors as indicated. Additional elements such as a semiconductor optical amplifier can be added as required in the forms of ELO, to provide sufficient modulator driving voltage. Prototypes of this system are now being put together at 60 GHz but they can be readily extended to 100s of GHz in integrated form.

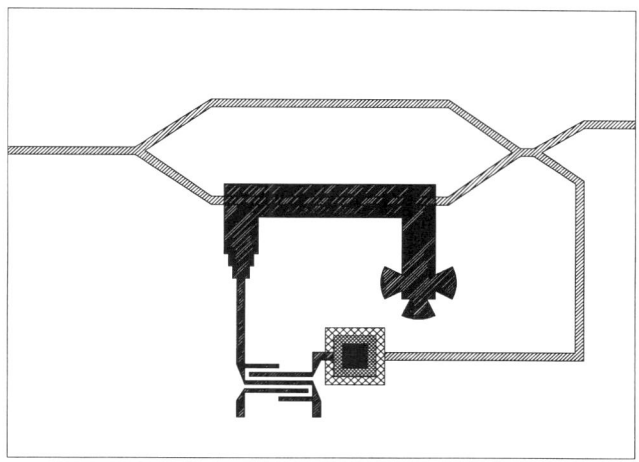

Figure 9. Optoelectronic oscillator: starting from the left, laser light goes into the waveguide, into the Mach Zehnder modulator and then into the HBT photodetector. The electrical signal from the detector is filtered and drives the MZ modulator. This oscillator is capable of extraordinary narrow linewidths at hundreds of GHz.

3.2 OPTICALLY-CONTROLLED SERIALLY-FED PHASED ARRAY SYSTEM

A new system suitable for phased array radar and imaging has been designed and demonstrated. This system is based on the time dependent nature of modern RF sensors and uses a novel optical serial feed which has all the advantages of true time delay and yet only requires one tunable laser, one optical modulator and one fiber optic grating delay. Initial results have shown true time delay for a 2-element array from 7 GHz through 12 GHz. The system uses high frequency polymer modulators and is very viable at millimeter and submillimeter frequencies. It many ways it represents the potential of future Terahertz systems which can perform practical functions and yet have a reasonable degree of complexity.

The transmit mode of this system can be best described in terms of a timing unit and a distribution network as shown in Figure 10. The desired delays for a given millimeter wave beam direction are generated sequentially by the timing unit and then transformed into concurrent signals to feed the radiators via the distribution network. The timing unit uses an electrically tuned DBR laser which is AM modulated with the desired microwave pulses using an optical modulator. After the modulator the laser light is directed through an optical circulator to a fiber optic grating. The grating reflects light back at different positions along the fiber depending upon the optical wavelength. Each serially fed optical pulse has a chosen wavelength and therefore a chosen time delay relative to the RF pulse gate, to produce a given pointing direction.

Returned light from the third port of the circulator enters the distribution network formed by the tapped optical delay line. Once the series of optical pulses arrive at the correct delay line taps for each antenna element, the microwave modulation is

352

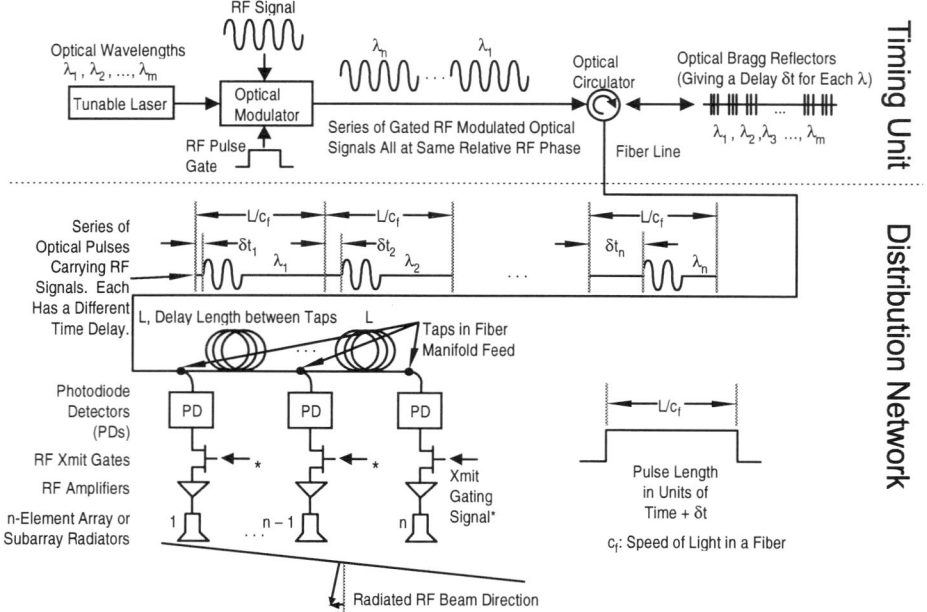

Figure 10. Basic XMIT Mode Implementation for Array of n Elements.

obtained using photodetectors. The microwave signals from these detectors are simultaneously gated on with microwave switches (Xmit gates) when the tapped line is fully loaded. After the signal is radiated the switches are turned off and the line is reloaded. To establish the viability of this approach, we have built a two-element transmitter with two optical wavelengths. The tunable laser was emulated by switching two fixed wavelength laser diodes. The RF signals feeding the array were monitored on a digital sampling oscilloscope. By gating the signals coming out from the photodetectors correctly, CH1 received the second pulse and CH2 received the first pulse. The pulse width was 6 ns at a repetition rate of $\approx$ 25 MHz which corresponded to the delay length $L \approx 4$ m. The true time delay measured from the two channels of the oscilloscope was 35 ps from 7 GHz through 12 GHz, as shown in Figure 11 and Figure 12. These two pulses were carried by two different wavelengths and were generated sequentially. The first pulse was then delayed by length L to match up with the second one. The Xmit gating ensures that only the correct pulse reaches each radiating element.

This new serially-fed true time delay system has been designed and the basic concept experimentally verified. It uses only one tunable laser, one optical modulator and one delay element to achieve beam steering. The system is quite versatile and can be used to control both one and two dimensional arrays. Furthermore, a natural extension to the receive mode can also be implemented for passive imaging. Essentially, by using the natural timing of our pulsed radar we are able to simplify greatly and increase the flexibility of optically controlled, high frequency systems.

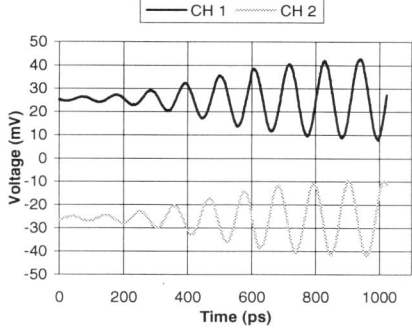

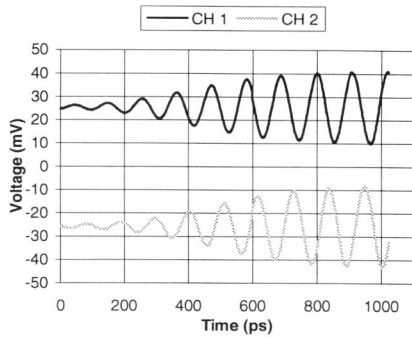

Figure 11. Signals feeding the radiators monitored on a digital sampling oscilloscope. The λ_1 pulse (CH2) was generated before the λ_2 (CH1) pulse. CH2 was leading CH1 by 35 ps. The RF frequency was 9.040 GHz.

Figure 12. λ_1 generated after λ_2. CH1 was leading CH2 in this case, a reverse of Figure 11. Using this delay the RF beam was directed to the opposite side of boresight direction.

The use of these systems for Terahertz frequencies depends on the optical components which, with minor exceptions, will easily have the required bandwidth. Another related example is our phase conjugation project discussed in the following section. In the same way as the phased array radar system, it also depends on the availability of fast modulators, optical on wafer guides and appropriate mixers and detectors.

3.3. MILLIMETER/SUBMILLIMETER WAVE PHASE CONJUGATION USING ARRAYS OF NONLINEAR OPTICALLY PUMPED DEVICES

Phase conjugation is a technique for reversing both the direction of propagation and the overall phase factor of an incoming wave. To date, most of the development has been concentrated on the optical regime due to the nonlinearity and power density constraints at longer wavelengths. The effort discussed here extends phase conjugation techniques to millimeter/submillimeter wave frequencies using arrays of mixing elements with antennas to form nonlinear surfaces [20]. In this work, we report the use of optical signals as carriers of the millimeter wave pump signal. Because of the size, losses and sheer complexity of these systems the optical injection technique is the key enabling technology to making simple, viable, low cost millimeter/submillimeter wave phase conjugation arrays.

To demonstrate the concept of using electronic devices with three wave mixing (TWM) to generate conjugate phases, a mixer can be used as the nonlinear element. The mixer output signal can be written as:

$$V_{IF} = a(V_{LO} + V_{RF})^2 \tag{1}$$

The conjugate phase arises from the term with the difference frequency: $\omega_{LO} - \omega_{RF}$. The incoming signal into the RF port of the mixer carries a phase factor

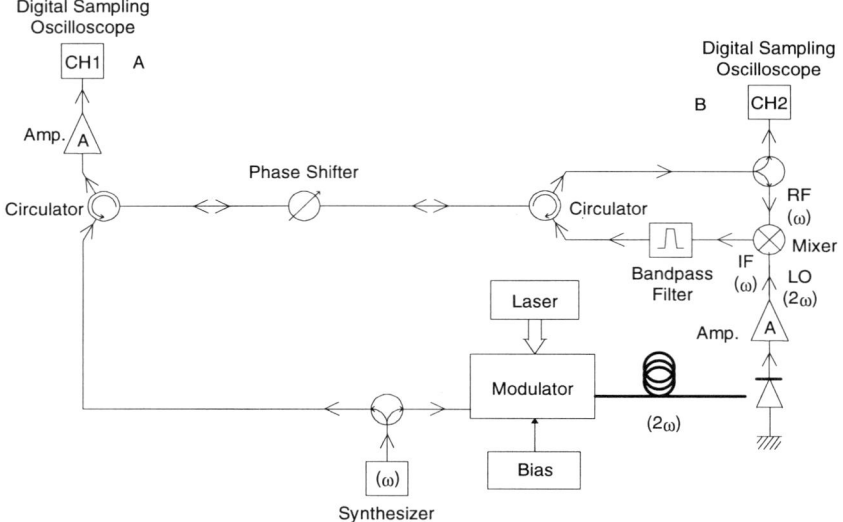

Figure 13. Initial experimental setup of optically injected conjugate phase generation.

$\omega t + \varphi$. It is mixed with $2\omega t$ (LO) to generate the conjugate phase: $\omega t - \varphi$ (IF). Of course, the other second-order terms can be filtered out by a bandpass filter. The initial efforts to demonstrate optical injection used fiber optics feeds and eliminated the need of a frequency doubling amplifier for generating the pump signal (2ω), as shown in Figure 13.

An electro-optical modulator was biased at its transfer function extremum, therefore producing modulation at 2ω (20 GHz) of the 1.3 μm laser beam. After this the light was routed via optical fibers to a PIN diode and the microwave pump signal 2ω was extracted. The 2ω signal was then amplified and sent to the mixer to generate the phase conjugate signal. The results are shown in Figure 14. To prove the signal received at A in Figure 13 is the phase conjugate signal of the input signal, a phase shifter was inserted in the microwave path. After the incoming signal passed through

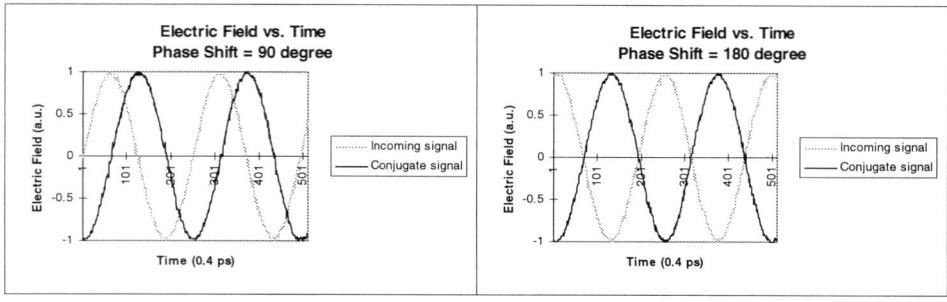

Figure 14. Incoming and conjugate signals of optically injected phase conjugation generation, monitored on the digital sampling oscilloscope. The incoming signals were recorded at B in Figure 13 and the conjugate signals were recorded at A in Figure 13.

the phase shifter, it carried a phase shift $+\varphi$. TWM reversed this phase shift to $-\varphi$ and then sent the signal back. As this $-\varphi$ phase-shifted signal passed through the phase shifter, it picked up an additional phase shift $+\varphi$. Therefore at A the overall phase shift caused by the phase shifter was zero: $-\varphi + \varphi = 0$. It is clearly shown in Figure 14 that the conjugate signal was virtually unchanged while the incoming signal was shifted due to the phase shifter change. Since this setup used only one element it was a one dimensional experiment and could not be applied to reversing the wavefront of a radiating field. In order to extend to three dimensional phase conjugation, a two dimensional planar array must be used.

In a microwave feed system, the field distribution poses a significant design challenge. Optical injection of the pump signal can virtually eliminate such problems. Another advantage of optical injection is that it removes the requirement of providing a high power, high frequency (2ω) microwave source. Currently, our study of the optical injection configuration concentrates on using a single HBT [14] as the phototransistor and the mixing device.

Up to this point, our discussion has concentrated on generating phase conjugate signal at a single element. These one dimensional demonstrations would have little practical use if one could not extend this phase conjugate generation technique to higher dimensions. It can be proven that if the generated phase factor is conjugate to that of the incoming wave on a plane, it will be conjugate everywhere. Therefore an array of conjugate generation elements can provide the ability of generating phase conjugate waves. To demonstrate this concept, an eight element array was built to show the directivity originating from phase conjugation. The setup is shown in Figure 15. Both the transmitting horn and the receiving horn are mounted on rotational stages. The

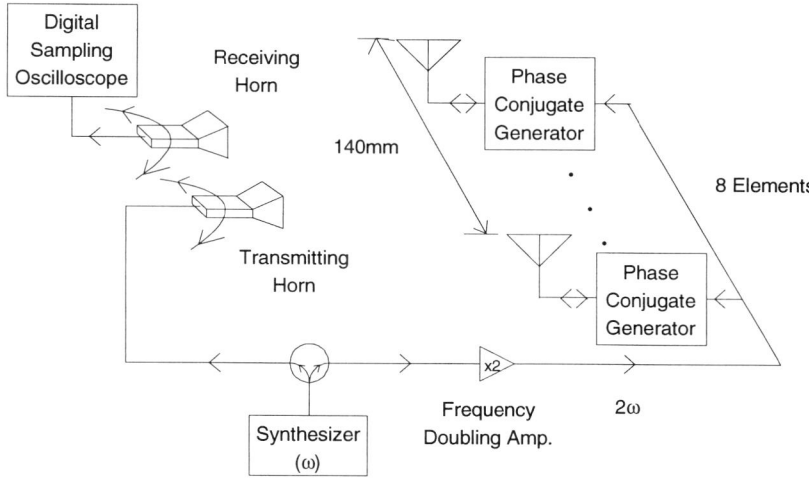

Figure 15. Experimental setup of the eight-element array phase conjugation system. The phase conjugate generators shown in the drawing are similar to the mixer circuit in Figure 13.

receiving horn is slightly higher and behind the transmitting one. After setting up the transmitting horn at an arbitrary angle, the receiving horn was moved around until the received signal was maximized on the sampling oscilloscope. The receiving horn was always able to follow the transmitting horn. To prove that this directivity was caused by phase conjugation rather than reflection, the pump signal was turned off at one element and the directivity disappeared. A dielectric material was also inserted into the microwave path and the received signal showed no phase shift.

The results obtained from this initial measurement confirmed our simulations as shown in Figure 16. This simulation show the basic self focusing property of this array as well as the limitations of diffraction. In terms of going to higher frequencies and more elements we look toward our new edge coupled HBTs fabricated with an optical waveguide in the base collector depletion regions. The passive waveguide is designed to maximize input light coupling while the HBT is configured in a traveling wave structure. These HBTs can then be integrated to form quasi-monolithic millimeter/submillimeter wave nonlinear optically pumped surfaces (MNOPS) as shown in Figure 17. These surfaces can be thought as an artificial nonlinear material which can operate at Terahertz frequencies. This is an excellent example of the integrated type of Lightwave/Terahertz structure that we project will be the building block of a new

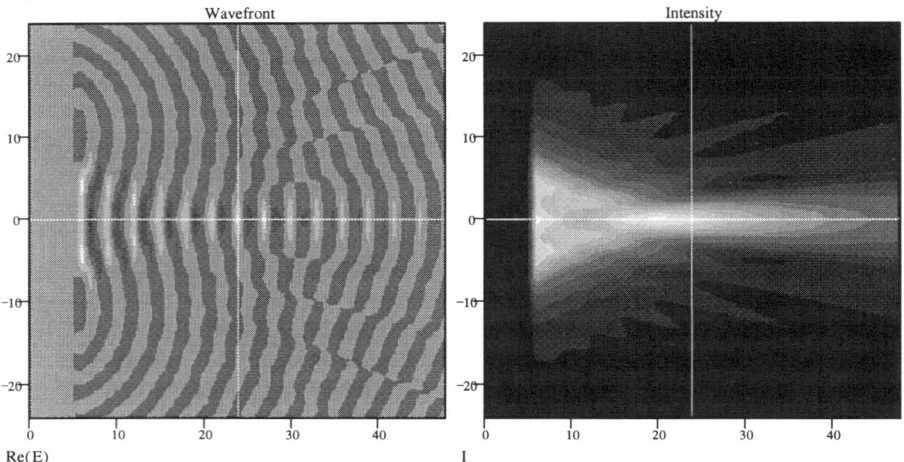

Figure 16. Computer calculated wavefront and intensity patterns of the conjugate signal of a 10 GHz point source at 30 cm from the conjugator array. The array is assumed to have 10 dipole elements with 2 cm spacing. The horizontal and vertical units are in centimeters measured from the center of the array.

generation of submillimeter systems.

4.   Conclusion

We have demonstrated several new optical devices that have impressive capabilities for millimeter/submillimeter wave performance. These include traveling wave phototransistors, ELO devices and electro-optic polymer modulators. Using these and similar building blocks new and novel forms of Terahertz systems become feasible. Examples of an optoelectronic oscillator, optically controlled phased array transmitters and optically controlled phase conjugators are discussed. The use of optics in these systems offers the hope of extraordinary performance, convenient transmission and practical, viable packages. This will be the vital element in making this portion of the spectrum useful and practical.

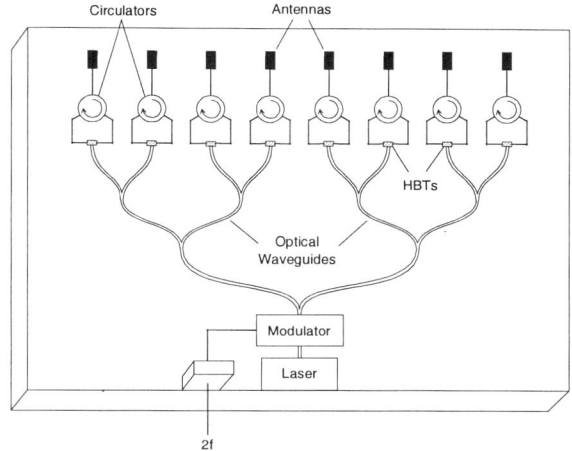

Figure 17. Microwaves nonlinear optically pumped surface for two dimensional phase conjugation generation.

Acknowledgment. I would like to thank my students: Yian Chang, D. Scott, D. Prakask, D. Chen, W. Wang, D. Bhattacharya, M. Ali and D. Cohen. Also, I appreciate the contributions of my colleagues: Professors Levi, Jalali and Steier and Drs Tsap and Newberg. This research is supported in part by AFOSR (under the direction of Dr. H. R. Schlossberg), NCIPT (DARPA) and ONR.

References

1. Froberg, N.M., Hu, B.B., Zhang, X.-C., and Auston, D.H. (1992) Terahertz radiation from a photoconducting antenna array, *IEEE J. of Quantum Electronics* **28**, 2291-2301.
2. Li, M., Sun, F.G., Wagoner, G.A., Alexander, M., and Zhang, X.-C. (1995) Measurement and analysis of Terahertz radiation from bulk semiconductors, *Appl. Physics Lett.* **67**, 25-27.
3. Luo, M.S.C., Chuang, S.L., Planken, P.C.M., Brener, I., Roskos, H.G., and Nuss, M.C. (1994) Generation of Terahertz electromagnetic pulses from quantum-well structures, *IEEE J. of Quantum Electronics* **30**, 1478-1488.
4. Kawase, K., Sato, M., Taniuchi, T., and Ito, H. (1996) Coherent tunable THz-wave generation from LiNbO3 with monolithic grating coupler, *Appl. Physics Lett.* **68**, 2483-2485.
5. Paiella, R., and Vahala, K.J. (1996) Four-wave mixing and generation of Terahertz radiation in an alternating-strain coupled quantum-well structure, *IEEE J. of Quantum Electronics* **32**, 721-728.
6. Carrig, T.J., Rodriguez, G., Clement, T.S., Taylor, A.J., and Stewart, K.R. (1995) Scaling of Terahertz radiation via optical rectification in electro-optic crystals, *Appl. Physics Lett.* **66**, 121-123.
7. Nahata, A., Auston, D.H., Wu, C., and Yardley, J.T. (1995) Generation of Terahertz radiation from a poled polymer, *Appl. Physics Lett.* **67**, 1358-1360.

358

8. Martins-Filho, J.F., and Ironside, C.N. (1994) Multiple colliding pulse mode-locked operation of a semiconductor laser, *Appl. Physics Lett.* **65**, 1894-1896.
9. Brown, E.R., McIntosh, K.A., Smith, F.W., Manfra, M.J., and Dennis, C.L. (1992) Measurements of optical-heterodyne conversion in low-temperature-grown GaAs, *J. of Appl. Physics* **73**, 1206-1208.
10. Chen, Y.K., Wu, M.C., Tubun-Ek, T., Logan, R.A., and Chin, M.A. (1991) Multicolor single-wavelength sources generated by a monolithic colliding pulse mode-locked quantum well laser, *Photonics Tech. Lett.* **3**, 971-973.
11. Arahira, S., Oshiba, S., Matsui, Y., Kunii, T., and Ogawa, Y. (1994) Terahertz-rate optical pulse generator from a passively mode-locked semiconductor laser diode, *Optics Lett.* **19**, 834-835.
12. McIntosh, K.A., Brown, E.R., Nichols, K.B., McMahon, O.B., DiNatale, W.F., and Lyszczarz, T.M. (1995) Terahertz photomixing with diode lasers in low-temperature-grown GaAs, *Appl. Physics Lett.* **67**, 3844-3846.
13. Fetterman, H.R., Prakash, D.P., Scott, D.C., and Wang, W. (1994) Integrated optically driven microwave/millimeter wave structures 3 *MTT-S IEEE Int. Microwave Sym. Digest*, 1493-1496.
14. Bhattacharya, D., Bal P.S., Fetterman H.R., and Streit, D. (1995) Optical mixing in epitaxial lift-off pseudomorphic HEMTS, *Photonics Tech. Lett.* **7**, 1171-1173.
15. Frankel, M.Y., Carruthers, T.F., and Kyono, C.S. (1995) Analysis of ultrafast photocarrier transport in AlInAs-GaInAs heterojunction bipolar transistors, *IEEE J. of Quantum Electronics* **31**, 278-285.
16. Beuhler, A.J., Wargowski, D.A., Singer, K.D., and Kowalczyk, T. (1995) Fabrication of low loss polyimide optical waveguides using thin-film multichip module process technology, *IEEE Trans. on Components, Packaging, and Manufacturing Tech.* **18**, 232-234.
17. Wang, W., Chen, D., and Fetterman, H.R. (1996) Optical heterodyne detection of 60 GHz electro-optic modulation from polymer waveguide modulators, *Appl. Physic Lett.* **67**, 1806-1808.
18. Ferm, P.M., Knapp, C.W., Wu, C., Yardley, J.T., Hu, B.B., Zhang, X.-C., and Auston, D.H. (1991) Femtosecond response of electro-optic poled polymers, *Appl. Physics Lett.* **59**, 2651-2653.
19. Yao, X.S., and Maleki, L. (1996) Converting light to spectrally pure microwave signals, *Optics Lett.* **21**, 483-485.
20. Cutler, C.C., Kompfner, R., and Tillotson, L.C., (1963) A self-steering array repeater, *Bell Syst. Tech. J.* **42**, 2013-2032.
   Scott, D.C., Plant, D.V., and Fetterman, H.R. (1992) 60 GHz sources using optically driven heterojunction bipolar transistors, *Appl. Physic Lett.* **61**, 1-3.

ELECTROOPTIC AND PHOTOCONDUCTIVE TECHNIQUES FOR PROBING AND IMAGING OF THZ ELECTRIC SIGNALS

H.G.ROSKOS, T.PFEIFER, H.–M.HEILIGER,
T. LÖFFLER and H.KURZ
*Institut für Halbleitertechnik II, Rheinisch–Westfälische
Technische Hochschule (RWTH) Aachen,
Sommerfeldstr. 24, D–52056 Aachen, Germany*

1. Introduction

Since femtosecond lasers have become available, they have been are applied to measure, in the time domain, the electric response of ultrahigh–speed devices with bandwidths in the GHz and THz frequency range. Such laser–based measurement systems exhibit a superior time resolution (bandwidth coverage) owing to the short duration of the optical pulses. Because of their complexity, femtosecond optical techniques remained confined to a small community working in ultrahigh–speed–device development where measurement alternatives are sparse. In recent years, however, optical probing has become more and more attractive because it has developed into a flexible test tool that allows us to choose from a variety of methods to solve a specific measurement problem. Here, we describe a modular measurement system that offers a number of alternatives to synchronize to, generate, and detect high-frequency electric signals in microelectronic devices and circuits [1]. A frequency range spanning more than three orders of magnitudes from 1 GHz to 4 THz is covered. For stroboscopic measurements on circuits driven by electronically generated clock signals, the system is configured to lock onto periodic signals of arbitrary frequency. Alternatively, impulsive time–domain mesurements can be performed on chip by signal injection with freely positionable photoconductive probes or by direct optical excitation of active devices. With respect to detection, the following approaches are available: sampling with freely positionable electrooptic and photoconductive probe tips, and probe–tip–free testing based on the field–dependent optical nonlinearity of the circuit's substrate material. The combination of the various modules allows us to optimize the approach to a specific testing problem. Measurements of the linear and nonlinear behavior of active and passive devices under test (DUTs) can be performed. The electric field, respectively potential, is measured locally (point measurements) or in its spatial distribution (mapping) both in the near and far field.

J.M. Chamberlain and R.E. Miles (eds.), New Directions in Terahertz Technology, 359–367.

2. Impulsive measurements

The time-resolved optoelectronic measurements are performed with a self–modelocked Ti:sapphire laser producing optical pulses of 150 fs duration (FWHM) over the wavelength range 0.7–1.0 μm. For synchronization of the laser pulse repetition frequency (f_L) to external oscillators, the laser cavity length is adjustable [2, 3]. The output mirror is mounted on a mechanical translation stage for varying f_L from 75.1 MHz to 75.7 MHz. Additionally, for fine tuning over a 400 Hz range, one of the mirrors of the folded cavity is attached to a piezoelectric transducer driven by a high–voltage amplifier.

Impulsive characterization is performed in pump/probe configuration. The lower part of Figure 1 illustrates a measurement on a coplanar waveguide. The laser beam is split in to two parts, a pump beam for generation of electric pulses and a time–delayed probe beam for detection. Electric pulse generation is achieved, in this case, with a freely positionable photoconductive switch (discussed below). The pulse is launched onto the DUT via the metallic needle tip of the switch that is in electric contact with the center conductor of the DUT. The signal is detected by the probe laser pulse in an electrooptic crystal positioned above the waveguide. The probe pulse is time–delayed with respect to the pump laser pulse by a mechanical translation stage. By variation of the time delay, the temporal waveform of the pump–pulse–induced electric signal is recorded. The pump/probe scheme avoids timing jitter, because pump and probe pulses are derived from the same laser pulse and are therefore inherently synchronized. Lock–in detection serves to reduce the signal noise. Spectral information is derived from the time–domain data by Fourier analysis.

While electrooptic probes are commercially available, photoconductive probes at present are not. Indeed, the concept of free positioning of such probes at any electrically accessible point of a DUT has only been introduced recently [4] and is under continuing development [5, 6, 7, 8, 9, 10]. The concept has received much attention for the following reasons: (i) photoconductive probes can be employed both for generation and detection of electric signals; (ii) absolute voltage calibration of generated and detected signals is straightforward (it is more of a problem in electrooptic probing); (iii) the sensitivity is extraordinarily high (potential differences of 1 μV can be measured); and (iv) the spatial resolution can be deep in the submicron regime. For the last two reasons in particular, photoconductive probes are of great interest also for scanning–tunneling microscopy [11, 12].

Two types of photoconductive probe that are routinely employed in our laboratory are depicted in the scanning–electron micrographs (SEMs) in the upper part of Figure 1 [5, 6, 7]. The switch and the contact tips are separated by just a few tens of μm from each other in order to limit propagation broadening of the generated electric pulses. The switch consists of a metal–semiconductor–metal electrode structure with a high–dark–resistance semiconductor. The switch is activated by optical carrier excitation in the semiconductor, and turned off after less than 1 ps by carrier trapping at defect sites in the semiconductor. The displayed probes are fabricated from silicon–on–sapphire (SOS). The subpicosecond photoconductance lifetime in the Si is achieved by ion implantation. The sapphire substrate is transparent and allows optically controlled positioning of the probe on the DUT and backside illumination of the switch by the laser pulses. Contact tips are fabricated in two ways: (i) A Pt/Au–coated, 5–μm–high Ti tip (left side in Figure 1) is realized by e–beam evaporation followed by a lift–off process. The tip diameter is 1 μm, but can be further

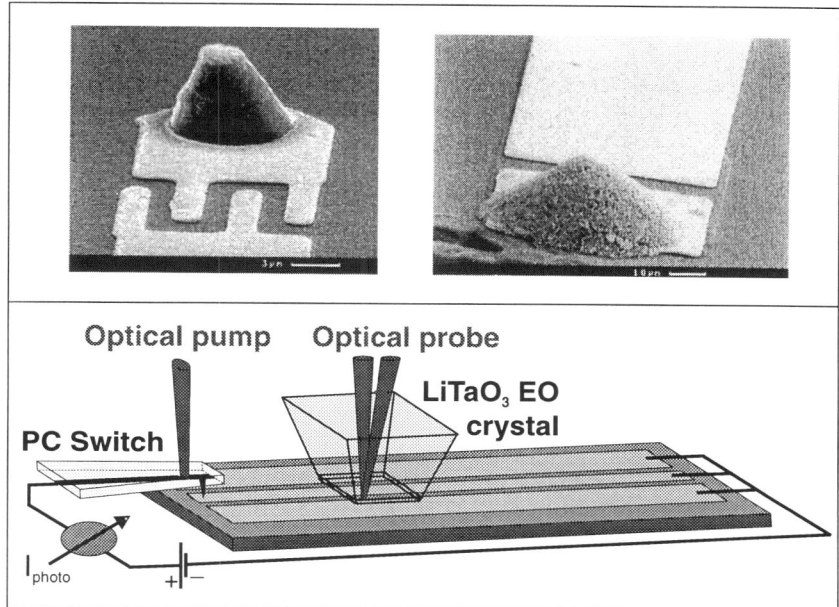

Figure 1. Upper part: SEM micrographs of two types of freely positionable photoconductive probe tips for generation and detection of electric pulses. Lower part: Pump/probe configuration for impulsive sampling. Electric pulses are generated with a probe tip, launched onto the DUT (here: a coplanar waveguide), and detected by electrooptic sampling.

reduced for better spatial resolution. (ii) A low–cost type of tip (right side of Figure 1) is fabricated from a drop of electrically conductive epoxy placed onto the switch electrode with the help of a sharpened glass fiber. The tip diameter is on the order of 10–30 μm and fits well to contact pads of standard mm–wave devices. The tip is relatively soft and flattens when pressed onto a surface for the first few times. If the epoxy has been cured properly, however, the initial change of shape is not severe, and the probe can be employed routinely.

Recently, probes have also been fabricated with another ultrafast photoconductor material, GaAs grown at low temperature (LT–GaAs) [1, 8, 9, 10]. These probes exhibit better time resolution and a higher switching efficiency than SOS probes. Electric signals with several volts amplitude can be generated allowing, e.g., the investigation of nonlinear device responses. Lift–off techniques are used to transfer LT–GaAs thin films from the growth to other substrates facilitating the realization of low–dielectric–load probes and of complex probe structures for complete optical S–parameter measurements.

With respect to signal detection, photoconductive sampling is different from electrooptic testing in two significant aspects: The first difference is that photoconductive probes measure potential difference, whereas electrooptic probes are electric–field sensors. Hence, only the latter are polarization sensitive. Both techniques measure not only absolute quantities but aslo sign changes of the signal: both amplitude and phase are

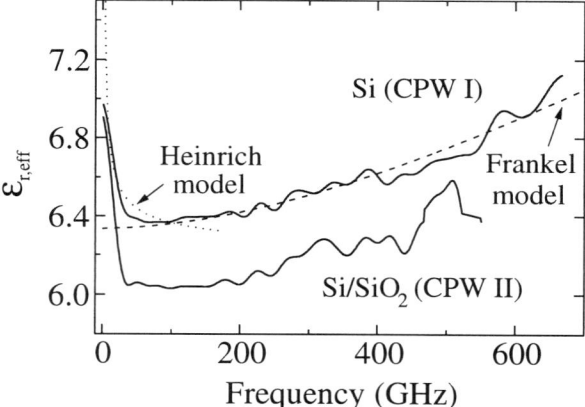

Figure 2. Frequency dependence of the effective permittivity of a coplanar waveguide on Si (CPW I) and Si/SiO$_2$ (CPW II). Full lines: measured data. Dashed line: simulation excluding the skin effect of the metallization (model by Frankel), dotted line: TEM simulation with the skin effect (model by Heinrich).

determined. This fact enables us to calculate the real and imaginary parts of the Fourier transforms and, from there, to extract successively the complex propagation constants of the signal and the complex device characteristics of the DUT. The second difference between the two techniques concerns the time resolution. With electrooptic probes, we achieve a time resolution of better than 300 fs. Photoconductive probing with SOS–based probes gives 2–ps resolution, whereas probes made from LT–GaAs can reach just below 1 ps [1]. Photoconductive testing is hence suitable for measurements with bandwidth requirements up to 1 THz, whereas electrooptic probing covers a range of several THz. The high sensitivity of photoconductive testing allows us to employ it readily for far–field measurements. The switch is then integrated into an antenna structure [1]. In this application, frequencies up to 4 THz are detectable. Far–field sensing via electrooptic sampling is more difficult, but recent research has shown ways to improve the sensitivity sufficiently to make it interesting [13].

An example of measurements with a SOS signal generator and electrooptic detection in the geometry of Figure 1 is shown in Figure 2 [14, 15]. The DUTs are coplanar waveguides on highly resistive Si, with and without a 200–nm–thick thermally grown SiO$_2$ insulator layer between the substrate and the waveguide metallization. The figure displays the waveguides' effective permittivity extracted from time–domain waveform measurements at various propagation distances. One can clearly identify the reduction of the permittivity by the insulator layer. The point to be made in this context is that useful spectral information is extracted from close to dc up to about 600 GHz. With LT–GaAs probes, this bandwidth has been increased to greater than 1 THz [1].

3. Measurements on free–running circuits

For stroboscopic optical probing of a circuit operating at an arbitrary, internally or externally generated frequency f_S, the repetition rate of the laser is locked to that frequency.

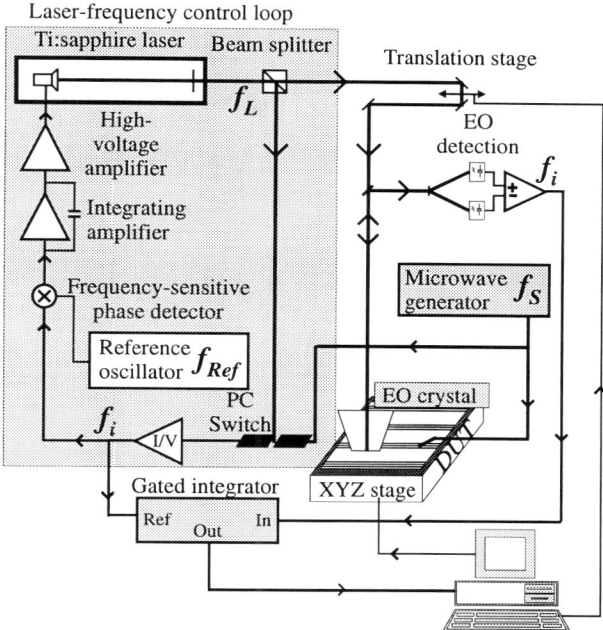

Figure 3. Scheme for stroboscopic measurements on free-running circuits (drive signal generated independently of the laser, here: by an external oscillator). The laser–pulse repetition rate is synchronized to a harmonic of the drive by adjusting the cavity length of the laser via mechanical and piezo–electric translation of one of the mirrors. A photoconductive switch serves as mixer to generate a signal at the intermediate frequency that is compared to a reference at fixed frequency.

Figure 3 displays the synchronization scheme [2, 3]. The laser pulses are mixed with the circuit's drive signal creating a replica at intermediate frequencies $f_i = f_S - n \times f_L$, n being an integer. The mixer (reference detector) is a photoconductive switch. The down-converted signal at the lowest frequency f_i is compared in a phase detector to the signal at f_{Ref} from a reference oscillator (f_{Ref}: typically 20 kHz). The error signal is minimized by adjusting the laser pulse repetition rate f_L via tuning of the laser cavity. We have demonstrated stable locking to microwave signals up to 75 GHz (the 995th harmonic of f_L), this frequency being limited by the available microwave source [2].

The measurements on a DUT are performed either with a second photoconductive receiver or (as shown in the figure) with an electrooptic probe. Two types of measurement are possible, (i) time–resolved waveform detection at discrete points on the cicuit (application as an optoelectronic sampling oscilloscope), or (ii) spatially resolved recording of electric fields at the fundamental or a higher harmonic of f_S (application as a field mapper). Both options are illustrated with measurements on an externally driven GaAs nonlinear transmission line (NLTL, see Figure 5) [3].

In the application as an optoelectronic sampling oscilloscope, the signal from the probe is processed by a time–gated integrator triggered by the mixer signal. The time evolution of the signal on the probe spot is sampled by mechanically delaying the probe

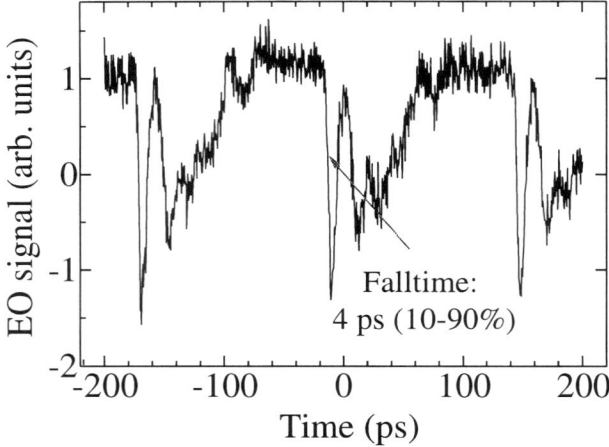

Figure 4. Example of a stroboscopic measurement with synchronized laser. A 6.310 GHz sine wave is fed into a NLTL. The electrooptic signal recorded as a function of time at the output port of the NLTL is displayed.

laser pulses relative to the laser pulses sent to the mixer. A sine wave at 6.310 GHz is fed into the circuit. Figure 4 displays the waveform measured at the output port of the NLTL. The nonlinear operation of the device is obvious as the shape of the waveform strongly deviates from that of a sine wave. The falling edge has steepened considerably, with a fall time (90 % – 10 %) of 4 ps. The strong inverted reflection 10 ps after each minimum is likely to result from circuit components behind the NLTL output port. The 4–ps feature does reflect the limit of the time resolution of the detection system. For the conditions of this particular measurement, the resolution is estimated to be 1 ps [3]. Fourier analysis of the data reveals that frequencies up to the 25th harmonic (150 GHz) of the drive signal can be resolved.

Figure 5 presents an example of a field–mapping experiment on the NLTL. The magnitude of the y–component of the electric field over the NLTL is displayed at the fundamental frequency (here: 7.512 GHz). Additionally, the phase is detected (not shown). The measurements are performed with a two–phase lock–in amplifier instead of the gated integrator. The DUT is moved under the electrooptic detector by a motorized x–y–z translation stage (distance between probe tip and DUT: 5 μm). The scan area is 2.5×2 mm^2 covering the whole NLTL section on the chip. The field on both sides of the center conductor of the waveguide is clearly visible. As expected, there is a 180° phase offset between the signal above the conductor and that below, because the field direction is opposite. A slight modulation of the signal along the NLTL results from the diodes. Signal attenuation on the NLTL is obvious from the decrease of the magnitude. The time needed for the measurement is 1 h indicating the long–time stability of the synchronization. Together with the phase information, the effective permittivity and the attenuation constant of the NLTL can be evaluated to be 22.5 ± 0.2 and (0.092 ± 0.007) mm^{-1}, respectively. Furthermore, the availability of both the amplitude and phase information allows to display the data to be displayed in the form of a movie to visualize signal propagation on the NLTL.

Spatial field mapping is of great interest for failure analysis of circuits. Conventional

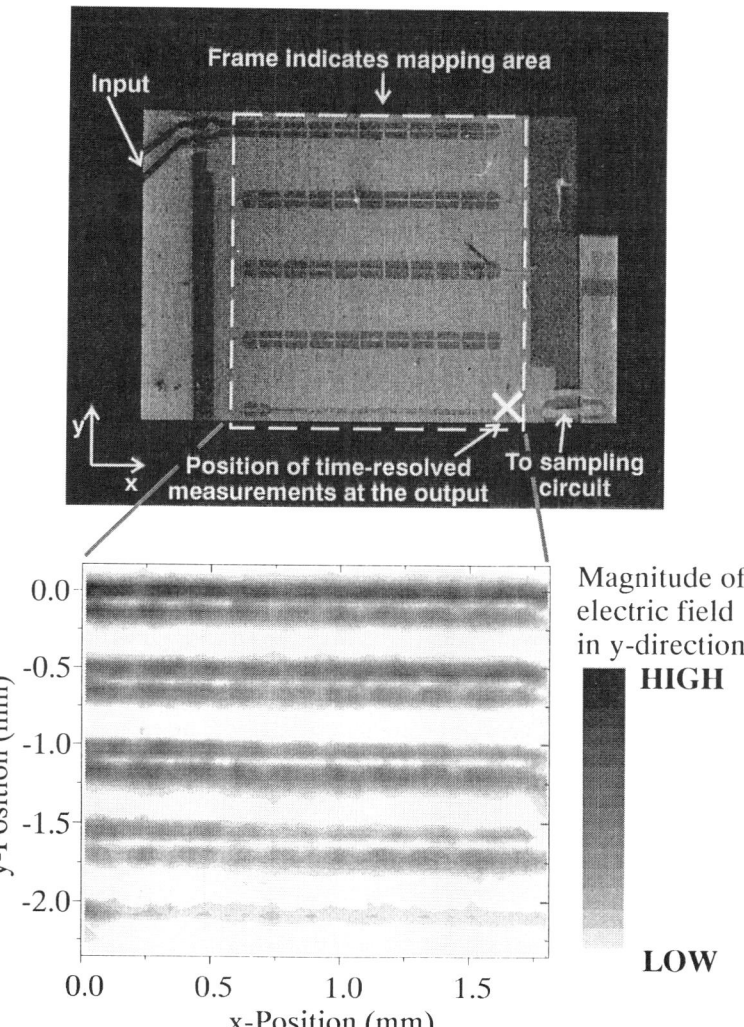

Figure 5. Upper part: SEM picture of the NLTL chip. The NLTL is a coplanar waveguide loaded periodically with Schottky diodes. It is folded into a meander with five horizontal sections (the fifth is narrower than the four others). The vertical sections are too narrow to be visible. The waveguide metallization appears bright, the dark stripes inbetween are GaAs. The barely visible bright stripes appearing periodically in the GaAs region are the electric contacts to the diodes. Lower part: Measured magnitude of the y–component of the electric field at the fundamental frequency of 7.512 GHz. 50 × 50 points are sampled.

GHz and THz electronic testing allows us to measure at designated circuits ports or in the far field only. From this information it can be hard to pinpoint the reasons for problems in a circuit. The detailed information on frequency–dependent local fields, provided by field–

mapping, facilitates the identification of the circuit components causing maloperation. We have demonstrated this recently with measurements on a planar 7–GHz antenna structure [16, 17] where parasitic modes could be localized. There exists an electronic alternative to optical field–mapping in the form of e–beam testing. This technique, however, appears to be fundamentally handicapped in frequency bandwidth limiting its application to the MHz and low GHz regime. In contrast, we have performed electrooptic near–field and photoconductive far–field mapping on Si–based resonator structures at a fundamental frequency of 76 GHz [18] which does not yet represent the fundamental limit of the detection scheme.

We finally mention recent work that opens the possibility of probe–free all–optical testing on Si circuits. On III/V–based circuits this has, with practical limitations, been possible for a long time by taking advantage of the electrooptic properties of the substrate itself. Si has a centrosymmetric crystal structure and is not electrooptically active. One can, however, take advantage of optical second harmonic generation as the optical nonlinearity depends on the external field. For further information on the application of this effect, the reader is referred to Refs. [1, 19, 20].

4. Summary

A brief overview has been given of our advances in optical probing of GHz and THz electronic devices. Building on the inherent advantages of light pulses — the high time resolution, good amplitude stability, the possibility of optical switching, and the avoidance of timing jitter in pump/probe measurements — the technique has been optimized for the needs of circuit testing. Of special interest for industrial applications is the field–mapping capability because of its potential impact in circuit failure analysis.

Given the versatility of optical testing, it is surprising that optical testing equipment is not yet offered commercially. This certainly has something to do with the existence of good electronic alternatives at least for the MHz and low GHz regime. Higher in the GHz range, however, alternatives are rare. There, the complexity and costs of the required femtosecond laser system are major obstacles. With the increasing availability of compact, reliable and inexpensive laser systems, the commercial situation will change.

5. Acknowledgements

We are grateful to D. van der Weide for supplying the NLTL and to the Daimler Benz Research Center at Ulm for their support. This work has been funded by BMBF (contract 01 M 2938 B) and DFG (Ku 540/22–1).

References

1. The present paper is a summary of a quite comprehensive description of our advances in optical testing: Pfeifer, T., Heiliger, H.–M., Löffler, T., Ohlhoff, C., Meyer, C., Lüpke, G., Roskos, H. G., and Kurz, H. (in press) Optoelectronic on–chip characterization of ultrafast electric devices: measurement techniques and applications, *IEEE J. Selected Topics in Quantum Electronics*.
2. Löffler, T., Pfeifer, T., Roskos, H. G., and Kurz, H. (1995) Detection of free–running electric signals up to 75 GHz using a femtosecond–pulse laser, *IEEE Photon. Techn. Lett.* **7**, 1189–1191.

3. Löffler, T., Pfeifer, T., Roskos, H. G., Kurz, H., and van der Weide, D. W. (1996) Stable optoelectronic detection of free–running microwave signals with 150 GHz bandwidth, *Microelectronic Engineering* **31**, 397–408.

4. Kim, J., Williamson, S., Nees, J., Wakana, S.–i., and Whitaker, J. (1993) Photoconductive sampling probe with 2.3–ps temporal resolution and 4–μV sensitivity, *Appl. Phys. Lett.* **62**, 2268–2270.

5. Pfeifer, T., Heiliger, H.–M., Stein von Kamienski, E., Roskos, H. G., and Kurz, H. (1994) Fabrication and characterization of freely positionable silicon–on–sapphire photoconductive probes, *J. Opt. Soc. Am. B* **11**, 2547–2552.

6. Pfeifer, T., Heiliger, H.–M., Roskos, H. G., and Kurz, H. (1995) Generation and detection of picosecond electric pulses with freely positionable photoconductive probes, *IEEE Trans. Microwave Theory Tech.* **43**, 2856–2862.

7. Heiliger, H.–M., Pfeifer, T., Roskos, H. G., Kurz (1996) External photoconductive switches as generators and detectors of picosecond transients, *Microelectronic Engineering* **31**, 415–426.

8. Hwang, J.–R., Cheng, H.–J., Whitaker, J. F., and Rudd, J.V. (1996) Photoconductive sampling with an integrated source follower/amplifier, *Appl. Phys. Lett.* **68**, 1464–1466.

9. Lai, R. K., Hwang, J.–R., Nees, J., Norris, T. B., Whitaker, J. F. (1996, in press) A fiber–mounted, micro–machined photoconductive probe with 15 nV/Hz$^{1/2}$ sensitivity, *Appl. Phys. Lett.*

10. Heiliger, H.–M., Vossebürger, M., Roskos, H. G., Kurz, H., Hey, R. and Ploog, K. (1996, in press) Application of lift–off low–temperature–grown GaAs on transparent substrates for THz signal generation, *Appl. Phys. Lett.*

11. Weiss, S., Botkin, D., Ogletree, D. F., Salmeron, M., and Chemla, D. S., (1995) The ultrafast response of a scanning tunneling microscope, *Phys. Stat. Solidi B* **188**, 343.

12. Groeneveld, R. H. M., Rasing, T., Kaufmann, L. M. F., Smalbrugge, E., Wolter, J. H., Melloch, M. R., and van Kempen, H. (1996) New optoelectronic design for ultrafast scanning tunneling microscopy, *J. Vac. Sci. Technol. B* **14**, 861–863.

13. Wu, Q., Hewitt, T. D., and Zhang, X.–C. (1996) Two–dimensional electro–optic imaging of THz beams, *Appl. Phys. Lett.* **69**, 1026–1028.

14. Pfeifer, T. Heiliger, H.–M., Stein von Kamienski, E., Roskos, H. G., Kurz, H. (1995) Charge accumulation effects and microwave absorption of coplanar waveguides fabricated on high–resistivity Si with SiO$_2$ insulation layer, *Appl. Phys. Lett.* **67**, 2624–2626.

15. Pfeifer, T., Heiliger, H.–M., Löffler, T., Roskos, H. G., and Kurz, H. (1996) Picosecond optoelectronic on-wafer characterization of coplanar waveguides on high–resistivity Si and Si/SiO$_2$ substrates, *Micro-electronic Engineering* **31**, 385–195.

16. Pfeifer, T., Löffler, T., Roskos, H. G., Kurz, H., Singer, M., and Biebl, E. M. (1996) Electrooptic measurement of the electric near field distribution of a 7 GHz planar resonator, *Electron. Lett.* **32** 1305–1307.

17. Pfeifer, T., Löffler, T., Roskos, H. G., Kurz, H., Singer, M., and Biebl, E. M. (submitted) Electro–optic near field mapping of planar resonators, *IEEE Trans. Antenna and Propagat.*

18. Pfeifer, T., Roskos, H. G., Kurz, H., Strohm, K. M., and Luy, J.F. (submitted) Electric near–field radiation of a 76–GHz SIMMWIC dipole resonator for automotive applications, *IEEE Microwave and Guided Wave Lett.*

19. Lüpke, G., Meyer, C., Ohlhoff, C., Kurz, H., Lehmann, S., and Marowsky, G. (1995) Optical second–harmonic generation as a probe of electric–field–induced perturbation of centrosymmetric media, *Opt. Lett.* **20**, 1997–1999.

20. Ohlhoff, C., Meyer, C., Lüpke, G., Löffler, T., Pfeifer, T., Roskos, H. G., and Kurz, H. (1996) Optical second–harmonic probe for silicon millimeter–wave circuits, *Appl. Phys. Lett.* **68**, 1699–1701.

TUNABLE COHERENT THZ RADIATION PULSES FROM OPTICALLY EXCITED BLOCH OSCILLATIONS

H. G. ROSKOS, T. PFEIFER, H.–M. HEILIGER,
T. LÖFFLER, AND H. KURZ
Institut für Halbleitertechnik II, Rheinisch–Westfälische
Technische Hochschule (RWTH) Aachen,
Sommerfeldstr. 24, D–52056 Aachen, Germany

1. Introduction

Quantum–mechanical tunneling is attractive for the realization of THz radiation sources both in superconducting and semiconducting structures. This paper deals with the phenomenon of *coherent tunneling* of electrons in semiconductor superlattice structures. Charge carrier tunneling through any potential barrier is called "coherent" if three conditions are fulfilled: (i) states of the same energy are involved on both sides of the barrier, (ii) the wave functions of the charge carriers populating the states have a fixed phase relationship, and (iii) the dynamics of the tunneling process is governed by a time–dependent phase difference between the wave functions. The concept of coherent tunneling is better known in the field of superconductivity where it has long been applied in Josephson junctions consisting of two superconductors separated by a thin insulating layer [1]. Application of a dc bias voltage induces a phase oscillation with voltage–dependent frequency between the ensembles of coherent Cooper pairs on both sides of the barrier. The periodic phase change results in a periodic motion of charge through the barrier which is the source for tunable electromagnetic radiation.

In the case of semiconductors, the concept of submillimeter–wave generation by coherent tunneling was first investigated with double–well potentials [2, 3] but it turned out that they allow only a limited tunability of the radiation frequency. Subsequent studies on Bloch oscillations, the quantum interference of states within a subband of a superlattice under dc electrical bias, revealed much wider tunability with continuous coverage from 0.3 THz to at least 4 THz (the detection limit of the receiver antennas) [4, 5]. In contrast to Josephson junctions, where superconductivity provides for the phase coherence of the charge carriers, it takes additional measures to synchronize the phases of the electronic wave functions in semiconductor quantum structures. This is accomplished by impulsive optical excitation. Coherence is maintained after excitation until it is destroyed by scattering processes of the carriers with phonons and other carriers. Phonon scattering is least efficient at cryogenic temperatures allowing for dephasing times of up to several picoseconds at low carrier densities and temperatures not exceeding several 10 K.

J.M. Chamberlain and R.E. Miles (eds.), New Directions in Terahertz Technology, 369–375.

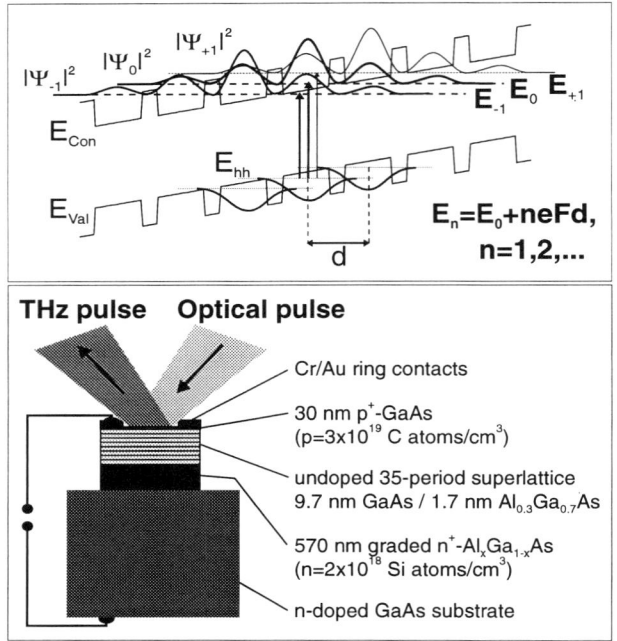

Figure 1. Upper panel: Scheme for optical excitation of Bloch oscillations. A laser pulse excites simultaneously several transitions from the valence band into the conduction band. Only transitions from one hole state are plotted. The populated conduction–band states form a wave packet performing Bloch oscillations. Lower panel: Cross section through the sample, and geometry of optical excitation and THz emission. Only the THz beam in reflection direction is shown. The beam in transmission direction is absorbed in the n–doped regions.

2. Superradiant emission from optically excited Bloch oscillations

The band–structure diagram in the upper part of Fig. 1 illustrates the excitation of Bloch oscillations with an ultrashort laser pulse. Valence–to–conduction–band transitions within one period of the superlattice (labelled as "0") are indicated. The superlattice is biased such that the energy difference $\Delta E = E_0 - E_{-1} = E_1 - E_0$ between the lowest conduction–band state in well "0" and the lowest states in the adjacent wells corresponds to the chosen frequency ν_B of the oscillation according to $h\nu_B = \Delta E = e\,F\,d$, d being the spatial period of the superlattice and F the electric field across it. Under the chosen bias conditions, the wave functions in the conduction band extend over several periods of the superlattice. In contrast, the heavy–hole wave functions are strongly localized in the respective wells because of the larger effective mass. An ultrashort laser pulse can simultaneously populate several (at least two) conduction–band states by transitions from the same heavy–hole state provided the temporal duration of the pulse is small enough that its spectrum covers these transitions. The diagram of Fig. 1 illustrates a typical experimental situation where the spectrum of the excitation covers the transitions "0", "-1", and "+1". The line thickness indicates the spectral weight; transition "+1" is weakly excited. The superposition of the excited wave functions represents a wave packet. After the optical pulse has passed, the different temporal evolution of the wave functions results in temporal and spatial oscillations of the wave packet with frequency ν_B. These are what are known as "Bloch oscillations". The dipole moment associated with the oscillations leads to emission of electromagnetic radiation. The energy of the radiation is extracted by transitions between the conduction–band states [6, 7].

When the optical pulse hits the sample, the same process is induced across the illuminated spot and in each well of the superlattice. Hence, the charge oscillations initially are fully phase–locked, and the emission can be treated as that of a phased array of dipoles [6, 7]. Provided the diameter of the excited spot is larger than $c/(2\pi\nu_B)$, the radiation is directed. It fulfills generalized Fresnel's laws, i.e., emission into free space occurs in the directions of the reflected and transmitted optical beams. Semiclassical theory predicts that the power of the radiation should scale quadratically with the density n_{exc} of photoexcited electron–hole pairs. Thus, one should be able to increase the emission efficiency (the ratio of emitted power to energy available for emission) by increasing the number n_{exc} of the dipoles. This property of coherent dipole ensembles allows us to adopt the expression of "superradiant" emission, better known from emitter ensembles with inversion between quantum states. As in inverted systems, fully coherent systems principally allow amplification of radiation when it is coupled back in a resonator geometry [6]. However, such amplification still remains to be demonstrated experimentally for Bloch oscillators.

In the following, we examine the excitation–density dependence of the emission [8]. The experiments are performed with a MBE–grown pin sample whose structure is displayed in the lower part of Fig. 1. The undoped 35–period GaAs/Al$_{0.3}$Ga$_{0.7}$As superlattice is sandwiched between a semi–transparent p$^+$–doped top contact and a n$^+$–doped bottom contact on the n–doped substrate. The contacts are needed for application of the dc bias voltage. Their influence on the THz–emission process is negligible as the Bloch oscillations are effectively decoupled from the capacitive influence of the contacts. The structure is patterned into 450–μm–wide mesas to limit current flow. Both, optical excitation and THz emission occur through the top contact of one of the mesas. As the dipole moment associated with Bloch oscillations is oriented perpendicular to the sample surface, the emission efficiency increases with the angle of the incident (and emitted) beam relative to the sample normal.

The experiments are performed in a standard setup for time–resolved THz–emission spectroscopy [5]. The sample is mounted in a closed–cycle cryostat and held at a temperature of 20 K. Bloch oscillations are excited by 150–fs pulses from a Kerr–lens–modelocked Ti:sapphire laser (pulse repetition time $\tau_{Rep} = 13$ ns). The optical beam hits the sample at an angle of 45° degree. Coherent THz pulses emitted in the direction of the reflected optical beam are detected as a function of time with a photoconductive antenna made from low–temperature–grown GaAs.

The left side of Fig. 2 displays Bloch transients measured with a Schottky–contact (instead of a pin) version of the superlattice structure for various applied bias voltages [5, 9]. The laser photon energy is centered at 1.561 eV. On the right side, the Fourier transforms are displayed after correction for the spectral sensitivity of the detection system (see full line at the bottom). The bias field increases with reverse voltage beginning at flatband at +0.7 V. Bloch oscillations are observed for a reverse voltage above -1.2 V. After a transition region between -1.2 -2.0 V, the frequency increases linearly with reverse voltage, and continuous tunability of the peak frequency of the emitted THz pulses is

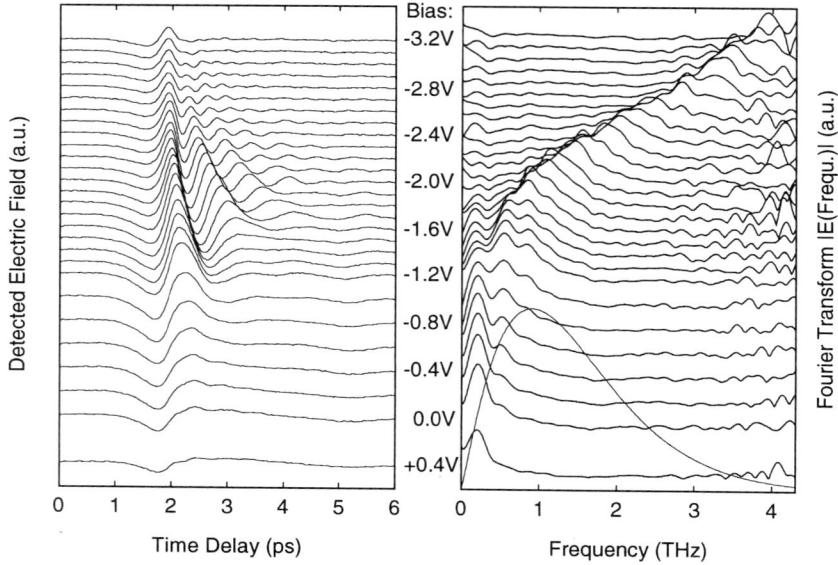

Figure 2. Left side: THz transients from the superlattice described in the text (width of first electron miniband: 18 meV). The applied dc field is varied from flatband conditions (at +0.7 V bias) to 14 kV/cm. Optical excitation occurs slightly before t = 2 ps. Right side: Fourier spectra of the transients, and sensitivity curve of the detection system.

achieved from several hundred GHz up to 4 THz, the detection limit of our system.

Figures 3 and 4 display results of measurements on the *pin* structure for various excitation powers but with fixed bias voltage [8]. The frequency of the Bloch oscillations is in the range from 1.7 THz at the lowest excitation power to 2.1 THz at the highest power, the frequency shift being a consequence of screening effects. Frequency ω, amplitude A, and decay time τ of the THz transients (not shown) are extracted by fitting the data with the function $A \exp(-(t + \phi)/\tau) \cos(\omega(t + \phi))$. Sets of data are presented in Figs. 3 and 4 for two different photon energies of the excitation pulses: 1.542 eV, for predominant excitation of excitons (squares), and 1.557 eV, mainly exciting uncorrelated electron/hole pairs (full circles).

The amplitude A of the fit to the Bloch transients is displayed in Fig. 3 as a function of n_{exc} estimated from the excitation power by taking reflection losses and an estimated absorption of 60 % at 1.557 eV (30 % at 1.542 eV) within the superlattice into account. The peak amplitude of the THz transients rises linearly with the carrier density as expected from simple superradiance theory. The linearity is maintained up to the highest carrier density of 2.2×10^{10} cm^{-2} per well. This density is still below the Mott density characteristic for the disappearance of the exciton resonances. It is noteworthy that no significant difference is found between the two data sets (circles and stars) of Fig. 3. Obviously, it matters little whether excitons or uncorrelated electron/hole pairs close to the bandgap are excited [8]. This is an indication that the dipole moment of the Bloch oscillations responsible for THz emission is similar in strength for the two cases.

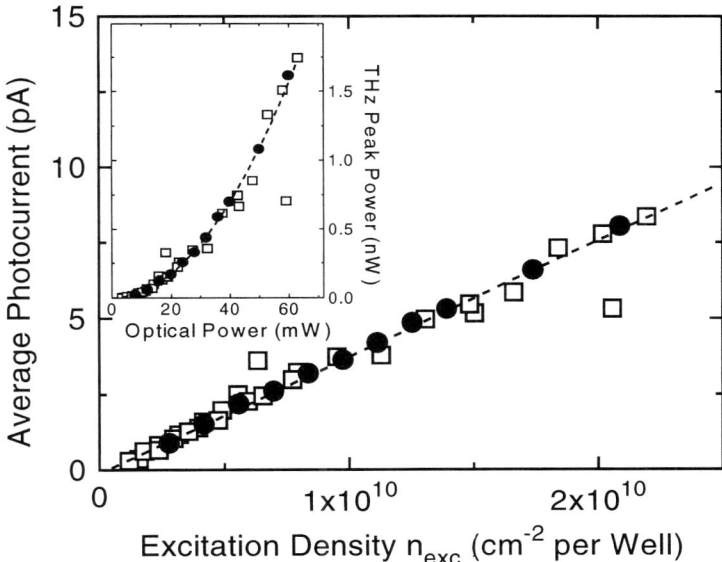

Figure 3. Peak amplitude of the detected electric THz field as a function of the density of photogenerated electron/hole pairs. The amplitude is given in terms of the measured photocurrent through the antenna. Inset: Time–averaged peak power of the THz pulses as a function of the time–averaged power of the illumination.

According to Ref. [10], one can estimate the power of the THz radiation from the measured antenna photocurrent if the sensitivity of the antenna has been quantified with a calibrated emitter. The inset of Fig. 3 displays the peak power ($\sim |A|^2$) of the THz radiation, time–averaged over τ_{Rep}, as a function of the optical input power. The THz peak power depends quadratically on the input power. At the highest input power of 60 mW, the THz power is about 2 nW (external conversion efficiency: 3×10^{-8}). This is an improvement by a factor of 20 over the previously reported values [6].

To relate the power measured outside of the sample to that generated within the superlattice, one has to take into account losses by free–carrier absorption and reflection at the highly doped top contact. For THz beam transmission through a conductive layer, one estimates that only 5 % of the incident power is transmitted through the top contact, 37 % is lost by Drude absorption in the contact, 58 % by reflection into the substrate where it is absorbed. Reabsorption in the superlattice region is not significant at the carrier densities of our experiments.

For practical purposes, the time–averaged power of the THz radiation $\langle P \rangle$ (i.e., the pulse energy divided by τ_{Rep}) is a more significant quantity than the peak power. In contrast to the latter, $\langle P \rangle$ depends strongly on the decay time τ of the THz transients. As τ decreases when the excitation density is raised, $\langle P \rangle$ does not scale quadratically with the input power.

Figure 4 displays the fit values of τ (inset) and the corresponding homogeneous linewidth $\Gamma = 2\hbar/\tau$ (main part of the figure) as a function of n_{exc} for excitation photon

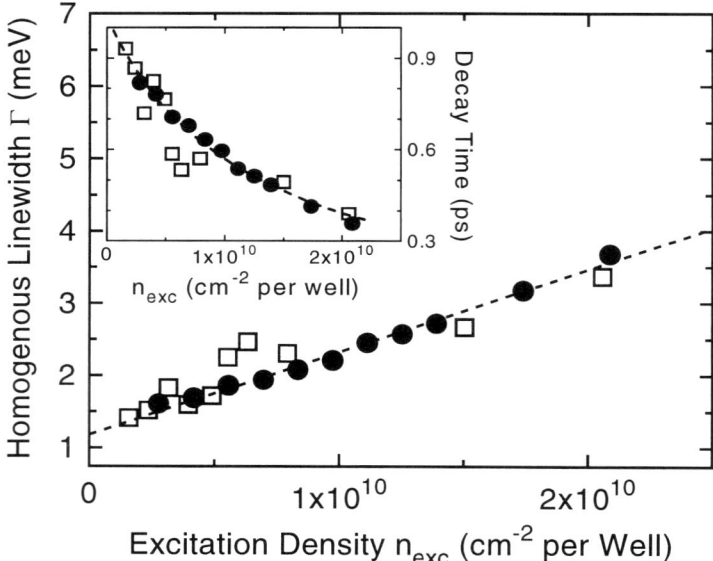

Figure 4. Density dependence of the dephasing of the Bloch oscillations (T = 20 K). Both the decay time τ (inset) and the homogeneous linewidth $\Gamma = 2\hbar/\tau$ are displayed.

energies of 1.557 eV (circles) and 1.542 eV (stars). For both photon energies, τ decreases continuously from 1 ps at the lowest density of 1.5×10^9 carriers per cm^2 in a quantum well to below 0.4 ps at 2.2×10^{10} carriers per cm^2 and well. Again, no significant difference is observed between the predominant excitation of exciton transitions versus one of continuum states. The analysis of the density dependence of τ reveals that the decay of the THz transients at T = 20 K results from carrier–carrier scattering [8].

Temperature–dependent measurements show that scattering with phonons becomes the dominant dephasing mechanism at elevated temperatures [5]. At room temperature, phonon–induced dephasing is so rapid that only Bloch oscillations with frequencies above 4.5 THz are observed [11]. Interestingly, Bloch oscillations at low temperatures are surprisingly robust against emission of LO phonons. Bloch oscillations are still observed when the electron wave packet is excited so high above the bandedge that the carriers relax via emission of up to three LO–phonons [12, 13].

3. Outlook

Enhancement of the emission efficiency suffers from the faster decay of the THz transients at higher excitation density resulting from more efficient carrier–carrier scattering. Efficiency enhancement by more than one order of magnitude without faster signal decay should be readily achieveable by better output coupling (now only 5 %) and excitation and emission at more oblique angles. In the long term, single–pass emitters should be replaced by systems with feed–back in a resonator structure in order to take advantage of the potential for superradiant amplification. To reduce the size of the THz system, the optical pump should then be a compact all–solid–state laser. Even integration of femtosecond semiconductor lasers with the superlattice on a single chip appears feasible. For ultimate miniaturization, however, schemes must be develop to drive the Bloch oscillations electrically instead of optically.

4. Acknowledgements

We acknowledge contributions by C. Waschke. This work has been funded by BMBF (contract 13N6526) and DFG (Ro 770/4–2). The superlattices have been grown by K. Köhler (Fraunhofer–Institut für Angewandte Festkörperphysik, Freiburg) and R. Hey (Paul–Drude–Institut, Berlin).

References

1. Kleiner, R., Steinmeyer, F., Müller, P., Kohlstedt, H., Pedersen, N. F., and Sakai, S. (1994) Dynamic behaviour of Josephson coupled layered structures, *Phys. Rev. B* **50**, 3942–3952.
2. Roskos, H. G., Nuss, M. C., Shah, J., Leo, K., Miller, D. A. B., Fox, A. M., Schmitt-Rink, S., and Köhler, K. (1992) Coherent submillimeter–wave emission from charge oscillations in a double–well potential, *Phys. Rev. Lett.* **68**, 2216–2219.
3. Nuss, M. C., Planken, P. C. M., Brener, I., Roskos, H. G., Luo, M. S. C., and Chuang, S. L. (1994) Terahertz electromagnetic radiation from quantum wells, *Appl. Phys. B* **58**, 249–259.
4. Waschke, C., Roskos, H. G., Schwedler, R., Leo, K., Kurz, H., and Köhler, K. (1993) Coherent submillimeter–wave emission from Bloch oscillations in a semiconductor superlattice, *Phys. Rev. Lett.* **70**, 3319–3322.
5. Roskos, H. G. (1994) Coherent emission of electromagnetic pulses from Bloch oscillations in semiconductor superlattices, in *Festkörperprobleme / Advances in Solid State Physics*, ed. Helbig, R., Vieweg, Braunschweig, Vol. **34**, 297–315.
6. Victor, K., Roskos, H. G., and Waschke, C. (1994) Efficiency of submillimeter–wave generation and amplification by coherent wave–packet oscillations in semiconductor structures, *J. Opt. Soc. Am. B* **11**, 2470–2479.
7. Roskos, H. G., Waschke, C., Victor, K., Köhler, K., and Kurz, H. (1995) Bloch oscillations in semiconductor superlattices, *Jpn. J. Appl. Phys.* **34**, 1370–1375.
8. Martini, R., Klose, G., Roskos, H. G., Kurz, H., Grahn, H. T., and Hey, R. (in press) Superradiant emission from Bloch oscillations in semiconductor superlattices, *Phys. Rev. B.*
9. Waschke, C., Leisching, P., Haring Bolivar, P., Schwedler, R., Brüggemann, F., Roskos, H. G., Leo, K., Kurz, H., and Köhler, K. (1994) Detection of Bloch oscillations in a semiconductor superlattice by time–resolved terahertz spectroscopy and degenerate four–wave mixing, *Solid State Electron.* **37**, 1321–1326.
10. van Exter, M. and Grischkowsky, D. (1990) Characterization of an optoelectronic terahertz beam system, *IEEE Trans. Microwave Theory Techn.* **38**, 1684–1691.
11. Dekorsy, T., Ott, R., Kurz, H., and Köhler, K. (1995) Bloch oscillations at room temperature, *Phys. Rev. B* **51**, 17275–17278.
12. Roskos, H. G., Waschke, C., Schwedler, R., Leisching, P., Dhaibi, Y., Kurz, H., and Köhler, K. (1994) Bloch oscillations in GaAs/AlGaAs superlattices after excitation well above the bandgap, *Superlatt. & Microstruct.* **15**, 281–285.
13. Meier, T., Rossi, F., Thomas, P., and Koch, S. W. (1995) Dynamic localization in anisotropic Coulomb systems: field induced crossover of the exciton dimension, *Phys. Rev. Lett.* **75**, 2558–2561.

MULTI-GIGAHERTZ OPTOELECTRONIC DEVICES

Part 1: Resonant Tunneling Diode (RTD) for optical modulation.

C. N. IRONSIDE and S. G. M^cMEEKIN*
The Department of Electronics and Electrical Engineering,
University of Glasgow,Glasgow, G12 8LT, United Kingdom.
* Cardiff School of Engineering, University of Wales Cardiff, PO Box
917, Newport Rd Cardiff, NP2 1XH, United Kingdom*

Abstract

A resonant tunneling diode (RTD) incorporated in the centre of an optical waveguide can achieve optoelectronic modulation by high-speed switching of an electric field that alters the semiconductor band-edge via Franz-Kelydsh effect.

1. Introduction.

The key objective of optoelectronic integration research is to develop technologies for optoelectronics which emulate the success that silicon-based electronics has achieved through integration: increasing speed, reliability and functionality. Optoelectronic integration has been developed largely for applications in optical communications. However, high-speed integrated optoelectronic devices in particular can find application in microwave technology and here in two articles we discuss two new multi-multigigahertz optoelectronic devices, a resonant tunneling diode optical moduoltor and a monolithic modelocked semiconductor laser. Both of these devices represent first-level simple integration but are capable of very high speed operation.

Part 1 reports work on the device potential of embedding a resonant tunneling diode (RTD) in an optical waveguide to achieve optoelectronic modulation at microwave frequencies. [1,2]

Since the discovery [3] that the RTD has sufficient negative differential resistance for practical devices most work has concentrated on entirely electronic devices for microwave applications. There has also been some work on the optoelectronic applications of RTDs [4-10]. In this part we report on a simple direct integration scheme to achieve optoelectronic modulation at microwave frequencies from the electric field associated with the RTD, by embedding the RTD directly in an optical waveguide. There are two key advantages of introducing an RTD into an optical

377

J.M. Chamberlain and R.E. Miles (eds.), New Directions in Terahertz Technology, 377–383.
© 1997 *Kluwer Academic Publishers. Printed in the Netherlands.*

waveguide. Firstly the electric field distribution across the waveguide is strongly dependent on the bias voltage, this allows small changes in the biasing voltage close to resonance to give a large change in the electric field distribution and therefore optical characteristics of the guide. Secondly, under the correct conditions, the RTD can introduce instabilities into the current-voltage characteristics of the guide which can enable the electric field across the waveguide to self-oscillate at microwave frequencies. The device has considerable potential as a microwave/optical interface by removing the need for large drive voltages to produce a significant level of modulation. It has also been shown that it is possible to lock the oscillation of an RTD device to an optical pulse train [9].

2. Device Description.

The RTD is incorporated in the centre of a 1 μm deep GaAs waveguide and consists of a 7 nm GaAs quantum well with two 1.7 nm AlAs barriers. Rib waveguides were etched to provide lateral confinement of the electrical and optical fields.

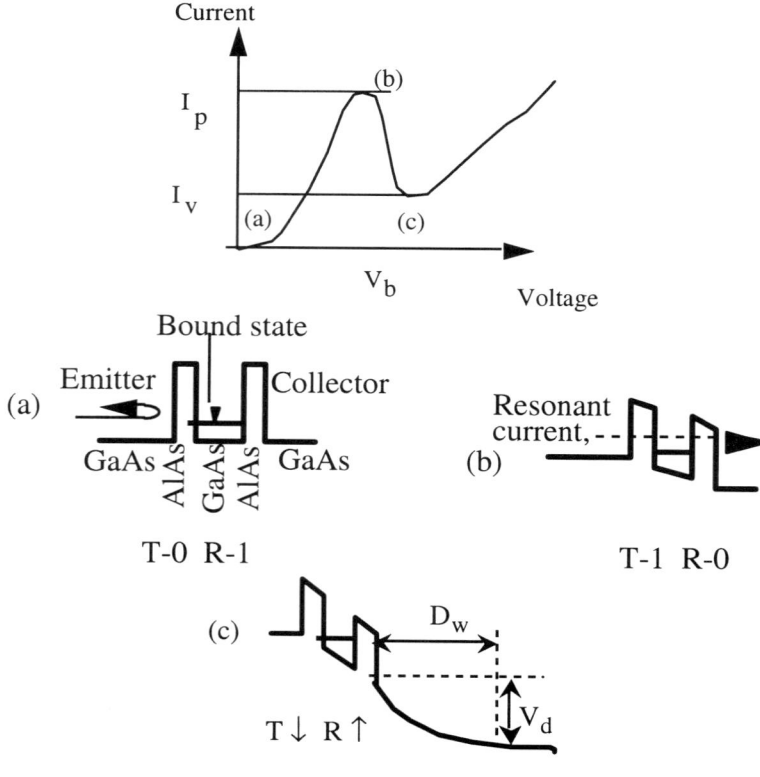

Figure 1.

Figure 1 shows a typical current voltage characteristic associated with a RTD at room temperature. To understand the dependence of the electric field distribution on the bias voltage we must examine the conduction band of the central section of the waveguide containing the RTD as a function of the bias voltage. At low applied bias voltages, below resonance, the current flow is due to electrons with enough energy to tunnel through the bound state. As the bias is increased, the bound state is pulled closer to the conduction band of the emitter and more electrons become available to tunnel through the bound state. When the applied bias is equal to V_b the bound state of the GaAs quantum well is aligned to the conduction band and the current increases to a maximum. At this point the maximum number of electrons can access the bound state and can easily tunnel through the barriers. A further increase in the bias across the RTD will increase the bottom of the conduction band above the bound state and current through the device drops to a minimum. The RTD now appears as a hetrojunction with a depletion region of width D_w formed on the collector side of the RTD. It is the voltage dropped across this region, V_d, that is responsible for the modulation of the optical field in the guide. A further increase in the bias voltage will only produce a gradual change in the electric field distribution as the depletion region is increased.

The value of V_d can be found from the dc current-voltage characteristics for the RTD, shown in Figure 1. The peak current at point b, I_p, is given by V_b/R_s where R_s is the total series resistance of the device including contact resistance and material resistivity. At point c the current drops sharply to I_v with no change in the series resistance and an incremental increase the external biasing voltage. The difference in current can be accounted for by the internal voltage dropped across the depletion region.

From the above analysis we have :-

$$I_p R_s = V_b = I_\upsilon R_s + V_d \tag{1}$$

$$\tag{1}$$

or

$$V_d = V_b \left(1 - \frac{I\upsilon}{I_p}\right) = V_b \left(1 - (PVR)^{-1}\right) \tag{2}$$

where V_d is the voltage dropped across the depletion region formed on the collector side of the RTD and PVR is the peak to valley ratio. As PVR increases V_d approaches V_b. The width of the depletion region is given by [10]:

$$D_w = \left(w^2 + \frac{2\varepsilon V_d}{eN_d}\right)^{\frac{1}{2}} \tag{3}$$

N_d is the doping concentration in the waveguiding region, W is the width of the barriers and well, $\varepsilon = \varepsilon_r \varepsilon_0$ is the dielectric constant. Assuming a uniform electric field, F, across the depletion region we obtain:-

$$F = \frac{V_d}{D_w} = \frac{V_b\left(1 - (PVR)^{-1}\right)}{\sqrt{\left(w^2 + \frac{2\varepsilon}{eN_d}\right)}} \qquad (4)$$

This field appears when the voltage is greater than V_b and gives rise to electroabsorption through the Franz-Keldysh shift in the band-edge due to the electric field across the RTD and the depletion region. When the applied bias is increased to the point where it is slightly greater than V_b, the device switches from point (b) to point (c) in I-V curve (Figure 1) and an electric field, F is induced within the waveguide. The induced electric field results in a reduction in the band-gap across the depletion region. The expected shift in energy of band-edge is given by:-

$$S = -\left(\frac{(eF\hbar)^2}{2m*}\right)^{\frac{1}{3}} \qquad (5)$$

m* is the effective reduced mass given by $(m_e^{1} + m_h^{-1})^{-1}$.

If we consider a lumped element analysis of the operation of the RTD then to a rough approximation the maximum operating frequency is limited by the intrinsic circuit elements and can be calculated from [11]:

$$f_{max} = \frac{1}{2\pi R_n C_w}\sqrt{(R_n / R_s - 1)} \qquad (6)$$

where R_n (all values in equation (6) are per unit area) is the negative differential resistance and is approximately the inverse of the slope of the I-V curve (Figure 1) between the points (b) and (c), R_s is the series resistance associated with the device. C_w is the capacitance per unit area across the RTD and the depletion region. Other factors affecting the maximum frequency, such as depletion region transit time, can be ignored since they will have a higher limit for the material parameters used.

Switching times are approximately $5R_nC_w$ when $R_d/R_n< 5$ where R_d is the resistance in the region beyond resonance. For some RTD devices this can be as short as 2.2 ps. For our devices which have an active area of 600 μm^2 we have estimated R_n from the IV characteristics to be less than 20 $\mu\frac{1}{2}.cm^2$ and C_w, calculated from equation(6), to be approximately 70 nF.cm^{-2}, corresponding to a calculated switching time less than 7 ps.

3. Results.

The high frequency electrical characteristics of the device were measured using a Wiltron network analyser. Peaks in the S_{11} reflection coefficient due to oscillations were observed at around 56 GHz from a 600 μm^2 RTD. This is considerably higher than the maximum frequency of 35 GHz predicted by the equation 6.

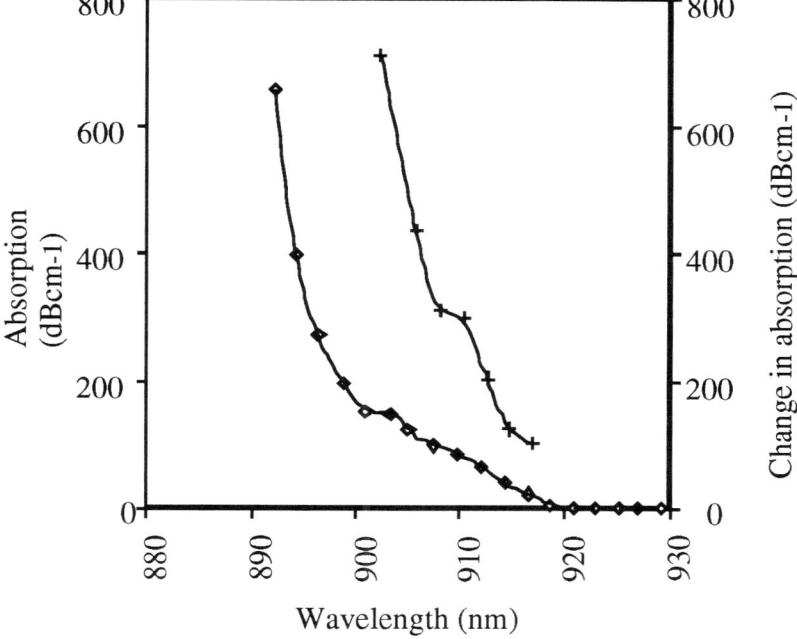

Change in Absorption induced by
the RTD switching.

◇   Measured bandedge

Figure 2.

      The change in the optical absorption spectrum induced by the Franz-Kelydsh effect when the device switched was measured using larger area devices which exhibited the electrically bistable characteristic. All the optical characterisation results used TE polarised light from a Ti:Sapphire laser which was tuneable in the wavelength region across the absorption edge of the GaAs waveguide i.e. from 850 nm.to 950 nm, TM polarised light was found to be very lossy due to the heavy doping of the cladding layers. Approximately 1 mW of light from the laser was coupled into the waveguide by an end-fire arrangement. The RTD was biased using electrical pulses of around 2 μs duration and period of 100 μs. The pulses were employed to minimise thermal effects. Typically, when the pulse amplitude exceeded the peak voltage of the RTD, V_b, the current drawn from the pulsed power supply falls rapidly, the voltage across the device then falls below V_b and the current increases until the voltage exceeded V_b causing the

382

device to switch again. A high-speed detector (2 GHz) was used to measure any change in the intensity of the transmitted light when the device was biased just above V_b. The time scale of the biasing pulse ~2 µs minimises the possibility of thermal effects inducing any electroabsorption.

A decrease in the intensity of the transmitted light associated with an increase in the absorption within the active region of the waveguide was observed when the RTD was switched above its resonance condition. The results are shown in Figure (2) where the change in the absorption coefficient associated with the bistable switching of the device is plotted against wavelength along with the bandedge measured for the waveguide. There is a shift in the bandedge of approximately 12 nm from the measured bandedge. For a 300 µm long device, a maximum modulation depth of 7 dB at approximately 900 nm was observed. From the typical I-V characteristic the voltage V_d and the depletion region was calculated from equations (2) and (3) to be 0.34 V and

0.157 µm respectively giving a field of 2.2 $MV.m^{-1}$ (22 kV/cm). This gives rise to a shift in the band-edge calculated from equation (5) of 9 nm which is in reasonable agreement with the 12 nm shift observed in Figure 8 considering the approximations made.

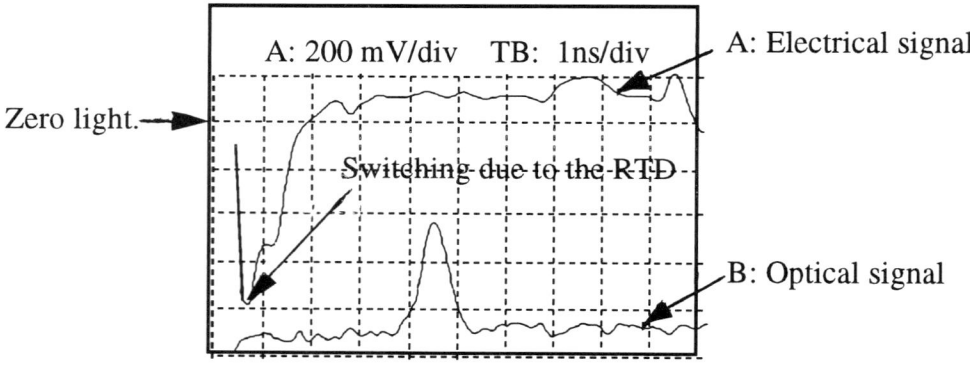

Figure 3

To investigate the high frequency optical response of the RTD optoelectronic modulator a pulsed and a rf biasing source was used to switch the RTD. Figure 3 shows a 7 dB change in optical intensity at 900 nm detected when a 0.95 V, 2 ns pulse was applied across a 100 µm long device. The rise and fall time of the optical pulse was approximately 500 ps which corresponds to the bandwidth of the oscilloscope. This observation is consistent with the calculated switching time which for our device is less than 7 ps. With direct electrical measurements it is not possible to have sufficient time resolution to measure this predicted switching time however ultrafast optical techniques such pump-probe and electro-optic sampling can access this timescale.

4. Conclusions

We have successfully integrated a RTD with an optical waveguide to provide optoelectronic modulation through the Franz-Keldysh effect due to the voltage drop across the depletion region of the RTD. The Franz-Kelydsh change in band-edge was measured to be approximately 9 nm with a maximum of 7 dB of modulation at a wavelength of 900 nm. Optoelectronic modulation was observed from 900 MHz cw signal and a 2 ns pulse signal with an optical response time of 500 ps. The use of a dc bias level and a bias tee enabled the pulse modulating signal to be reduced to 0.1 V. In addition we have demonstrated that the same device is capable of self-oscillating at frequencies up to 56 GHz.

References

1   Mc Meekin,S. G., Taylor, M. R. S. and Ironside,C. N. (1996) Optical modulation with a resonant tunnelling diode, *IEE Proc. Optoelectronics* **143** 12-16.

2   McMeekin, S. G., Taylor, M. R. S., Vögele, B., Stanley, C. R. and Ironside, C. N. (1994) Franz-Keldysh effect in an optical waveguide containing a resonant tunneling diode, *Appl. Phys. Lett.* **65** 1076-1078.

3   Sollner, T.C.L.G., Goodhue, W.D., Tannenwald, P.E., Parker, C.D. and Peck, D.D. (1983) *Appl Phys Lett* **43**,588.

4   Grave, I., Kan, S. C., Griffel, G., Wu, S. W., Sa'ar, A. and Yariv, A. (1991) Monolithic integration of a resonant tunneling diode and a quantum well semiconductor laser, *Appl. Phys Letters* **58**, 110-112.

5   Kurata, H., Tsuchiya, M. and Sakaki, H., (1990) A novel optical bistability device consisting of resonant tunneling diode and a quantum stark modulator: Experimental demonstration, *Surface Science,* **228** 468-471.

6   England, P., Golub, J.E., Florez, L.T. and Harbison, J.P., (1991) Optical switching in a resonant tunneling structure, *Appl Phys Lett.,* **58** 887-889.

7   Moise, T.S., Kao, Y. C., Garrett, L. D., and Campbell, J. C., (1995) Optically switched resonant tunneling diodes, *Appl. Phys. Lett.,* **66** 1104-1106.

8   Chen, L., Kapre, R. M., Hu, K. and Madhukar, A., (1991) High-contrast optically bistable optoelectronic switch based on InGaAs/GaAs (100) asymmetric Fabry-Perot modulator, detector, and resonant tunneling diode, *Appl. Phys. Lett.,* **59** 1523-1525.

9   Lann, A.F., Grumann, E., Gabai, A., Golub, J.E. and England, P., (1993) Phase locking between light pulses and resonant tunneling diode oscillator, *Appl. Phys. Letts.* **62** 13-15.

10 Van Hoof, C., Genoe, J., Raymond, S. and Borghs, G., (1993) Giant optical bistable behaviour using triple barrier resonant tunneling light-emitting diodes, *Appl. Phys. Lett.,* **63** 2390-2392.

11 Brown, E. R, Goodhue, W. D. and Sollner, T. C. L. G. J., (1988) Fundamental oscillations up to 200 GHz in resonant tunneling diodes and new estimates of their maximum oscillation frequency from stationary-state tunneling theory, *Appl. Phys.* **64** (3) 1519-1529.

MULTI-GIGAHERTZ OPTOELECTRONIC DEVICES

Part 2: Monolithic, harmonic modelocking of semiconductor lasers

C. N. IRONSIDE, S.D. McDOUGALL, E. A. AVRUTIN
AND J. F. MARTINS-FILHO*
*The Department of Electronics and Electrical Engineering, University of
Glasgow, Glasgow, G12 8LT, United Kingdom.
*Departamento de Fisica, Universidade Federal de Pernambuco
(UFPE), Cidade Universitaria, Recife - PE 50.610-901, Brazil*

Abstract

*By employing a harmonic modelocking technique called multiple colliding pulse
modelocking (MCPM) it is possible for a monolithic modelocked semiconductor laser to
produce a pulse train with a repetition rate of up to 375 GHz. These devices are
compact, robust and efficient sources of signals modulated at multi-gigahertz
frequencies.*

1. Introduction

Semiconductor lasers are compact and efficient sources of picosecond and
subpicosecond pulses [1,2]. Due to their small size, low pumping power requirement,
low cost and robustness they are the most suitable laser device for integration in optical
circuits for ultrafast photonic applications, such as high speed communication systems,
ultrafast data processing, optical computing and opto-microwave-electronic interfacing
[3]. Among the different methods of short pulse generation in semiconductor lasers
mode-locking is able to produce shorter pulses, with better spectral characteristics at
higher repetition rates. The most common approaches to mode-lock semiconductor
lasers are passive, active and hybrid mode-locking, in monolithic or external extended
cavity configuration. Monolithic mode-locked semiconductor lasers devices offer the
possibility of very low cost sources of ultrashort pulses which are compact, reliable,
robust, efficient and can be mass-produced. The monolithic passive mode-locking
scheme is attractive for achieving high repetition rates, above 100 GHz at average
powers of around tens of mW. Furthermore, these devices do not require the injection
of a high frequency RF signal. This scheme can be very simply implemented by
integrating a saturable absorber in the laser cavity. The fast dynamics of the nonlinear

J.M. Chamberlain and R.E. Miles (eds.), New Directions in Terahertz Technology, 385–391.
© *1997 Kluwer Academic Publishers. Printed in the Netherlands.*

behaviour of the saturation and recovery of both absorber and gain sections of the laser are responsible for its ultrafast characteristics.

More recently a growing interest has been observed in the generation of ultrashort pulses at ultrahigh repetition rates, reaching terahertz rates. Such high speed is achieved by inducing the laser to operate at harmonics of the fundamental inverse round-trip time of the cavity [1],[5]. In this paper we present the progress on generation of short pulses at high repetition rates by multiple colliding pulse mode-locked (MCPM) quantum well lasers.

2. Description of device

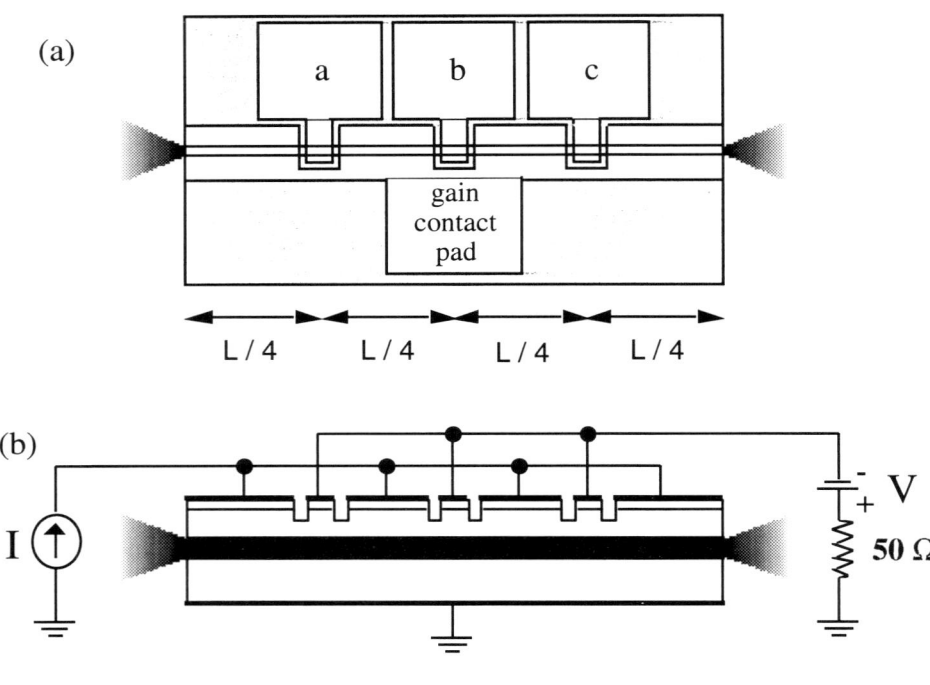

Figure 1.

By employing a harmonic modelocking technique called multiple colliding pulse modelocking (MCPM) it is possible for a monolithic modelocked semiconductor laser to produce a pulse train with a repetition rate of up to 375 GHz. Figure 1-a shows the top view diagram of the MCPM laser and figure 1-b shows its longitudinal cross-section with the electrical connections for forward and reverse biasing. The laser has a contiguously electrically connected gain section and 3 separated sections placed at every quarter of the cavity length. Each of the 3 sections (labelled a, b and c on the diagram) is separately electrically addressable and when reverse biased it behaves as a saturable absorber and when forward biased as a gain section. Therefore, by selectively biasing, one can choose the number and position of saturable absorbers in the laser cavity, which causes the laser to have 1, 2, 3 or 4 pulses circulating in the cavity, giving first to fourth harmonic of the repetition rate, respectively.

The MCPM laser is fabricated on a GaAs/AlGaAs 4 quantum well material grown by atmospheric pressure metal organic vapour phase epitaxy (APMOVPE). The semiconductor material structure consists of 0.2 μm heavily p-doped ($2x10^{19}$ cm^{-3}) GaAs cap layer followed by 1 μm p-type carbon doped ($2.2x10^{17}$ cm^{-3}) $Al_{0.43}Ga_{0.57}As$ upper cladding layer. The quantum well structure consists of four 10 nm GaAs wells spaced by 10 nm $Al_{0.20}Ga_{0.80}As$ barriers and they are surrounded by two 0.1 μm $Al_{0.20}Ga_{0.80}As$ layers in a separate confinement structure. The lower cladding layer is formed by $Al_{0.43}Ga_{0.57}As$ silicon n-doped ($1.4x10^{17}$ cm^{-3}) layer 1.7 μm thick.

3. Results

If the central saturable absorber section is reversed biased the laser operates with two pulses circulating in the resonator which collide in the saturable absorber. With are two pulses circulating in the resonator we find that in the frequency domain the spacing between the longitudinal modes of the laser doubles and the laser produces a pulse stream with a time between pulses of nl/c (i.e half a resonator round trip time).

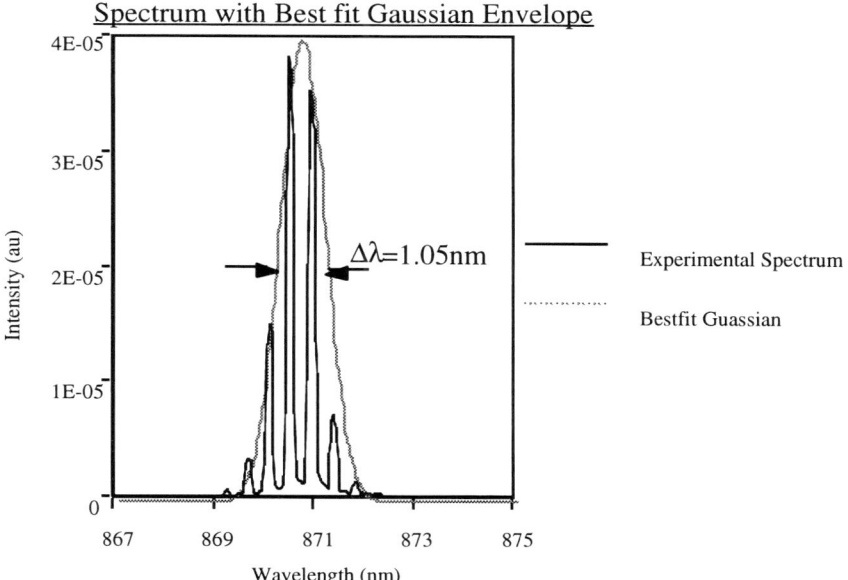

Figure 2.

388

Figure 2 shows the spectrum of the laser operating with 2 pulse colliding pulse modelocked and Figure 3 shows an autocorrelation of the laser output. The relationship between the frequency and time domain measurements indicates that the pulses are transform limited with no chirp.

If the laser is operated with all saturable absorbers turned on, that is with all the sections a,b and c reversed biased, then the mode spacing doubles again and the laser has 4 pulses circulating in the resonator. The laser produces a pulse stream with a time between pulses of nl/c2 (i.e quarter of a resonator round trip time). The pulse circulation scheme is illustrated in Figure 4. In four pulse MCPM operation the laser has operated with repetition rate of up to 375 GHz [1].

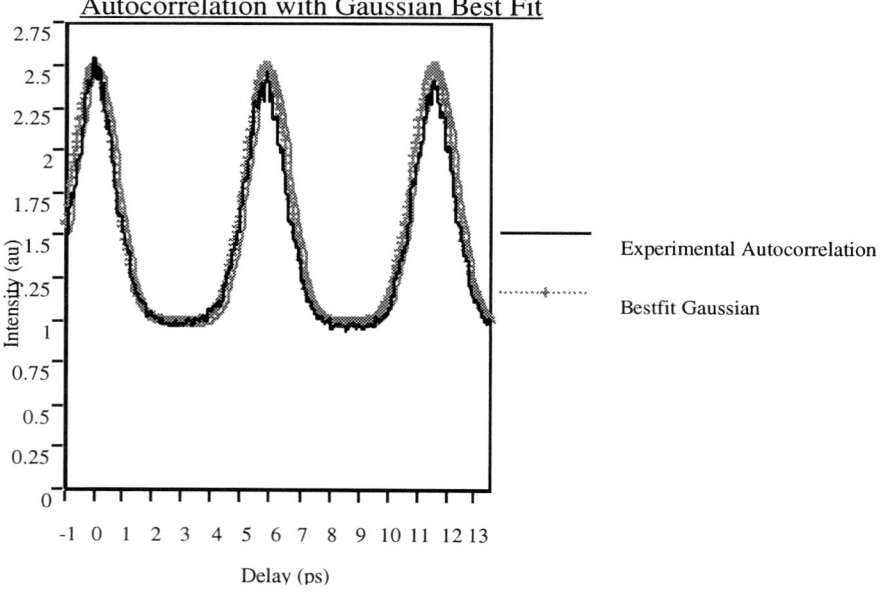

Figure 3

4. Theory

The theory of the device operation can be understood from a frequency domain approach [4]. In multimode operation the longitudinal modes of the laser build up independently from noise and the output from the laser is noisy. With modelocked operation the longitudinal modes of the laser have a fixed phase relationship. This requires a mechanism which couples the phase of longitudinal modes, the saturable absorber sections provide this mechanism. However, in the multisectioned lasers the spatial configuration of the saturable absorber sections is important because the spatial overlap of the longitudinal modes in the various sections can be utilised to alter the coupling between modes. For example with the saturable absorber in the centre it turns out that the spatial overlap and thereby coupling in the saturable absorber between every second mode is maximised and with the all three saturable absorbers turned on the coupling between every fourth mode is maximised. The spatial overlap factors are defined by the equation :-

Start

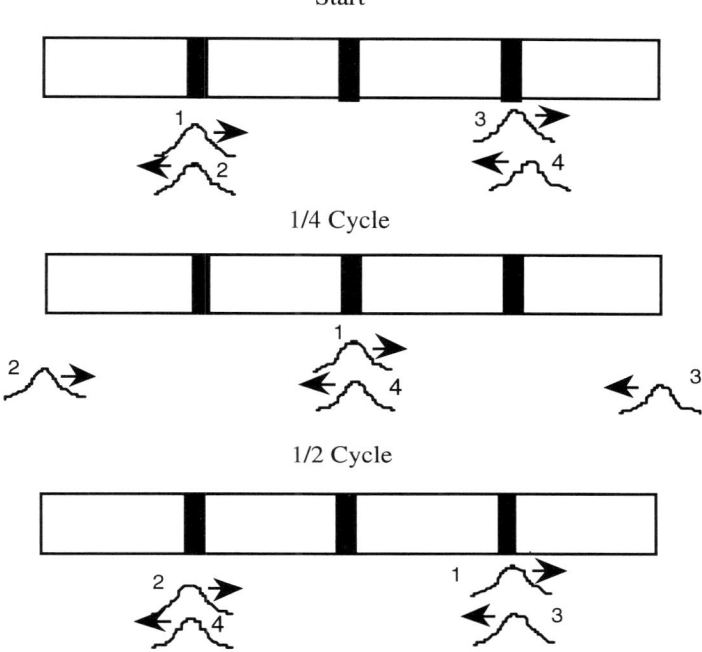

Figure 4. Four pulse operation

$$\xi_m^{(g,a)} = \int_{g,a} dz \cdot (u_k u_{k+m}^*) \cos\left(\frac{m\pi}{L}(z + \frac{L}{2})\right) \tag{1}$$

In equation (1) the integration is over the fraction of the laser length occupied by the corresponding section, the zero of z is assumed in the centre of the cavity, u_k are the longitudinal mode profiles (wave functions) of the cavity.

Figure 5 shows the spatial overlap factors for gain and absorption sections for the two pulse and four pulse operation.

The theoretical results from the frequency domain model, [1], strongly suggest that for short monolithic cavities, fast nonlinearities play a major role in mode-locking operation and that the device geometry is highly favourable to MCPM operation. Numerical simulations with a time domain model were performed, using saturation coefficients and fast recovery times estimated from the frequency domain model. Good agreement between the models and the experimental results for MCPM operation was obtained, although there are still some features which require further investigation for a complete understanding of the laser. In particular the nature of the very rapid recovery of the saturable absorber, which we postulate here (the time domain simulations use a recovery time of around 0.5 ps), has not been observed in pump-probe experiments on reverse biased multiple quantum well material.

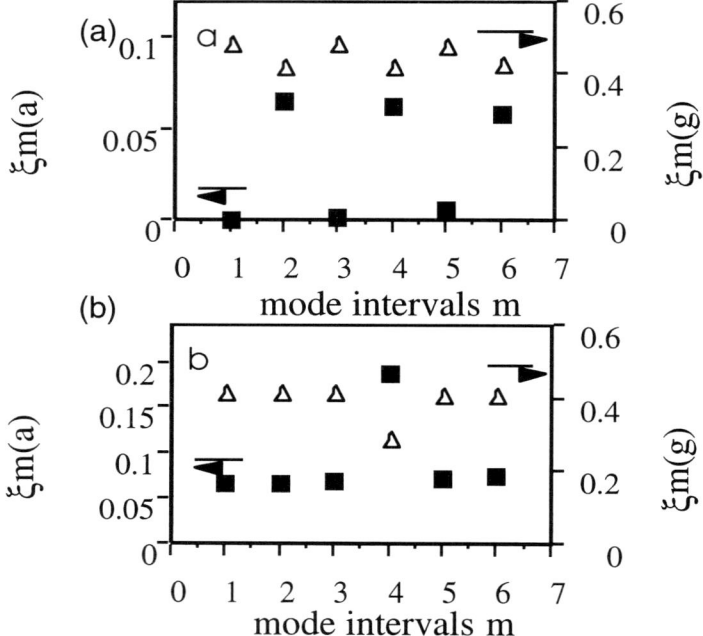

Figure 5. Spatial overlap factors: (a) only central saturable absorber on, (b) all three saturable absorbers on.

5. Conclusions

The operation of a monolithic multi-sectioned semiconductor laser has been described. The laser can be operated as either single mode, multimode, self pulsation (also called Q-switched) and mode-locked in either 1, 2, 3 or 4 pulse operation. This device is a very versatile source of high repetition rated (up to 375 GHz) ultrashort pulses (around 1-3 ps) of light. The various modes of operation of the device have been experimentally investigated and we have found experimental indications of the presence of excitonic nonlinearities in its operation. The device is a monolithic semiconductor chip which can be mass produced; it is low-cost, robust, efficient, and reliable. These are crucial features if ultrashort pulses of light are to find wide-spread application outwith the laboratory.

Multigigahertz up to -1500 GHz modulation can be generated using optical pulses from ultrashort pulsed semiconductor lasers. If we compare self pulsation with modelocked source we can note the following general points; self pulsation gives a wider repetition frequency range but has broadband phase noise. Modelocking gives shorter pulses, lower phase noise and higher maximum repetition frequency [3]. The next challenge in this work is to synchronise the devices at these very high frequencies to either an electrical or an optical signal.

References

1   Martins-Filho, J. F., Avrutin, E. A., Ironside, C. N. and Roberts, J. S., (1995) Monolithic Multiple Colliding Pulse Mode-locked Quantum-Well lasers:Experiment and Theory, *IEEE Journal of Selected Topics in Quantum Electronics,* **1** 539-551.

2   Vasil'ev, P. P., (1992) Ultrashort pulse generation in diode lasers, *Optical and Quantum Electron.,*. **24**, 801-824.

3   Helekey, R. J., Derickson, D. J., Mar, A., Wasserbauer, J. G. and Bowers, J. E., (1993) Millimeter-wave signal generation using semiconductor diode lasers, *Microwave and optical technology letters,* **6** 1-5.

4   Siegman, A. E.*Lasers*, University Science Books.

5   Arahira, S., Oshiba, S., Matsui, Y., Kunii T. and Ogawa, Y., (1994) Terahertz-rate optical pulse generation from a passively mode-locked semiconductor laser diode, Opt. Lett.,. **19**, 834-836.

LIST OF SPEAKERS
NATO ASI: New Directions in Terahertz Technology

Prof G Beaudin
Lab. de Radioastronomie Millimetrique
Observatoire de Paris/ Meudon
5 Place Jules Janssen
F-92195
FRANCE

Dr J Bowen
Department of Cybernetics
University of Reading
Whiteknights P O Box 225
Reading RG6 2AY
GREAT BRITAIN

Professor B Carli
Instituto di Recerca Sulle Onde
Ellettromagenetiche del CNR
50217 Firenze
Via Panciatichi 64
ITALY

Dr J M Chamberlain,
Department of Physics
The University of Nottingham
Nottingham NG7 2RD
GREAT BRITAIN

Dr N J Cronin
School of Physics
The University of Bath
Claverton Down
Bath BA2 7AY
GREAT BRITAIN

Professor H R Fetterman
UCLA Electrical Engineering
Department
Los Angeles
CA 90024
USA

Dr P Goy
AB Millimetre
52 Rue Lhomond
75005 Paris
FRANCE

Dr W J Hall
MMS Space Systems Ltd
FPC320, PO Box 16
Filton
Bristol BS12 7YB
GREAT BRITAIN

Dr Paul Harrison
Department of Electronic and Electrical
Engineering
The University of Leeds
Leeds LS2 9JT
GREAT BRITAIN

Prof T. de Graauw
University of Groeningen FDL
Nijenborgh 4
9747 AG Groeningen
The Netherlands

Dr ir Chris van Hoof
IMEC - VZW
Kapeldreef 75
3030 Leuven
BELGIUM

Dr C Ironside
Department of Electronics and
Electrical Engineering
The University of Glasgow
Glasgow G12 8LT
GREAT BRITAIN

Professor T Itoh
UCLA Department of Electrical
Engineering
405 Hilgard Avenue
Los Angeles
CA 90024-1594
USA

Professor E Kollberg
Department of Microwave Technology
Chalmers University of Technology
54196 Gothenburg
SWEDEN

Professor J Leotin
SNCMP-INSA
Complexe Scientifique de Rangueil
31077 Toulouse
Cedex
FRANCE

Professor D Lippens
Departement Hyperfrequences et
Semiconducteurs
Institut d'Electronique et de
Microelectronique du Nord
Cite Scientifique, Avenue Poincare
BP 69 59652 Villeneuve d'Ascq Cedex
FRANCE

Dr C Mann
Rutherford Appleton Laboratory
Chilton
Didcot OX11 OQX
GREAT BRITAIN

Dr R E Miles
Department of Electronic & Electrical
Engineering
The University of Leeds
Leeds LS2 9JT
GREAT BRITAIN

Prof R D Pollard
Department of Electronic & Electrical
Engineering
The University of Leeds
Leeds LS2 9JT
GREAT BRITAIN

Professor Dr H P Roser
DLR Institut Fur Weltraumsensorik
Rudower Chaussee 5
12489 Berlin
GERMANY

Dr Hartmut Roskos
Institut fur Halbleitertechnik II
RWTH Aachen
Sommerfeldstr. 24
D-52074 Aachen
GERMANY

Prof D Rutledge
Department of Electrical Engineering
MS 136-93
Caltech
Pasadena CA 91125
USA

Dr D Rytting
Research and Technology Manager
Hewlett-Packard Company
Santa Rosa Systems Division
1400 Fountaingrove Parkway
Santa Rosa, CA 95403-1799
USA

Dr H Sigg
PSI Zurich
Badenerstrasse 569
CH 8048 Zurich
SWITZERLAND

Dr W S Truscott
Electrical Engineering and Electronics
Department
PO Box 88
Manchester M60 1QD
GREAT BRITAIN

Dr D Wake
BT Networks and Systems
BT Laboratories
Martlesham Heath
Ipswich IP5 7RE
GREAT BRITAIN

Professor Dr T Wenckebach
Faculty of Applied Physics
Delft University of Technology
P O Box 5046
2600 GA Delft
THE NETHERLANDS

Dr R J Wylde
Thomas Keating Ltd
Station Mills
Billinghurst
West Sussex, RH14 9SH
GREAT BRITAIN

LIST OF PARTICIPANTS
NATO ASI: New Directions in Terahertz Technology

BELGIUM

S Brebels
IMEC VZW
Kapeldreef 75
B-3001
Leuven
BELGIUM

P Pieters
IMEC VZW
Kapeldreef 75
B-3001
Leuven
BELGIUM

CANADA

M Davies
Building M50
Institute for Microstructural Sciences
National Research Council of Canada
Montreal Road
Ottowa K1A OR6, Ontario
CANADA

DENMARK

M Jensen
Department of Physics
University of Essex
Wivenhoe Park
Colchester, Essex CO4 3SQ
GREAT BRITAIN

EIRE

W Lanigan
Department of Physics
St Patricks College
Maynooth
Co. Kildare
EIRE

FRANCE

L Castaing
SNCMP
INSA - Complexe Scientifique de
Rangueil
31077 Toulouse Cedex
FRANCE

M Gross
AB MillimÉtre
52 Rue Lhomond
75005 Paris
FRANCE

R Havart
Institut d'Electronique et de
Microelectronique du Nord
Cite Scientifique, Avenue Poincare
BP69
59652 Villeneuve d'Ascq Cedex,
FRANCE

R K Kupka
Lure B,timent 209D
Centre Universitaire
Paris-Sud
91405 Orsay Cedex
FRANCE

S Megtert
Lure Batiment 209D
Centre Universitaire
Paris - Sud
91405 Orsay Cedex
FRANCE

C Meny
SNCMP, INSA - Complexe
Scientifique de Rangueil
31077 Toulouse Cedex
FRANCE

S Pasquier-Puech
SNCMP
INSA - Complexe Scientifique de
Rangueil
31077 Toulouse Cedex
FRANCE

GERMANY

R Feineaugle
DLR Berlin
Institut fur Welstraumsensorik
Rudower Chaussee 5
12498 Berlin,
GERMANY

V Krozer
TU Chemnitz
Fak. f.Elektrotech 8
Informationstech
Reichenhainerstrasse
D-09126 Chemnitz
GERMANY

A Linhart
Institut fur Weltraumsensorik
Rudower Chaussee 5, Geb 16 16
12489 Berlin
GERMANY

C Preis
Universitat Regensburg
Institut fur Experimentelle und
Angewandte Physik
93040 Regensburg
GERMANY

U Rauschenbach
Universit%otsstr 31
D-93053 Regensburg
LS Prof. Renk Institute for Applied
Physics
GERMANY

GREAT BRITAIN

P Buckle
Electrical Engineering and Electronics
Department
P O Box 88
Manchester M60 1QD
GREAT BRITAIN

J Digby
Physics Department
University of Nottingham
Nottingham NG7 2RD
GREAT BRITAIN

S Feiven
Department of Physics
University of Essex
Wivenhoe Park
Colchester CO4 3SQ
GREAT BRITAIN

P Gunning
PP B55/125a
BT Laboratories
Martlesham Heath
Ipswich IP5 7RE
GREAT BRITAIN

C Kuo
Electrical Engineering and Electronics
Department
P O Box 88
Manchester M60 1QD
GREAT BRITAIN

M Lynch
Electrical Engineering and Electronics
Department
P O Box 88
Manchester M60 1QD
GREAT BRITAIN

G Parkhurst
Physics Department
University of Nottingham
Nottingham NG7 2RD
GREAT BRITAIN

A Saher
Department of Electronics and
Electrical Engineering
University of Glasgow
Glasgow G12 8QQ
GREAT BRITAIN

R Stone
Clarendon Laboratory,
University of Oxford,
Parks Road
Oxford OX1 3PU
GREAT BRITAIN

D Thompson
Microwave and Terahertz Technology
Group
Department of Electronic and Electrical
Engineering
University of Leeds
Leeds LS2 9JT
GREAT BRITAIN

J Thorpe
Department of Electronic and Electrical
Engineering
University of Leeds
Leeds LS2 9JT
GREAT BRITAIN

B Towlson
Department of Cybernetics
University of Reading, P O Box 225
Whiteknights
Reading RG6 6AY
GREAT BRITAIN

S Wootton
School of Physics
The University of Bath
Claverton Down
Bath BA2 7AY
GREAT BRITAIN

GREECE

L Karatzas
Department of Cybernetics
The University of Reading
P O Box 225
Whiteknights
Reading RG6 6AY
GREAT BRITAIN

ITALY

G Modugno
Scuola Normane Superiore
Piazza dei Pisa Cavalieri
56100
ITALY

MALTA

C Sammut
Department of Physics
University of Malta
Msida MSD 04
MALTA

NETHERLANDS

P Dieleman
University of Groningen FDL
Nijenborgh 4
9747 AG Groningen
THE NETHERLANDS

F Ghianni
Faculty of Applied Physics
Delft University of Technology
P O Box 5046
2600 GA Delft
THE NETHERLANDS

H Pellemans
Faculty of Applied Physics
Delft University of Technology
P O Box 5046
2600 GA Delft
THE NETHERLANDS

J Weijtmans
University of Groningen FDL
Nijenborgh 4
9747 AG Groningen
THE NETHERLANDS

NORWAY

M Nawaz
Department of Physical Electronics
Norwegian University of Science and
Technology
NTH, N-7034
Trondheim,
NORWAY

PORTUGAL

Jose Maria L Figueiredo
Lab. de Fisica- Fac. De Ciencias
Universidade do Porto
Rua Campo Alegre, 687
P-4150 Porto
PORTUGAL

SPAIN

J Carbonell
Institut d'Electronique et de
Microelectronique
Du Nord (IEMN)
Cite Scientifique
Avenue Poincare BP 69
F-59652 Villeneuve d'Ascq, Cedex
FRANCE

SWEDEN

L Dillner
Department of Microwave Technology
Chalmers University of Technology
S-412 96 Gothenburg
SWEDEN

J Stake
Chalmers University of Technology
Department of Microwave Technology
S-41296 Goteborg
SWEDEN

SWITZERLAND

D Erni
Institut fuer Feldtheorie
und Hochfrequentztechnik ETHZ
Gloriastrasse 35
CH-8092 Zurich
SWITZERLAND

S Graf
Paul Scherrer Institut
Badenerstr 569
CH-8048
SWITZERLAND

A Magun
Institute of Applied Physics
Sidleustr 5
Univ. of Bern
CH-3012 Bern
SWITZERLAND

USA

O Boric-Lubecke
Jet Propulsion Laboratory
Mail Stop 246-101
4800 Oak Grove Drive
Pasadena
CA 91109-8099
USA

D Bhattacharya
UCLA Department of Elec. Eng.
405 Hilgard Avenue
64-147 Engineering IV
Los Angeles
CA 90095-1594
USA

C Church
Director
Electronics Division
US Army Research Office
PO Box 12211
Research Triangle Park
NC 27709-2211
USA

I Galin
Areojet
PO Box 296 (Dept.8661, Bld. 170)
1100 West Hollyvale Street
Azuga
CA 91702
USA

V Mitin
Department of Electrical and Computer
Engineering
Wayne State University
Detroit
Michigan 48202
USA

D Rutledge
Department of Electrical Engineering
MS 136-93
Caltech
Pasadena CA 91125
USA

D Scott
UCLA Department of Elec. Eng.
405 Hilgard Avenue
64-147 Engineering IV
Los Angeles, CA 90095-1594
USA

G Sirmain
Lawrence Berkely Laboratory
University of California
Bldg 2, Room 200, 1 Cyclotron Road
Berkely
CA 94720
USA

NATO ASI: New Directions in Terahertz Technology
Chateau de Bonas, Castera- Verduzan, France, June 30 - July 11, 1996.
Photographer: *Mr. Deliniere, 32190 Vic Fezensac, France*

405

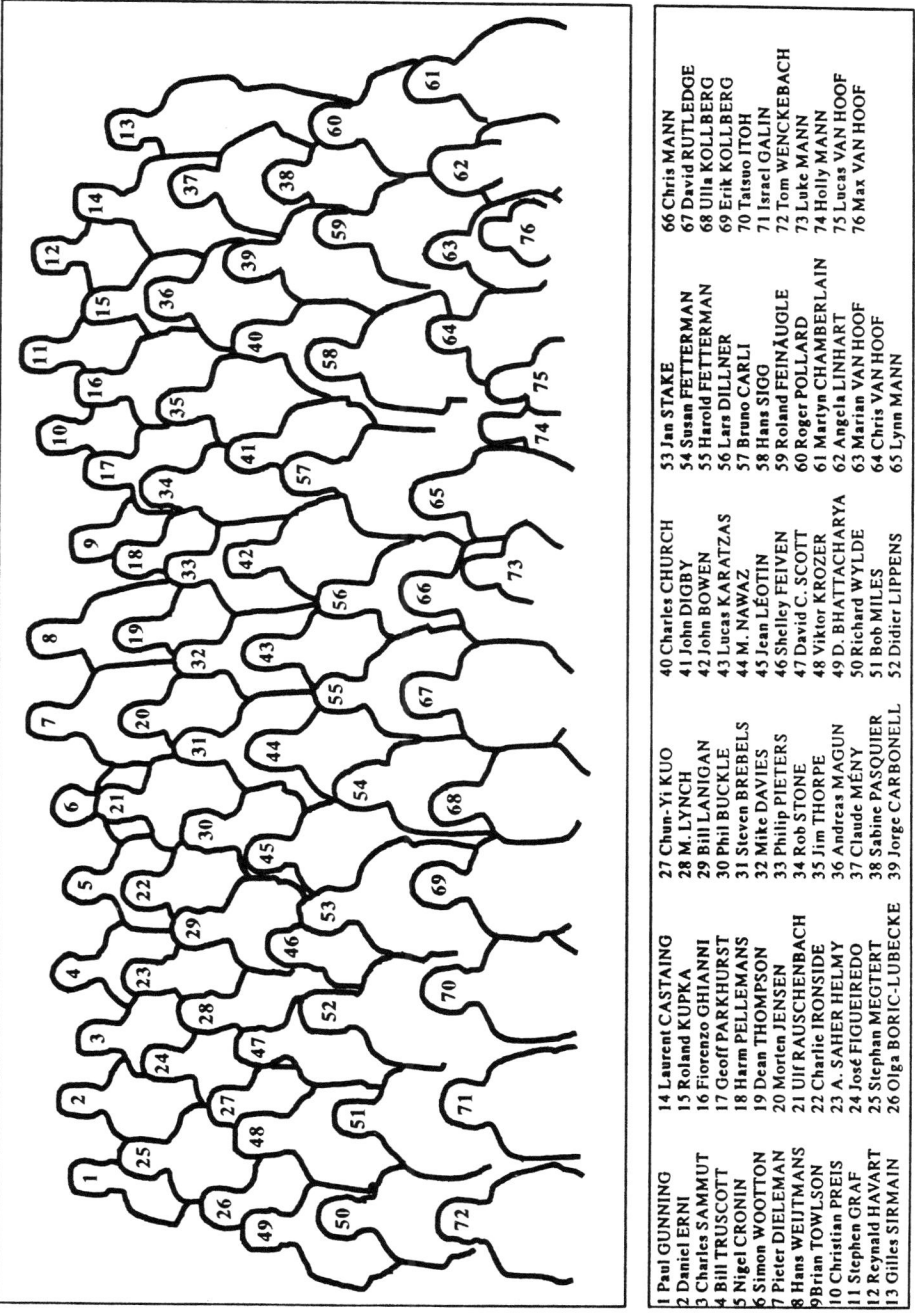

1 Paul GUNNING	27 Chun-Yi KUO	53 Jan STAKE
2 Daniel ERNI	28 M. LYNCH	54 Susan FETTERMAN
3 Charles SAMMUT	29 Bill LANIGAN	55 Harold FETTERMAN
4 Bill TRUSCOTT	30 Phil BUCKLE	56 Lars DILLNER
5 Nigel CRONIN	31 Steven BREBELS	57 Bruno CARLI
6 Simon WOOTTON	32 Mike DAVIES	58 Hans SIGG
7 Pieter DIELEMAN	33 Philip PIETERS	59 Roland FEINÄUGLE
8 Hans WEIJTMANS	34 Rob STONE	60 Roger POLLARD
9 Brian TOWLSON	35 Jim THORPE	61 Martyn CHAMBERLAIN
10 Christian PREIS	36 Andreas MAGUN	62 Angela LINHART
11 Stephen GRAF	37 Claude MÉNY	63 Marian VAN HOOF
12 Reynald HAVART	38 Sabine PASQUIER	64 Chris VAN HOOF
13 Gilles SIRMAIN	39 Jorge CARBONELL	65 Lynn MANN
14 Laurent CASTAING	40 Charles CHURCH	66 Chris MANN
15 Roland KUPKA	41 John DIGBY	67 David RUTLEDGE
16 Fiorenzo GHIANNI	42 John BOWEN	68 Ulla KOLLBERG
17 Geoff PARKHURST	43 Lucas KARATZAS	69 Erik KOLLBERG
18 Harm PELLEMANS	44 M. NAWAZ	70 Tatsuo ITOH
19 Dean THOMPSON	45 Jean LÉOTIN	71 Israel GALIN
20 Morten JENSEN	46 Shelley FEIVEN	72 Tom WENCKEBACH
21 Ulf RAUSCHENBACH	47 David C. SCOTT	73 Luke MANN
22 Charlie IRONSIDE	48 Viktor KROZER	74 Holly MANN
23 A. SAHER HELMY	49 D. BHATTACHARYA	75 Lucas VAN HOOF
24 José FIGUEIREDO	50 Richard WYLDE	76 Max VAN HOOF
25 Stephan MEGTERT	51 Bob MILES	
26 Olga BORIC-LUBECKE	52 Didier LIPPENS	

Index